Human Heredity
and Society

Copyediting: June Gomez
Composition: Graphic World, Inc.
Text and Cover Design: John Rokusek/Rokusek Design
Illustrations: George Barile/Accurate Art
Production, Prepress, Printing, and Binding by West Publishing Company.

WEST'S COMMITMENT TO THE ENVIRONMENT

In 1906, West Publishing Company began recycling materials left over from the production of books. This began a tradition of efficient and responsible use of resources. Today, up to 95 percent of our legal books and 70 percent of our college texts are printed on recycled, acid-free stock. West also recycles nearly 22 million pounds of scrap paper annually—the equivalent of 181,717 trees. Since the 1960s, West has devised ways to capture and recycle waste inks, solvents, oils, and vapors created in the printing process. We also recycle plastics of all kinds, wood, glass, corrugated cardboard, and batteries, and have eliminated the use of styrofoam book packaging. We at West are proud of the longevity and the scope of our commitment to our environment.

Photo Credits

8 Dian Fossey: AP/Wide World Photos, Charles Darwin: American Museum of Natural History, Mountain Gorilla: Russ Kinne, Science Source/Photo Researchers, Female Chimpanzee: Tom McHugh, Science Source/Photo Researchers, Jane Goodall: Charles Knoblock, AP/Wide World Photos; 11 Courtesy of Michael Cummings; 15 Jean Boughton, Stock, Boston; 138 David Parker, Science Photo Library/Photo Researchers; 161 (top) Courtesy of Michael Cummings, (bottom) Catherine Ursillo, Science Source/Photo Researchers 162 (top) Courtesy of Michael Cummings, (bottom) Courtesy of Dr. Ira Rosenthal; 164 (top) Courtesy of Michael Cummings, (bottom) Courtesy of Dr. Ira Rosenthal; 165 Courtesy of Michael Cummings; 190 Courtesy of Dr. Ira Rosenthal; 262 Supplied by Carolina Biological Supply Company.

COPYRIGHT ©1992   By WEST PUBLISHING COMPANY
50 W. Kellogg Boulevard
P.O. Box 64526
St. Paul, MN 55164-0526

Printed in the United States of America

99 98 97 96 95 94 93 92   8 7 6 5 4 3 2 1 0

Library of Congress Cataloging in Publication Data

Woodward, Val W.
    Human heredity and society / Val Woodward.
        p.   cm.
    Includes bibliographical references and index.
    ISBN 0-314-93390-5
    1. Human genetics.   I. Title
QH431.W68   1992
573.2´1—dc20   91-39537
                                                    CIP

ISBN: 0-314-93390-5

Human Heredity and Society

Val Woodward
Department of Genetics and Cell Biology
University of Minnesota

West Publishing Company

St. Paul   New York   Los Angeles   San Francisco

To Dean and Carol, Jim and Terry, Kathryn and Laurence, and "my" eight grandchildren, all of whom are doing quite well even without a genetic explanation why.

Contents

3 Genes, Sex Chromosomes, and Sexual Dimorphism, 121

4 Evolution, Populations and Species, 176

Preface

When I enrolled in college nearly half a century ago it seemed clear to many people that the allies would win WWII; later it seemed equally clear that the United States would maintain a permanent peace with its atomic bomb hegemony. Few undergraduates appeared to worry about it one way or another, their concerns being riveted on sports and pranks, both of which were segregated by ethnicity and sex.

During the past fifty years many societal changes have taken place, and students' awareness has increased by orders of magnitude. Today undergraduates are bombarded with bad news — whaling and spear-fishing controversies, acid rain, toxic waste, holes in the ozone, apartheid, drug abuse, the homeless, overt racism and sexism, crime, dozens of wars that never cease, "death with dignity," skyrocketing medical costs, the AIDS epidemic, unemployment, terrorism, drinking pilots, crimes of the rich and famous, famines, and so on. As a teacher I am bombarded with questions from students who are searching for ways to reduce the magnitude of one or another of these life-threatening problems. I do not recall undergraduates fifty years ago carrying equivalent burdens of consciousness. A few must have, but most appeared to be innocent.

Since then it has become socially acceptable to discuss some of these "touchy" issues, but new problems are spawning a new kind of innocence. Generalists often refer to this new kind of innocence as *single-issue-itis*, intense focus on one issue and near total ignorance of related issues. To illustrate with a trivial example, well-meaning people who work hard soliciting funds to save the otters may feel competition from well-meaning people who work equally hard to save the whales, the spotted owls, or the rainforests. Feelings of competition among well-meaning groups often damps the desire (and mask the option) to discover underlying principles of species extinction, thus ensuring innocence, and the innocent too seldom intuit that the only antidote for innocence is growing understanding.

The environments within which university students study and play have changed. Today women rarely wear bobby-sox and few men wear crew-cuts. A smaller percentage of women graduates become home-makers and mothers, fewer men become providers within nuclear families, and fewer students are Euro-american in origin; a smaller percentage are age 17-21; and importantly, too few graduates today can count on finding steady employment, especially in jobs that extend their university-acquired skills.

As life styles change and as new fears fend off tranquility, what case can be made for advising students to learn about human genetics? What has genetics to do with sexuality, safe sex, AIDS, getting a job, having a family, remaining single, integrating the schools, providing first class citizenship for ethnic minorities and non-nationals, women, the disabled and diseased, young and old, the homeless and the unemployed, single parents, homosexuals, and other victims of social inequities that went unrecognized back when most university women wore bobby-sox and men crew-cuts?

With or without knowledge of genetics, if we succeed in reducing the magnitude of any of these societal problems, and I think we will, it will become necessary to understand that there is no such thing as a quick fix — making one thing right while leaving related things wrong. Quick fixes to societal problems must be rejected out of hand. Indeed, the development and history of genetics illustrates this point, even though genetics is being advertised today to be a major provider of quick fixes in the fight against ill health.

At the turn of the century it was believed by many geneticists that the sterilization of "diseased" persons would ultimately eliminate not only individual "diseases," such as feeblemindedness and poverty, but also the social blights alleged to be brought on by such diseases. There was no scientific basis for this belief, only the ideological desire to preserve the planet for "Nordic peoples." Today tax payers are footing the bill for a three billion dollar project to locate every gene on every human chromosome, for the advertised purpose of leading toward the elimination of genetic diseases. The insertion of a normal gene into the chromosomes of persons carrying "bad genes" is a classical example of a quick fix, a magic bullet. However, knowledge of the technology of gene surgery and ignorance of the fact that poverty curtails access to gene surgery illustrates one of the consequences of this new kind of innocence.

One of the most difficult concepts for nonscience students to grasp is that science is a societal activity, that scientists are socialized in the same ways that non-scientists are socialized, and that as a result the same spectrum of political persuasions found among philosophers, auto mechanics, medical doctors, and department store clerks are found within the community of scientists. The knowledge required to develop a scientific magic bullet does not inform the developer whether, how, or when to use the bullet. Indeed, many of the arguments that take place within the scientific community are about the social uses of science, not about the laws, theories, and facts of science. Clearly the arguments about the so-called Star Wars initiative of the Reagan administration illuminated far more about the political posturing of the contending scientists than about their different interpretations of scientific data.

Genetics was sequestered into a political controversy the year it became a formal science (1900). In fact, that controversy (about which people and populations are biologically superior) had been raging hot for fifty years prior to the birth of genetics. In the early 1850's Herbert Spencer formulated a concept of human nature that later came to be called social darwinism, and from 1869 on, Francis Galton fathered and fueled the ideology of eugenics, which promised to improve upon human nature. Both social darwinism and eugenics preempted genetics decades before it became possible to formulate a definition of the gene. Social darwinism rationalized British colonialism on the grounds of biological superiority; the eugenics movement promised to create a more superior population by sterilizing the biologically unfit.

In 1907 the first sterilization law was enacted in the state of Indiana. A concept of the gene was beginning to emerge, and it was said that feeblemindedness, imbecility, sex perversion, poverty, drug addiction, and other forms of asocial behavior are caused by specific genes. By 1930 sterilization laws were passed in more than 30 states, and in July 1933 these same sterilization laws were imported by Hitler to become the core of the Nazi Hereditary Health Laws.

Eugenics laws were not premised upon genetics facts, but upon political and ideological ambition. Eugenics lost favor after WWII, in part because of the atrocities of the Nazi regime, and in part because it was a political, not a scientific concept. However, the political ideas that gave birth to eugenics did not die; they are alive and well today. Eugenics is hailed by prominent people throughout the world and is a rallying cry of neofascists everywhere. Extreme eugenics sees the science of genetics as a technological means for "improving the human stock." This is as gross a misuse of science as can be found in the annals of science history.

Eugenics was and is a political and idelogical outlook that has been and is supported by leading geneticists who affirm that genetics holds the key to improving the quality of human species. Enthusiasts have advocated that "high quality" people should have lots of babies and that "low quality" people should be sterilized. However, they still have not defined high and low quality, nor have the alleged genes been isolated—not to mention the fact that ever increasing numbers of "high quality" people are being caught committing low quality crimes.

After the eugenics movement had popularized a cause-effect relationship between genes and asocial behavior, it seemed logical to resurrect the social darwinist view that social classes are the result of an inequitable distribution of "good and bad" genes, not of society's resources. Clearly, they insisted, some people are ill because of mutant genes, so what is wrong with the assumption that some social classes are ill, i.e., prone to poverty, criminality, sex and drug abuse, and low intelligence because of mutant genes? This logic tells us that the way to resolve societal problems is to "wage war" on mutant genes. These are the kinds of claims and promises that contribute to our unawareness and to the continued search for the alleged mutations.

On the other hand, great strides were made quickly between 1944 and 1965 in our understanding of what genes are, what genes do, and how genes do what they do. Great strides were made even more quickly after 1970 in our understanding of how to clone genes, move genes from one organism to another, and manipulate the expressions of genes. But this knowledge of molecular genetics did not tell us how to use genetics to improve human societies. In part this is because we know so little about human societies and in part because we have exaggerated the roles played by genes during the evolution of human societies. This confusion will persist if we continue to remove genes from their biological contexts, and to call upon them to resolve problems that exist within societal contexts.

The caveats expressed here about extending the influence of genes beyond proteins, which are encoded by genes, to the characteristics of human societies, which are influenced by their histories, do not detract from the exquisite beauty of genes; rather, they are warning flags stationed proximal to the simplistic assumption that knowledge of human genes can double as knowledge of human cultures. Without doubt genetics holds its own among the other sciences such as mathematics, physics, and chemistry, and without doubt the more we learn about genes the more exciting the whole of biology becomes. Geneticists have every reason to take pride in their work, and every reason to be proud of their science. But genes in all of their exquisite beauty cannot explain the history of the human drama. I have learned from thousands of first year university students that beginners can understand what genes are and do, and how genes do what they do. They also can understand that to know the origins of music and literature it is necessary to look beyond the gene.

Indeed, to fully appreciate the central role of genes during the course of evolution, *it is as important to know what genes are not as it is to know what they are.* My attempts to call attention to this fact led to the making of this book. *Human Heredity and Society.*

To my view, knowledge of genes is not sufficient for minimizing the threat of any current life-threatening problem. To improve upon our quality of life we must possess (a) the skills needed for distinguishing between facts and pseudo facts, between self-serving ideologies and scientific theories, and (b) a desire to preserve the planet and ameliorate human suffering.

What better way for teachers to help students in a world as uncertain as ours than to help them learn how to distinguish between realities that can be verified by experiments and observations, and images of reality that require dogma, ritual, and authority.

This book is dedicated to students and teachers who learned and helped me learn to resist simplistic determinist explanations of ourselves, and to regard human phenotypes as emergent outcomes of our histories, of our genes whose influences upon us are continuously changed by those environments, and of our environments whose influences upon us are continuously changed by our genes.

Acknowledgements

First I am grateful to my busy colleagues who took time to review early drafts of this book. All of them made important suggestions but none is responsible for the book's weaknesses. The reviewers are: James F. Arnesen, William Rainey Harper College; L. Herbert Bruneau, Oklahoma State University; Frank C. Dukepoo, Northern Arizona University; Mark Fenlon, Jefferson Community College; Jane E. Ferguson, The University of Wisconsin-Oshkosh; Richard W. Frederickson, St. Joseph's University; Michael A. Gates, Cleveland State University; Eugenia Georges, Rice University; Janis Hutchinson, University of Houston; Judith M. James, Eastern Illinois University; Lyle W. Konigsberg, University of Tennessee-Knoxville; Melanie W. S. Loo, California State University-Sacramento; Charles R. Peebles, Michigan State University; Ronald W. Wilson, Michigan State University; Virginia Wolfenberger, Texas Chiropractic College.

Second I thank the people at West Publishing who helped me through a full range of moods — during illness, while proof reading and indexing, and in health. My many interactions with Cliff Kallemeyn have been exceedingly pleasant; I am most grateful to him for never letting his store of patience and helpful expertise run low. Jerry Westby is solely responsible for convincing me to transform my class notes into this book, promising not riches, but tennis lessons. Even though our tennis relationship resides in a future time, thanks is due him for many engaging discussions that rescued me every time an urge to quit began acting up. June Gomez is a most capable and understanding copy editor whose writing style and organizational suggestions are greatly appreciated.

Third, heartfelt thanks to Professor Clare Woodward who took time off from discovering how proteins fold to provide loving support and help during every phase of the book's development.

Yet even with all of this help mistakes remain; these are all mine.

Val Woodward

1

The First Method For Sighting Genes

Chapter Outline

Are We More Different than Alike? . . .

A Short Story About a Gene

About one of every 20,000 babies born in the United States will develop **Huntington's disease** (HD). Until the folk singer Woody Guthrie died of Huntington's disease in 1958, very little public concern was shown for this disease, even though persons with the disease experience a tragic deterioration of nerves and muscles before they die. After Guthrie's death, his widow Marjorie formed the Committee to Combat Huntington's Disease. The committee did not attract the attention of the federal government or the **medical-research** community, but it enjoyed some success collecting money for education about the disease and for helping the families of diseased patients; it was not a typical success story.

In 1968 Nancy Wexler, today a well-known geneticist, learned that her mother had begun showing symptoms of HD, which meant, she understood, that she and her sister Alice each had a 50/50 chance of developing the disease later in their lives. Wexler also learned, after she was 22 years of age, that all three of her mother's brothers had had Huntington's disease. Shortly after her father, Milton Wexler, had informed her of her mother's condition, the entire family helped to organize the Hereditary Disease Foundation, which for a while worked jointly with Marjorie Guthrie's Committee. At that time almost nothing was known of the disease except its tragic manifestations and that it is caused by a **gene.**

In 1972 at one of the first workshops of the Hereditary Disease Foundation, a Venezuelan physician, Ramon Avila Giron, showed a movie he had taken of a large number of people with Huntington's disease. All of the diseased persons belonged to the same extended family, and all lived in small villages along the shoreline of Lake Maracaibo, on the north coast of Venezuela. No one living there knew why so many people were ill with HD, but they knew enough of their history to pinpoint a time, around 1800, when the first German and Spanish sailors visited the area and sired more than a few babies. The disease, they speculate, may have been introduced into the indigenous population by one of the sailors.

Another location of an abnormally high frequency of HD is the island of Mauritius, east of Madagascar, in the Indian Ocean. Toward the end of the 18th century, the French aristocrat August de Bourbon moved his family to Mauritius, being afraid, as he was, of the social consequences of the French Revolution. At least one reproductive member of his entourage carried the gene for HD, since among the descendents (13,000 are living today), 20 have died of HD, 6 currently have the disease, and 25 are at **risk** (25 persons who have at least one parent with HD).

In both Venezuela and Mauritius, the gene that initiates the development of Huntington's disease was transported from Europe by one person to one or more children. None of the members of the resident populations in either location carried the disease gene prior to its intrusion from afar. In Chapter 4 we will discuss the significance of gene transport from one population to another in light of genetic diversity within and between modern human populations.

After 1968 Wexler decided to study genetics and to concentrate her research efforts on Huntington's disease. In 1979 she made her first visit to the villages surrounding Lake Maracaibo, where she learned that there were about 2,600 descendents of the first family to carry the HD gene; of these, about 100 persons had the disease and more than 1,000 were at risk for developing the disease. Wexler set out to obtain blood samples from as many of the descendents of the original family as she could find and, of course, persuade to donate blood.

Ten years earlier the proposal to collect blood samples to identify the causative gene of HD would have been met with disbelief. Except for a few genes that determine the characteristics of a few blood proteins, human genes could not have been "sighted" by examining samples of blood; the few human genes that had been reported by 1970 were "inferred" from family pedigrees or simply deduced from **phenotypes,** a flawed method for "sighting" genes, as we shall see shortly. But during the 1970s more than a few remarkable discoveries were made that permitted geneticists to locate human genes on human **chromosomes** as easily as they could locate mouse genes on mouse chromosomes. These discoveries will be discussed in Chapter 2 after an introduction to the chemical composition and physical structure of genes.

To continue the Huntington's disease saga, after two adventurous trips to the villages surrounding Lake Maracaibo, Nancy Wexler and those of her associates who had mastered the use of the new technology discovered the exact location of the HD gene. Moreover, they developed a technique for determining *whether persons at risk do or do not carry the gene.* This was one of the most celebrated discoveries in genetics during the 1980s, and there were many.

Problems That Transcend Genetics

The discovery of a method for identifying among individuals at risk those who do and do not carry the HD gene exposed another kind of problem, a personal, emotional one, not the kind that genetics technology can solve. Wexler felt this problem acutely—to wit, did she want to know if she carried a gene which, if present, would unquestionably destroy her life, as it had destroyed her mother's her three uncles', and her grandfather's lives? Arlo Guthrie, the famous son of Woody Guthrie, faced the same dilemma, as have less well-known people with an HD parent.

As you might expect, some people at risk choose to know and to plan their lives according to the most reliable information they can gain about themselves; others choose not to know, feeling more secure in coping with the hope that they do not carry the gene than with the certainty that they do. There are no easy solutions to these kinds of problems, and as we shall see, genetics discoveries sometimes expose these kinds of problems. (If you wish to read more about the adventures of Wexler that led to the discovery of the location of the HD gene and about the kinds of human problems that can and cannot be solved by way of the new genetic technology, see the book *Genome* by Bishop and Waldholz, described at the end of the chapter. Also, a new book on Huntington's disease is being written by Alice Wexler and will be published by *Times Books/Random House.*)

Another kind of problem exposed by discoveries in genetics has to do with the way we see and evaluate ourselves. All of us, to some degree, are faced with a desire to be different but not too different, to be unique but also normal. Given our cultural and religious histories, it isn't easy for any of us to be objective as we strive to answer questions about our *biological similarities and differences.* As individuals we may imagine that we possess qualities that separate us from the crowd; but as members of societies, in particular ethnic or national groupings, we may exaggerate and denounce the characteristics of *other* as we build a case for the legitimacy or superiority of *self.* These are not problems that the science of genetics can solve.

The societal questions raised by new genetic information often are left for the politicians to answer. But more and more, with each new discovery, biologists are taking a more aggressive stance for the position that science, not politics, is a better source of the information needed for making societal decisions.

Consider the case of Huntington's disease. First, no one denies that persons with advanced symptoms of Huntington's disease differ from the rest of us. George Huntington, the Long Island physician who first described the **syndrome,** wrote a short paper in 1872 describing his patients who had the disease. Early symptoms included involuntary, jerky movements; later symptoms included a chronic, downward slope of nerve and muscle degeneration. He correctly pointed out that disease onset is usually during the fourth decade of life and that death follows onset by about 15 years. During those 15 years, total mental deterioration takes place. Hungtington also presented evidence that the disease is hereditary. Remember the date, 1872.

There can be no doubt that Huntington's observations of HD persons were more astute than those present-day observers who have reported that persons with early symptoms of Huntington's disease are under the influence of alcohol. Indeed, Wexler's mother was accosted by a traffic cop for crossing a busy street, in broad daylight, under the influence of alcohol; this was before she lost all control of her movements. Nevertheless, it is true that alcohol in sufficient quantity may induce involuntary and jerky movements, and to the extent that it does, and in the absence of other information, it may be offered as an **explanation** of the early symptoms of HD, however wrong that explanation might be.

Explanations are necessary to the advancement of scientific understanding. Science is, in fact, an art and a technology of discovering which among contending explanations is the correct one. Correct explanations open doors to ever deeper and more profound explanations, while wrong explanations lead to blind alleys. For example, persons with HD are similar to *some* other persons by phenotype, but they differ from *all* other persons by **genotype.** Therefore, while persons under the influence of alcohol may appear phenotypically *similar* to persons with HD, inebriated persons are genetically *different* from persons with HD. In other words, there may be many causes of involuntary, jerky movements (phenotype) — alcohol, cerebral palsy, accidents, and the HD gene — but there is only one cause of HD, and that is the HD gene. In general, geneticists look for genotypic explanations of phenotypic differences, but *not all phenotypic differences are best explained by genotypes.*

If an observer adopts the alcohol explanation for HD phenotypes and thus enters a blind alley the next erroneous step may consist of uncomplimentary remarks about the character or upbringing of the person with involuntary, jerky movements. The first wrong explanation, alcohol, invites an additional wrong explanation, flawed character. And most certainly if one tries to improve this flawed character in order to improve the health of HD persons, the entire project will stumble in a blind alley.

On the other hand, if we discover that the HD phenotype and the HD gene are always bound by a **cause-effect relationship,** the next discovery may give rise to an explanation of how the gene initiates the development of the phenotype, and this to strategies for health betterment. As we begin to discuss genotype/phenotype relationships in greater depth, it will become more obvious that causal explanations are as necessary to scientists as they are anathema to candidates for political office. But for starters, consider that if HD persons differ from non-HD persons by one gene, then that one gene is a good place to begin the search for the cause of HD.

The science of genetics is an organized search for the causes of similarities and differences — among individuals within species and among all plants, animals, and microorganisms. In the early days of genetics the study of similarities (like begets like) was called **heredity.** Today the words heredity and genetics are used in much the same way.

Even Professionals Disagree

Geneticists disagree about the causes of similarities and differences among individuals, just as nongeneticists do, but their disagreements are couched in a more technical vocabulary. Even so, the debate is whether genes, experiences, competition, cooperation, or some combination of these is the root cause of phenotypes and phenotypic differences among members of the same species. The debates among geneticists differ from those among nongeneticists in that geneticists have the option to keep searching, by way of experiments and controlled observations, for new evidence and for better explanations.

It hasn't been easy for geneticists to explain cause-effect relations between human genes and human phenotypes. Until about 1980, while acknowledging that humans are among the most interesting animals on the planet, geneticists concluded that humans are the most difficult to study. Humans are interesting because of their complex biology — reproduction, growth, development, and evolution — and because of their complex behavior — individual and societal.

Human genetics was difficult to study prior to 1980, compared with plant, insect, and small mammal genetics, because with humans the choice of mates is made by those who do the mating, not by geneticists; because the human generation time is long, nearly half that of a geneticist's career, compared to 10 days for some insects and less than an hour for bacteria; and because the numbers of progeny resulting from a mating pair are low (1 to 3 mostly and rarely more than 10), making it difficult for the geneticist to reach general conclusions. In other words, geneticists cannot plan human genetic experiments, and if they could they wouldn't see many results from those experiments before they grew old and died.

During the decade of the 1980s, human genetics became easier to do. To illustrate, between 1900 and 1980 fewer than 400 human genes had been identified in such a way that genes, phenotypes, and the locations of genes on chromosomes could be demonstrated to share cause-effect relationships. Between 1980 and 1990 the number of verifiable relationships jumped from 400 to more than 2,500. Progress is so fast today that some geneticists predict that by the end of the 20th century, the detailed genetic structures and locations of all of the estimated 100,000 human genes will be known. The ever-accelerating pace at which advances are being made is exciting to some and frightening to others.

Excitement is high among the geneticists who are making the discoveries. A discovery does for a scientist what a win does for an athlete, what a performance does for a musician, what a Pulitzer Prize does for a writer. However, unlike athletes who share their triumphs with millions of nonathletes, scientists are pretty much left to share their triumphs with one another. As athletes become better at what they do, they attract the interest of more nonathletes. Millions of nonbasketball players admire and enjoy the skills of "Air" Jordan, "Magic" Johnson and Clyde "The Glide" Drexler. But as the numbers of scientific discoveries increase, the numbers of interested, nonscientist spectators increase slowly if at all (probably in line with the numbers of nonpoets who are passionate about the poetry of MacArthur genius award winner Mae Swenson).

For others, the response to rapid progress in genetics arouses fear rather than excitement, a response that appears to be unique to science. Anxiety about improved athletic skills and better poetry is rare. Why are public responses to science so different? The skills needed for excellence must become highly fine tuned; such skills are rare, admirable, and close to the edge of human potential. Why is it that skills in science are less appreciated than skills in athletics? Part of the answer may have to do with public access to the professions. Many nonprofessionals have played basketball; many nonpoets have written poetry; but few nonscientists have performed a scientific experiment. Newspapers devote entire sections to athletics and to explaining how the games are played, who the players are, and how each stacks up against peers and former heroes. The vocabulary is rehearsed seven days a week, year after year. And if a fan misses the newspaper, television is there to take up the slack. Not so with science.

Fear of science also has been a theme of fiction, as in the cases of Dr. Jekyll and Frankenstein. In the United States, the fictitious scientist usually speaks with a foreign accent and is doing research on a doomsday machine. This genre of bad press, thankfully, is dying out.

As the Dr. Jekylls are dying out, real science is emerging. Nonscientists have more access to science today than ever before. Advances in genetics make the news nearly every day. Even so, many nonscientists are wary of science and scientific discoveries, and as a result many scientists express the worry that modern society will become further divided as science becomes even more essential to societal well-being and more members of society fail to understand how or why.

No one knows to what extent societies will become further divided by a dependence upon science that only a few understand and control. Many citizens may feel that patching the hole in the ozone is beyond their responsibility and control. The greenhouse effect may or may not be a worry, "but I can't do anything about it." How many people worry about the destruction of the rain forests, the loss of species, or whether it is right or wrong to sacrifice chimpanzees for medical research? How many people blame science for creating these problems in the first place?

If life-threatening and socially important questions fail to motivate us to study and to act in the best interests of the world around us, what are the chances of our being motivated by abstract questions such as, Are human beings more alike than different or more different than alike? While this question may not grip everyone's curiosity, it *lies at the heart of the science of genetics.* There is no doubt that humans are like chimpanzees and canaries biologically, and there is no doubt that humans differ from chimps and canaries as judged by individual and social behavior.

It has been suggested by some biologists that humans possess a biological **nature** akin to that of other primates and that humans differ from other primates in that they possess a different social nature. From such a proposition it has been suggested that both natures influence our behavioral characteristics, but that our biological nature is the more important determinant of individual behavior and that our social nature is the major determinant of our societal behavior. A practical problem with such a proposition is that no one has, as yet, distinguished between these two natures, nor has anyone teased apart the two kinds of influence each nature is said to have upon us. While the controversies generated by questions of similarities and differences will not be resolved in this book, the relevance of the controversies will be highlighted alongside each of the major themes of the book.

For example, that a few people develop Huntington's disease is a biological fact; that the disease is initiated by a gene is a biological fact. But the approach taken toward individuals with the disease, people at risk, and the families affected is opinionated and complicated by our societal behavior. One society may operate on the premise that the greater good is best served by sterilizing all persons suspected of carrying the HD gene. Another society may operate on the premise that the greater good is best served by learning how to improve the quality of life for persons with the disease. This latter societal trajectory, unlike the first, calls for the development of a medical-psychological-sociological research infrastructure, and it will cost a lot of money.

It may be that you are not related to or even know a person with HD, and it may be that because of this you feel comfortable leaving the worry to others, as many of us have done with the hole in the ozone, diminishing rain forests, and endangered species. But on the other hand, in all likelihood you are related to or know someone with a genetic disease, possibly someone you care about. If this is true, you may have wondered why, in a rich country, so little is being done to study the disease and to develop means for minimizing the suffering of those with the disease. Now the question is whether such an experience will motivate you to transform your personal concerns into societal acts. If not you, then who?

As we learn more about heredity and human society, more of these kinds of questions will become apparent; and while we ponder the question, someone out there will be in the act of transforming his or her concerns into societal acts. All of the sciences are creating worries for human societies but probably no other science more so than genetics. While we can't know how human nature will be described at the end of the next century, it is safe to say that genetics will be included in that description.

Genetics is more than a way to solve some of our societal problems; it is a lens through which we see ourselves. What we see influences not only the way we solve problems but also the way we define ourselves.

An Introduction to the Vocabulary of Genetics

Geneticists use the word phenotype to describe an individual's physical, physiological, and biochemical characteristics. The word phenotype is used to describe one, a few, or all of an individual's characteristics. Noses may be narrow or broad, pointed or "pug," small or large; eyes may be blue or brown; skin colors range from light pink to dark brown; height ranges between 3 and 9 feet; and our bodies assume a variety of shapes. Noses, nose shapes, eye color, height, and skin color are aspects of phenotype (Figure 1.1).

Since a vocabulary does not exist with which to provide a complete description of any person's total phenotype (we do not know enough about human biology to do that), it is common practice for geneticists to partition total phenotype into phenotypic **traits** and each trait into **contrasting forms.** For example, the contrasting forms of eye color, a phenotypic trait, are brown and blue; of hair shape, a trait, curly and straight; and of earlobes, attached and unattached.

Dian Fossey

Charles Darwin

Mountain Gorilla

Female Chimpanzee

Jane Goodall

Figure 1.1 Aspects of Phenotype. Jane Goodall is famous throughout the world for her studies of chimpanzees. Dian Fossey studied highland gorillas for 17 years and was trying to save them when she was brutally murdered. Darwin is best known for explaining the origins of all species. Together they illustrate the concept that phenotypic variation is one of our species' two prominent collective attributes. The other is our commonality.

The sum of phenotypic differences among individuals within a defined population is called that population's **phenotypic variation.** Geneticists want to discover the causes of phenotypic variation, whether genes, environmental factors, interactions among genetic and nongenetic factors, or more complex processes. While it is true that eye color differences are influenced by genes and religious preferences by nongenetic factors, some aspects of phenotype (e.g., skin color) are influenced by genes and by nongenetic factors (e.g., ultraviolet light).

Most geneticists do not try to study the total phenotype of any organism nor the total phenotypic variation within any large population. Rather, they focus upon one or a few phenotypic traits (e.g., a particular genetic disease, the inability of some people to distinguish between the colors red and green, or the inability of some cells and tissues to recognize a hormone). Geneticists seek to learn how specific phenotypic traits develop and why it is that traits differ, for example, in color (blue versus brown eyes), in form (five fingers versus six fingers per hand), in size (5'3" versus 6'3"), and so on.

Since genes are known to influence the development of some phenotypic differences, it is obvious why geneticists are so very interested in what genes are, in what they do, and in how they do what they do.

But if the search for genetic causes of phenotypic differences becomes an obsession to the point of ignoring other causes of phenotypic differences, those in the search may overlook the vast interaction—indeed the interdependence—of genetic and nongenetic influences upon phenotypes. While intense focus upon a tiny part of the universe is one form of genius, seeing the big picture is another form. Therefore as we embark upon a detailed study of what genes are, what they do, and how they do what they do, we will examine evidence that phenotypes are more than the sums of the genes said to give rise to them. There are many nongenetic determinants of phenotypes. The environments within which genes do what they do are very complex and vastly influential upon genes and upon the processes of phenotypic **development.**

For example, some chemicals cause genes to **mutate (mutagens),** some induce the development of **cancers (carcinogens),** other chemicals induce developmental mistakes in embryos **(teratogens),** and still others are cell poisons that prevent cells from dividing **(toxins).** Food (amount and kind) and oxygen, temperature and atmospheric pressure, parental care and exercise, and many other aspects of environment influence the development of phenotypes. It is difficult for any scientist to keep an accurate and up-to-date account of all of the many influences upon the emergence of human phenotypes and to sort out from among them all the specific effects of one.

For example, a professional geneticist who is studying the chemical properties of a gene suspected of causing a genetic disease may be chided by a family doctor for ignoring the effects of hospital care upon children with the disease. The doctor may have observed that even though a gene initiates the disease, the severity of the disease ranges between near death to near normal health, depending upon the health care environment. Nongeneticists striving to understand basic genetics principles must acquire the combined knowledge of the family doctor and the geneticist because these basic principles include gene activities, nongenetic factors that influence phenotypes, and interactions among genes and between genes and nongenetic factors.

. . . Or More Alike Than Different?

Each of our individual, developmental histories dates back to the moment that one of our mother's **egg cells** was **fertilized** by one of our father's **sperm cells.** The fertilization event resulted in a cell (called a fertilized egg cell, or **zygote**) with 23 maternal and 23 paternal chromosomes (Figure 1.2). The 23 chromosomes that each of us received from our mothers carried one copy of about 100,000 genes; the 23 chromosomes that each of us received from our fathers also carried a copy of those same 100,000 genes. Each normal zygote, then, carries two copies of the 23 chromosomes and two copies of the 100,000 genes. Judged by these processes, the vast majority of individuals are more alike than different.

Genetic endowments are complete at the moment of fertilization. After that moment we receive no more genetic instructions from our parents; but beginning at the moment of fertilization, our genetic instructions and the environments within which the instructions are translated into biological activity participate in the processes that lead to the emergence of adults from zygotes. The emergence begins slowly. Zygotes remain zygotes for about 30 hours, after which they multiply (by dividing) to become two-cell **embryos;** about four hours later the two cells multiply, giving rise to four-cell embryos. Several days later a multicelled embryo, called a **blastocyst,** will become attached at the top of the mother's **uterus.** Even though a blastocyst contains about 100 cells at the time, it is no larger than the egg cell from which it developed.

About nine months later a baby is born, a small human being as biologically complex as its parents think it is beautiful. **Growth** during those first nine months includes an increase in cell number, from 1 to nearly 100 billion, and an increase in the size of some cells. **Differentiation** during the same time shows an increase in the number of cell types, from 1 to about 100 (muscle cells, skin cells, hair cells, and so on). A dozen or so years later an even more complex, adult human will emerge. One of the "hot" research frontiers in biology today is the study of how these complex transformations occur. The formal name given to the geneticist's corner of this field of biology is **developmental genetics.**

While humility is not a common feature of the scientist's behavior, most biologists will agree that we do not understand much about the processes of growth and differentiation. True, we are making headway, but we have a very long way to go. At the same time we are not totally ignorant about biological development.

Skipping the details for a moment, we know that *genes are inherited* from parents and that *phenotypes develop* after the zygote has received its two copies of 100,000 genes (i.e., each of us inherits genes, but our noses and thumb shapes, eyes and eye colors develop). We know that genes are necessary but not sufficient for development, just as nongenetic factors are necessary but not sufficient. In other words, we know that the processes of development cannot be predicted from knowledge of genes alone. One way to get a better grasp of these statements is to examine what are called **norms of reaction.**

Norms of Reaction

A norm of reaction is the range of phenotypes that may emerge from a single genotype if that genotype is allowed to initiate development in all possible

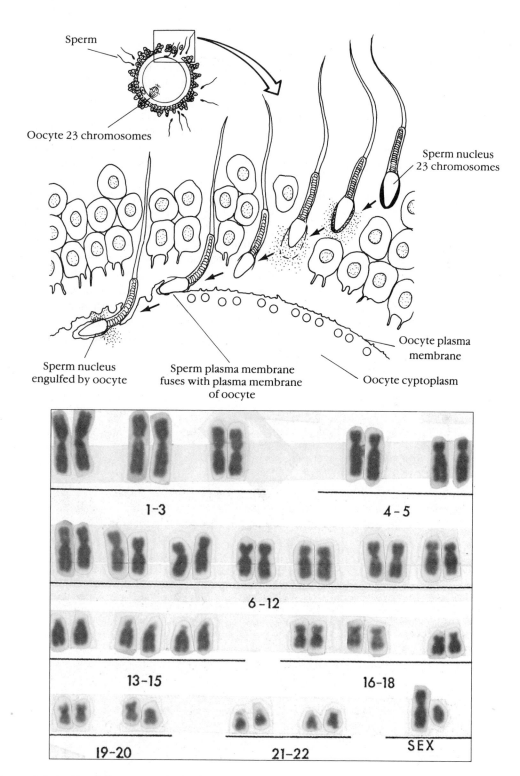

Sperm

Oocyte 23 chromosomes

Sperm nucleus
23 chromosomes

Sperm nucleus
engulfed by oocyte

Sperm plasma membrane
fuses with plasma membrane
of oocyte

Oocyte plasma
membrane

Oocyte cyptoplasm

1-3

4-5

6-12

13-15

16-18

19-20

21-22

SEX

Figure 1.2 An Egg Cell or *oocyte* being fertilized by a sperm cell (above). The figure follows the pathway of a single sperm cell into an oocyte. One of each pair of chromosomes (below) is contributed by the egg and the other by the sperm.

environments. Norms of reaction refer to the observation that genotypes do not "dictate" identical phenotypes in all environments.

A precise norm of reaction is difficult to measure, but its qualitative meaning is relatively easy to understand. Norm of reaction means that genotypes and phenotypes *are not locked into* precise cause-effect relationships. It means that we cannot predict the phenotypic outcomes of genotypes over wide ranges of environment. Some phenotypic traits are known to be more resistant to environmental influence than others (e.g., eye color is more resistant to change than is skin color), but the only way to know the "plasticity" of phenotypes is to observe their development in a wide range of environments because the information about the plasticity of phenotypes is not to be found in the properties of genes.

Development is influenced by genes and by environments (Figure 1.3). The three horizontal rows of plants represent three different elevations at which seven plants were grown. The top row of plants was cultivated at an elevation above 9,000 feet (3,050 m), the middle row at an elevation of about 4,500 feet (1,140 m), and the bottom row at an elevation very near sea level (30 m). Otherwise the plants at all three elevations were provided the same conditions for growth (e.g., soil and water). All of the plants arose as cuttings from seven

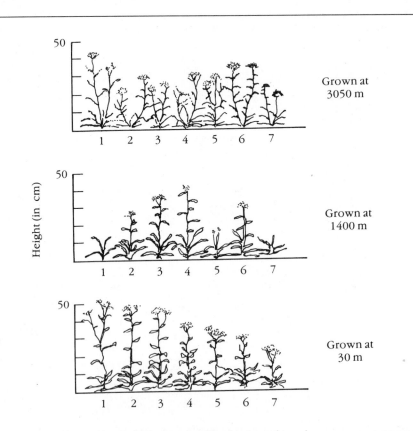

Figure 1.3 Identical genotypes with nonidentical phenotypes. The horizontal numbers represent seven genetically different plants of the same species. The three plants above number 1 are genetically identical, as are those above 2, 3, ..., 7. At different altitudes, genetically identical plants develop different phenotypes.

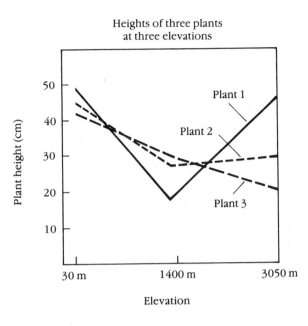

Heights of three plants
at three elevations

Figure 1.4 Phenotypic Expression is Influenced by Environment.

parent plants, permitting the observation of the development of seven
genotypes in three environments. The three plants above the number 1 are
genetically identical, and the same is true for the three plants above numbers 2
through 7.

Plant number 1 grows tall (50 cm) and flowers at the low elevation; it is short
and does not flower at 4,500 feet above sea level; it grows tall and flowers at the
high elevation, much as it does at sea level. The three plants, while genetically
identical, develop different phenotypes in different environments. Elevation
(including the physical changes that accompany change in elevation, such as
atmospheric pressure, oxygen, and sunlight) influences the development of
phenotypes. Each of the seven plants is phenotypically different from the others,
which is to say that genotypes influence the development of phenotypes.

These results demonstrate that genotypes *do not fix* plant height or
flowering; rather, genotypes establish the capacity for each plant to develop
a wide range of phenotypes, a range that is characteristic of the species. Within this range of phenotypes, environmental factors exert a demonstrable
influence.

It is the range of phenotypes that can be expressed by a single genotype that
is called that genotype's norm of reaction (Figure 1.4). As you can predict from
Figure 1.3, it is relatively easy to obtain a large number of plants of the same
genotype; all that is needed is to make cuttings from a single plant, that is, to
reproduce the plant **asexually.** This method for creating large numbers of
genetically identical individuals **(cloning)** simply doesn't work with humans,
horses, or hamsters.

As it turns out, we do not know the norm of reaction for any human genotype.
Figure 1.5 illustrates what we would have to do to get this information. As you
can see from the figure, it will be a long time indeed before we know the norm
of reaction for any human genotype. And since there are nearly 6 billion human

genotypes, we cannot expect to discover the full range of phenotypes that human genes are potentially capable of initiating.

As will be discussed in Chapter 4, it has been possible to gain some information about norms of reaction of a few human genotypes. Of every 1,000 human births, 2 to 3 are **identical twins** (Figure 1.6). Identical twins arise from one fertilization event (i.e., from one egg cell fertilized by one sperm cell). Such twin pairs are called **monozygotic** (MZ) twins. It is after the fertilized egg cell divides into two cells that such twin pairs are formed, that is, once in a while the two cells of a young embryo will separate, and each will develop into an adult. The two members of a twin pair will then have identical genotypes. To the extent

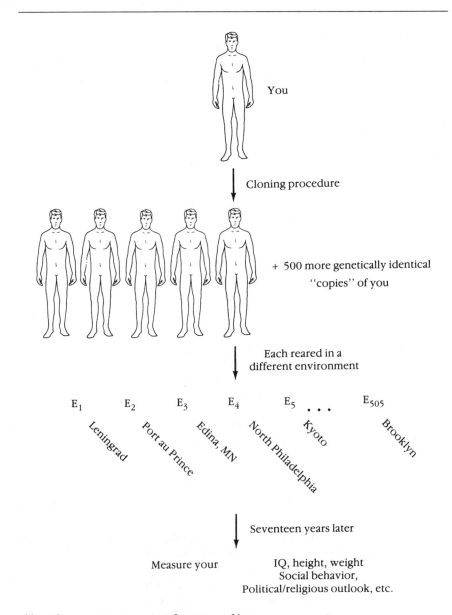

Figure 1.5 What would it take to measure norms of reaction of human genotypes?

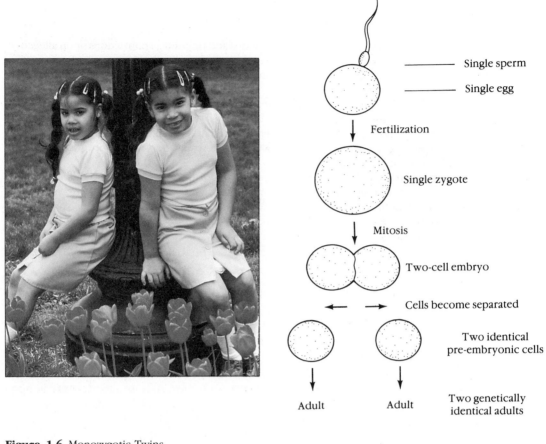

Figure 1.6 Monozygotic Twins.

that MZ twins differ in phenotype, the differences can be attributed to environmental differences experienced by them. But most MZ twins are reared together, which means that in addition to carrying identical genes they experience similar environments; a few twin pairs have been adopted into different families and reared in somewhat different environments.

The kinds of environment within which identical twin genotypes have developed have been limited to two for each genotype, and usually the two kinds of environment are similar (e.g., a similar religion, ethnic, and economic setting). Studies of MZ twins reared in different foster homes do not reveal norms of reaction of their genotypes, but their similarities and differences do provide some information about the plasticity of human phenotypes.

How Do We Know that Genes Influence Phenotypes?

Like sighting rare birds, sighting genes takes practice, and with practice sighting techniques get better. A few paragraphs back it was cautioned that the practice of drawing conclusions about genotypes from casual observations of phenotypes is a flawed method for sighting genes; it can be said of this practice that

An Aside

Biologists and nonbiologists who rationalize their religious and/or political convictions upon alleged biological facts and generalizations often take advantage of our collective ignorance of norms of reaction of human genotypes. These kinds of rationalizations have been popularized within every decade of this century, in many cases long before the results of relevant experiments had been published. For example, Harvard's Richard Herrnstein has argued that social success and failure are the outcomes of genotypes, and that therefore social classes are at least in part genetically determined (see Suggested Reading). Indeed, **eugenics** laws were enacted in most states of the Union to rid our species of "bad genes," genes alleged to cause alcoholism, sexual perversion, drug abuse, low intelligence, and antisocial behavior.

But if genes are inherited and if phenotypes develop, and if phenotypes develop differently in different environments, then cause-effect relationships among genes and human behaviors must be more complicated than these simplistic, biological determinist views would have us believe. For an exceedingly well-documented history of the misuses of genetics, see Suggested Reading for D. J. Kevles' book *In the Name of Eugenics*.

"sometimes the magic works and sometimes it doesn't." People who swear by this practice always feel certain, but they are not always correct.

The first reliable method for sighting genes was that of following the **inheritance patterns** of contrasting phenotypic traits generation after generation. However, during the first two decades of this century the study of inheritance patterns of contrasting phenotypic traits was done in parallel with the study of the physical partitioning of chromosomes into **gametes** that, upon fertilization, become zygotes. This became the first, standard criterion for identifying genes and the **methodological** foundation for the new science of genetics. But, good as the method was, genes were not observed; genes were inferred from inheritance patterns.

Curious women and men have searched for the causes of phenotypic variation for thousands of years, but not until Gregor Mendel's creative experiments was it possible to isolate specific influences of specific genes upon phenotypic variation. Mendel's experiments were published in 1865, but were neither understood nor confirmed by a second party until 1900 (recall that George Huntington described the hereditary nature of Huntington's disease in 1872, 28 years before the science of genetics "was born"). Mendel did not focus upon composite phenotypes of individual pea plants but only upon one contrasting trait at a time (e.g., tall versus short pea plants, green versus yellow pea seeds, smooth versus wrinkled pea seeds, and so on).

How did Mendel escape the fact that a range of phenotypes may emerge from a single genotype? The answer has two parts. First, Mendel studied peas within a narrow range of environments (the garden of a monastery in Central Europe). Second, the phenotypic traits selected by him happen to be stable over a wide range of environments. In other words, the norms of reaction of the contrasting phenotypes studied by Mendel did not overlap as they are shown to do in Figure 1.4. It is common knowledge that people with brown eyes have brown eyes whether they live in Helsinki or in Nairobi. It also is common knowledge that skin color darkens in the presence of ultraviolet light and that health worsens with dietary deficiencies.

What Did Mendel Do?

One thing that Mendel did that no biologist had done before was to select from a large population of plants a few that differed from the majority in a particular trait. He found a few short-stemmed pea plants within a large population of long-stemmed plants, a few with yellow seeds among the normal plants with green seeds, wrinkled seeds among mostly round seeds, and so on. In all, he studied seven such contrasting phenotypic traits (Table 1.1).

The next thing Mendel did was to **cross** plants exhibiting contrasting traits (e.g., tall plants were crossed with short plants, plants with green seed were crossed to plants with yellow seed, and so on). Then he waited for the **progeny** of each cross to develop, and he observed their phenotypes. The progeny plants were allowed to produce a second generation of progeny, and the phenotypic traits of these plants were observed. In some cases Mendel followed inheritance patterns through four or five generations. To understand the meaning of Mendel's experimental strategy for sighting genes, it is necessary to review the reproductive characteristics of peas.

Pea plants do not exist as male or female plants as dogs and cats are either male or female animals. Each pea plant is both male and female, that is, a **hermaphrodite.** Male and female sex organs are found on each plant, within each flower. The male sex organs produce pollen (male sex cells, or gametes), and the female sex organs produce eggs (female sex cells, or gametes). The petals on each flower shield the sex organs such that pollen within each flower never gets out. This ensures that fertilization events occur only between pollen grains and eggs of the same flower, a process called **self-fertilization** (Figure 1.7).

Self-fertilization in natural populations over many generations ensures that pea plants **breed true.** The progeny plants of a single parent plant will be genotypically uniform, which is to say that Mendel began his experiments with **pure lines** of peas. There was no problem of parental identity; there was no genetic variation among the progeny plants of pure-line parent plants used by him until he interfered with the natural process by creating **hybrid** plants.

Table 1.1 The seven pairs of contrasting phenotypic traits studied by Mendel

Cross	*F_1 Phenotypes*	*F_2 Phenotypes*	*F_2 Phenotypic Ratios*
1. Round × wrinkled seed	Round seed	5474 round 1850 wrinkled	2.96:1
2. Inflated × wrinkled pods	Inflated pods	882 inflated 229 wrinkled	2.95:1
3. Long × short stem	Long stem	787 long stem 277 short stem	2.84:1
4. Yellow × green seed	Yellow seed	6022 yellow 2001 green	3.01:1
5. Gray × white seedcoat	Gray seedcoat	705 gray 224 white	3.15:1
6. Axial vs terminal flowers	Axial flowers	651 axial 207 terminal	3.14:1
7. Green × yellow pods	Green pods	428 green 152 yellow	2.82:1

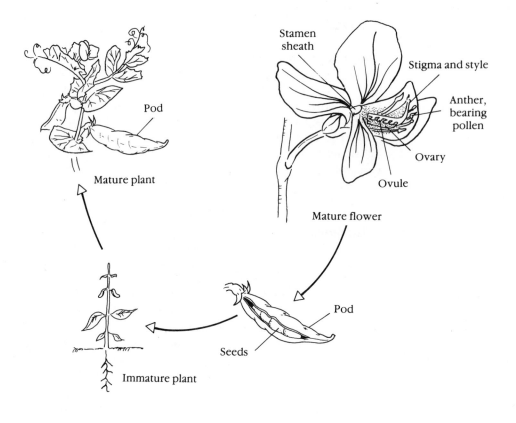

Figure 1.7 Self-fertilization of peas.

Mendel created hybrid plants by **cross-fertilization.** He opened flowers of plants from which he wished to harvest pollen; he gathered the pollen in tiny paper bags or simply placed the pollen on his thumbnail. Then he opened flowers on plants whose eggs he wished fertilized by the collected pollen, removed the male sex organs of the recipient flower, and dusted the pollen onto the female sex organs (Figure 1.8).

Mendel selected the original parent plants on the basis of their contrasting phenotypes. The progeny **(F_1)** plants always looked like one or the other of the parent plants; they were never intermediate in phenotype. For example, progeny plants from the tall-stem × short-stem crosses were tall; from the smooth seed × wrinkled seed crosses, smooth seed, and so on. However, when the F_1 plants were allowed to reproduce naturally by self-fertilization, the second generation progeny (**F_2 plants**) were of two contrasting phenotypes: some looked like one of the original parent plants (tall) and some looked like the other (short) (Table 1.1).

F_1 hybrid plants derived from tall and short parent plants were uniformly tall and indistinguishable from their tall parent plants, whether the tall parent was female or male. F_2 plants were of two types, tall and short. But what caught Mendel's attention was the fact that three fourths of the F_2 plants were tall, and one fourth was short, that is, the parental phenotypes always appeared in a ratio of $3:1$ in the F_2 generation (Figure 1.9), a *very specific inheritance pattern.*

Mendel then set out to explain the results of his breeding experiments. He postulated that each plant carries two **factors** (factors were renamed genes 40 years later) for each phenotypic trait and that one of the two factors is passed to the next generation of plants through each egg cell and one through each pollen grain. Mendel's model of inheritance is shown in Tables 1.2 and 1.3, and it is described as follows: The tall parent plants carry two copies of the T gene, both copies in the *T* form; the short parent plants carry two copies of the T gene, both in the *t* form. Both parental plants are **homozygous** for the gene that influences plant height, but the tall parent is homozygous for the *T,* and the short parent is homozygous for the *t* form of the T gene. Sex cells of tall plants carry one copy of the T gene, always in the *T* form, and sex cells of short plants carry one copy of the T gene, always in the *t* form. In short, Mendel proposed that *contrasting forms of genes explain contrasting forms of phenotypic traits.*

The fertilization events engineered by Mendel, then, united T and t sex cells into Tt zygotes, which developed into Tt plants. Plants possessing different forms of a gene are **heterozygous,** that is, all F$_1$ plants are heterozygous for the gene that influences plant height, and all are phenotypically tall. Sex cells of heterozygous plants are of two types: one half will carry the *T* form of the T gene, and one half will carry the *t* form.

Heterozygous plants carry one copy of the gene for height from each parent plant, but phenotypically they resemble only one of the parents (i.e., Tt plants are tall). Mendel introduced the terms **dominant** and **recessive** to describe the two forms of a gene. *T* is dominant to *t* in that when they reside together

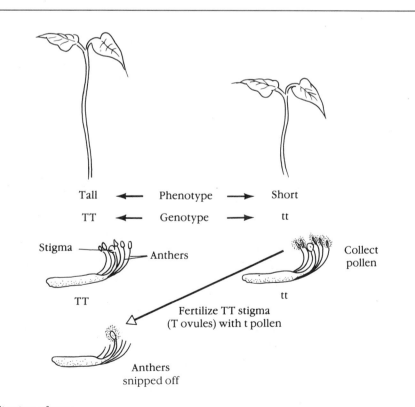

Figure 1.8 Cross-fertilization of peas.

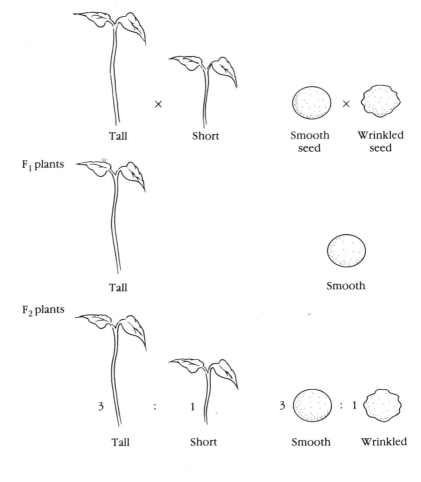

Figure 1.9 Phenotypic ratios among F_2 plants.

in the same plant, the plant is tall; the analogue of this is to describe t as being recessive to T.

Tables 1.2 and 1.3 outline in more detail Mendel's explanation of his results. Recall that the F_1 plants are heterozygous and that each egg cell and each pollen grain will carry one or the other of the two forms of the T gene. In formal terms this means that the probability of an egg cell carrying T is one half; the probability of an egg cell carrying t also is one half. The same is true of each pollen grain (i.e., the probability is one half that it will carry T and one half that it will carry t).

Therefore Mendel predicted that upon self-fertilization of F_1 plants, the probability of T eggs being fertilized by T pollen, giving rise to TT progeny, is $1/2 \times 1/2 = 1/4$; of t eggs being fertilized by t pollen, $1/2 \times 1/2 = 1/4$; and of T eggs being fertilized by t pollen (1/4) and t eggs being fertilized by T pollen (1/4), producing Tt zygotes, $1/4 + 1/4 = 1/2$. Thus the predicted genotypic ratio—TT:Tt:tt—is $1:2:1$ (Table 1.2). But since both TT and Tt plants are tall

(1/4 + 1/2 = 3/4), while only tt plants are short (1/4), the predicted phenotypic ratio is 3:1.

T and t are symbols for the two contrasting forms of the T gene; these two contrasting forms are called **alleles.** Homozygous plants carry two copies of the same allele, TT and tt, while heterozygous plants carry one copy of each allele, Tt. In all likelihood the t allele of the T gene is a mutant form of the gene. More on this later.

Table 1.2 Explanations of Mendel's results

Parent plants	Phenotype	tall × short
	Genotype	TT × tt
F$_1$ plants	Phenotype	tall
	Genotype	Tt
F$_2$ plants	Phenotype	tall and short
	Genotype	TT, Tt tt

$$\underbrace{TT, Tt}_{3/4} \quad \underbrace{tt}_{1/4}$$

(a) Phenotypic ratio ¾ Tall: ¼ short

Genotypic ratio ¼ TT:½ Tt:¼ tt

(b)

		♂ gametes	
		T	t
♀ gametes	T	¼ TT	¼ Tt
	t	¼ Tt	¼ tt

Table 1.3 A further explanation of Mendel's results

Generation	*Phenotype*	*Gametes*	*Genotype*
Parent	Tall	T	TT
	Short	t	tt
F$_1$	Tall	½ T	Tt
		½ t	
F$_2$	¾ Tall*		⅓TT
			⅔Tt
	¼ Short		tt
F$_2$ Genotypes	TT	Tt	tt
Phenotypes	Tall	Tall	Short
	↓	↓	↓
F$_3$	Tall	¾ Tall	Short
		¼ Short	

*T	T	t	
T	TT	Tt	⅓ of tall F$_2$ plants are TT
t	Tt		⅔ of tall F$_2$ plants are Tt

An Aside

More than 2,500 contrasting forms of human phenotypic traits are now known to be influenced by contrasting alleles of as many genes. In many of these cases the rule of thumb used for interpreting relationships between genes and phenotypes is first to *visualize* contrasting phenotypic traits, a simple example of which is the presence or absence of hair on the middle segment of the fingers. Second, assume that the trait is influenced by one gene; in this case call it the MDH gene that exists in at least two contrasting forms, *MDH* and *mdh*. Finally, assume that each of the contrasting forms of the trait, the presence or absence of hair, are influenced by one or the other of the two contrasting forms of the gene; in this case, *MDH* leads to the presence of mid-digital hair and *mdh* to its absence. All of these assumptions will be verified or nullified by observations of inheritance patterns within families that exhibit both forms of phenotype (Figure 1.10). Other, more significant human phenotypes are discussed below.

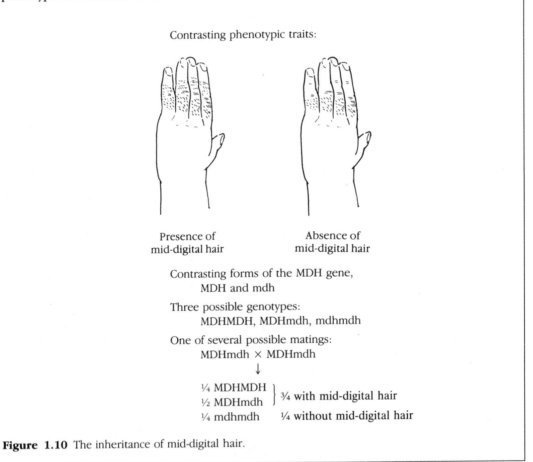

Contrasting phenotypic traits:

Presence of mid-digital hair   Absence of mid-digital hair

Contrasting forms of the MDH gene,
MDH and mdh

Three possible genotypes:
MDHMDH, MDHmdh, mdhmdh

One of several possible matings:
MDHmdh × MDHmdh
↓

¼ MDHMDH ⎱
½ MDHmdh ⎰ ¾ with mid-digital hair
¼ mdhmdh   ¼ without mid-digital hair

Figure 1.10 The inheritance of mid-digital hair.

Independent and Dependent Segregation

Mendel knew nothing about chromosomes; his breeding experiments had been terminated by the time chromosomes were discovered. However, he did suggest that genes exist in cells called **germ line** cells and that germ line cells somehow, "package" one copy of each pair of genes into each gamete. Gametes formed within germ lines that carry identical copies of one gene are always identical with respect to that gene; gametes formed within germ lines that are heterozygous at one gene **locus** will be of two types, and the two types of gamete will be produced in equal frequency.

The biological significance of independent segregation was amplified upon the discovery of chromosomes and especially upon the discovery of the fates of chromosomes during **meiosis,** the name given to the kind of cell division that gives rise to gametes (Figure 1.11a and b and Perspective 1.1). From the diagram in Figure 1.11a you see that chromosomes appear in pairs in the cells that participate in meiosis, and that only one member of each pair appears in each gamete, the end result of meiosis. Mendel did not know that *meiosis choreographs the movements of chromosomes into gametes.*

Mendel described the genetic consequences of meiosis after he discovered that each pair of alleles segregates independently of other pairs of alleles. He

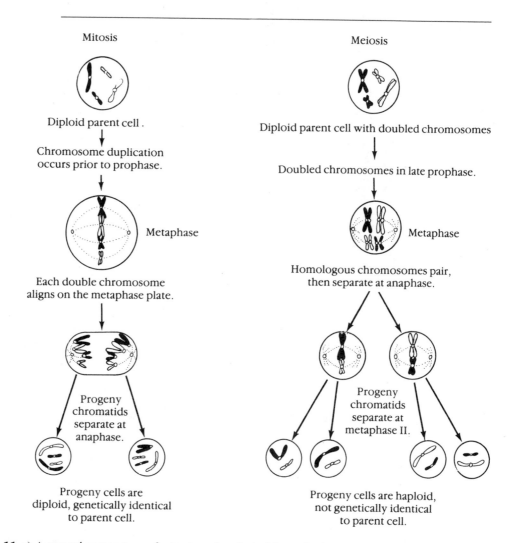

Figure 1.11 a) A general comparison of mitosis and meiosis. Mitosis leads to two cells identical to one another and to their parent cell. Meiosis is a two stage event during which four cells derive from one parent cell; however, the four cells are different from one another and from the parent cell.

(continued)

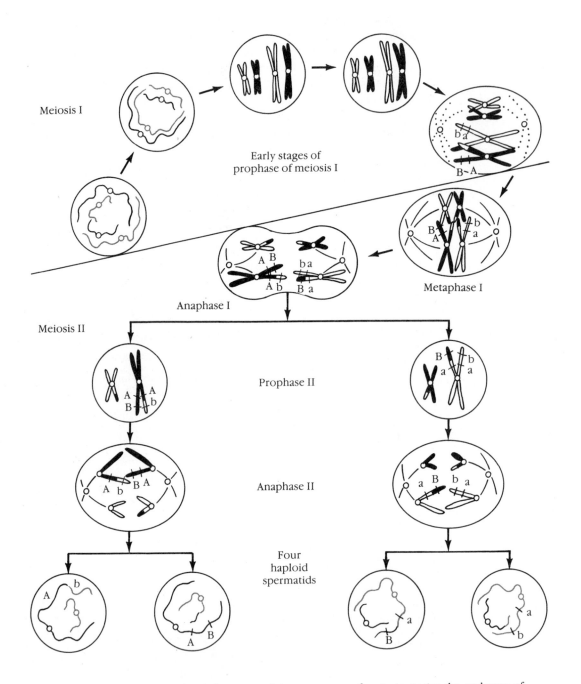

Figure 1.11 (continued) b) A detailed cartoon of the two stages of meiosis; notice the exchange of chromatid segments in Meiosis I.

made this discovery by following the inheritance patterns of two pairs of alleles within the same germ line. Parent plants with yellow and smooth seeds were crossed to parent plants with green and wrinkled seeds; the cross is symbolized as YYSS × yyss. F_1 plants had yellow/smooth seeds and were genotype YySs. F_2 plants were of four phenotypes: 9/16 had yellow/smooth seeds; 3/16 had

Perspective 1.1

If we could observe every step of the process by which a human zygote develops into a baby, and if we could focus clearly on objects as small as chromosomes inside human cells, we could become witnesses to two fascinating phenomena: the first is cell division, a process by which one cell becomes two cells; the second is cell differentiation, a process by which a cell of one type becomes a cell of a different type. The first phenomenon would show us how a single cell, a zygote, becomes a baby made up of billions of cells; the second would show us how one cell type, the zygote, gives rise to about one hundred cell types (cells as different in appearance as neurons, skin cells, muscle cells, and bone cells all carry chromosomes and genes identical to the chromosomes and genes carried by the zygote).

On close examination we would discover two kinds of cell division, one of which is called **mitosis** and the other **meiosis.** Mitosis is sometimes called *equational division* in that the progeny cells of mitosis are genetically identical (equal) to one another and to the parent cell. During mitosis each chromosome is duplicated exactly, and each progeny cell receives a copy of each chromosome. Chromosome duplication is followed by cell division. However, it is a figure of speech to say that progeny cells are identical to their parent cells, because mitosis is a process whereby *a parent cell becomes two progeny cells;* the parent cell no longer exists after the two progeny cells are formed. But still, the two progeny cells carry identical copies of the chromosomes that earlier were carried by the parent cell (Figure 1.11a).

Meiosis is a different kind of cell division. A parent cell does not become two identical progeny cells. During meiosis a parent cell, following chromosome duplication, will participate in a sequence of two cell divisions, which is to say that from one parent cell four progeny cells will emerge. The significant difference between mitosis and meiosis is that *after meiosis the progeny cells differ from one another and from the parent cell.* The most notable difference is that each progeny cell carries exactly one half the number of chromosomes carried by the parent cell.

The vast majority of contemporary geneticists who have become familiar with Mendelian genetics find it relatively easy to visualize Mendelian inheritance patterns after having learned to attach, in the mind's eye, a pair of alleles to a pair of **homologous chromosomes** and then to track the chromosomes through meiosis into gametes. Indeed, the alleles and the chromosomes form a mental-image amalgam so tight that it becomes easy to forget that genes and chromosomes were discovered by very different research strategies, by scientists who were searching for different kinds of answers to different kinds of questions. Mendel figured out the genetic consequences of meiosis without ever knowing that chromosomes exist or how gametes are formed; and later biologists, using microscopes, discovered chromosomes without having the foggiest idea about genes.

Using the terminology of human biology as opposed to pea biology, the parent cells of meiotic cell divisions give rise to egg cells in females and to sperm cells in males. Egg cells are progeny cells of cells called primary oocytes, and sperm cells are progeny cells of primary spermatocytes. Primary oocytes and primary spermatocytes have 46 chromosomes in their nuclei. These cells are **diploid** in that they contain two sets of chromosomes. Egg and sperm cells have 23 chromosomes in their nuclei and are called **haploid** cells. If you think about a complete human life cycle, the diploid-haploid cycle is easy to remember. All the cells in your body except the sex cells (egg and sperm) are diploid (two sets of chromosomes, for a total of 46). Sex cells are haploid (one set of chromosomes, for a total of 23). At the time an egg cell and a sperm cell unite to form a zygote, their combined chromosomes (23 each) restore the diploid number (46). By way of mitotic cell divisions zygotes become adults, ensuring that each of the billions of progeny cells carry 46 chromosomes, identical to those carried by the zygote, except for the sex cells, the end products of meiotic cell divisions: 46 to 23 via meiosis; 23 + 23 to 46 via fertilization; 46 to 46 via mitosis.

Meiosis includes a sequence of two cell divisions (see Figure 1.11b). The first of these, called Meiosis I, or simply M-I, begins with a primary oocyte or a primary spermatocyte; the events of M-I are about the same for both. At the beginning of the process each chromosome is a double structure, that is, each chromosome is composed of two **chromatids.** As the chromosomes

(continued)

prepare for the first cell division by aligning themselves in an orderly way along a plane at the center of the cell, homologous chromosomes line up alongside one another, one on each side of the center plane. While homologous chromosomes are so aligned, chromatid exchanges occur. Then, as the **homologues** begin to move in opposite directions, toward opposite sides of the cell, each will carry with it chromatid segments of the other. After the two sets of chromosomes have moved apart, the primary oocyte, or primary spermatocyte, begins to divide into two cells, each with 23 chromosomes.

These two cells are called, in one case, secondary oocytes and in the other case secondary spermatocytes. These are the cells that will initiate the second set of cell divisions, Meiosis II, or M-II. But M-II is different in the two sexes. In females one secondary oocyte is larger than the other. The small secondary oocyte is called a polar body, and when it divides it gives rise to two additional polar bodies, neither of which becomes an egg cell. Division of the larger secondary oocyte gives rise to one large egg cell and one small polar body. In short, meiosis gives rise to four haploid cells, one egg cell, and three nonfunctional polar bodies. The egg cell carries 23 chromosomes in its nucleus, exactly half the number found in the primary oocyte.

Secondary spermatocytes are equal in size. After each of these cells divide (i.e., after M-II) four **spermatids,** of equal size, each with 23 chromosomes, will have emerged from these meiotic events. The four spermatids soon undergo developmental changes that transform them into sperm cells.

To understand Mendelian genetics in the terms dictated by the choreography of chromosomes as they wend their ways into gametes, we begin with one gene and its two alleles and one pair of homologous chromosomes. The first point is to illustrate the chromosomal meaning of first-division segregation, that is, the segregation of alleles during M-I (Figure 1.11b).

The *A* allele of the A gene is located on one homologue, and the *a* allele is located on the other. Each homologue is composed of two chromatids, and on each chromosome the chromatids are identical. Therefore homologue 1 carries two *A* alleles, and homologue 2 carries two *a* alleles. As the two homologues pull apart and move toward opposite sides of the cell,

homologue 1 will carry two *A* alleles to one end and eventually into one secondary spermatocyte and homologue 2 will carry two *a* alleles to the other secondary spermatocyte. One secondary spermatocyte will give rise to two sperm cells, each carrying *a;* the other secondary spermatocyte will give rise to two sperm cells, each carrying *A.* Beginning with a heterozygous primary spermatocyte, the meiotic event leads to the segregation of *A* from *a* during M-I, and to two kinds of sex cell, *A* and *a,* in equal numbers.

The second point is to illustrate that an exchange of chromatid segments between homologous chromosomes delays the segregation of some alleles until M-II. Now we trace the same ground, through M-I and M-II, but with two genes on each homologue: *A* and *B* on homologue 1 and *a* and *b* on homologue 2. This time, as the homologues align themselves along the equatorial plane of the cell, a chromatid exchange occurs between two of the four chromatids and between genes A and B. As the homologues move apart after the exchange, homologue 1 will carry two *A* alleles but one *B* and one *b* allele; that is, after M-I, *A* will have segregated from *a* but *B* will not have segregated from *b.*

Homologue 2 will carry two *a* alleles and, like homologue 1, one *B* and one *b.* During M-II, the two chromatids of homologue 1 will become chromosomes, one of which will carry *A* and *B* and the other *A* and *b.* Likewise, homologue 2 will give rise to two chromosomes, one carrying *a* and *B,* the other *a* and *b.* The A gene alleles segregate during M-I, and the B gene alleles segregate at M-II. Mendel wrote part of the script when he said that individuals heterozygous for two genes will produce four kinds of gametes: AB, Ab, aB and ab (peas and humans are alike in this case); he didn't have a clue, however, about either meiosis, or genes located side by side on the same chromosome, or first- versus second-division segregation of alleles.

First- and second-division segregation of alleles cannot be demonstrated in humans, as shown here, but the phenomena have been demonstrated in organisms whose haploid cells are produced such that the geneticist can follow the process, step by step. No one doubts that the genetic outcomes of meiosis are pretty much the same in all organisms in which meiosis occurs.

yellow/wrinkled seeds; 3/16 had green/smooth seeds; and 1/16 had green/wrinkled seeds, producing a phenotypic ratio of 9:3:3:1 (Figure 1.12).

With one eye on Figure 1.12, then, and one on the text, notice among the F_2 progeny of **dihybrid** F_1 plants, that 3/4 are yellow seed and 1/4 are green seed and that 3/4 are smooth seed and 1/4 are wrinkled seed, the same ratios observed among the progeny of **monohybrid** F_1 plants (Table 1.1). In other words, Mendel observed that the two allelic pairs segregate in exactly the same way, whether they segregate in two different germ lines of monohybrids or in the same germ line of a dihybrid.

There are two things going on here: one is meiosis, the process overseeing half-chromosomes into gametes (Figure 1.11b), and the other is the fact that the

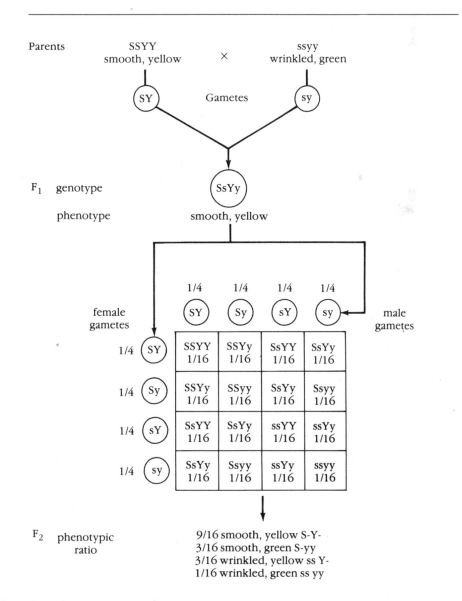

Figure 1.12 The independent segregation of two genes.

gamete into which an allele, or a chromatid, ultimately finds itself is a matter of chance. It may be easier to grasp the meaning of chance if you simulate segregation by flipping pennies into the air (Figure 1.13). Flip a penny 50 times, and count the number of times it lands heads. If the penny is not "loaded," the number will be close to 25. By chance the penny will land heads half of the times it is flipped. Now, flip two pennies at the same time 100 times, and count the number of times both land heads, both land tails, and one is heads and one is tails. The numbers will be close to 25, 25, and 50. Next, if you paint one penny red and the other blue, and then flip them together, you will find that the red

a) Pennies, alleles, and chance:

Flip one penny: $P^* = \frac{1}{2}$ that it will land heads (H)
                  $P = \frac{1}{2}$ that it will land tails (T)

Flip two pennies: $\frac{1}{2}$ H × $\frac{1}{2}$ H = $\frac{1}{4}$ HH
                  $\frac{1}{2}$ H × $\frac{1}{2}$ T = $\frac{1}{4}$ HT $\Big\}$ $\frac{1}{4} + \frac{1}{4} = \frac{1}{2}$ HT
                  $\frac{1}{2}$ T × $\frac{1}{2}$ H = $\frac{1}{4}$ HT
                  $\frac{1}{2}$ T × $\frac{1}{2}$ T = $\frac{1}{4}$ TT

A heterozygous germ line cell is Ss:
                  $P = \frac{1}{2}$ that the next gamete will be S
                  $P = \frac{1}{2}$ that the next gamete will be s

The cross, Ss × Ss, then, will produce
                  $\frac{1}{2}$ S × $\frac{1}{2}$ S = $\frac{1}{4}$ SS
                  $\frac{1}{2}$ S × $\frac{1}{2}$ s = $\frac{1}{4}$ Ss $\Big\}$ $\frac{1}{4} + \frac{1}{4} = \frac{1}{2}$ Ss
                  $\frac{1}{2}$ s × $\frac{1}{2}$ S = $\frac{1}{4}$ Ss
                  $\frac{1}{2}$ s × $\frac{1}{2}$ s = $\frac{1}{4}$ ss

b) Two genes and four alleles:

A heterozygous germ line cell is SsYy; the probability
for each kind of gamete will be:
                  $\frac{1}{2}$ S × $\frac{1}{2}$ Y = $\frac{1}{4}$ SY
                  $\frac{1}{2}$ S × $\frac{1}{2}$ y = $\frac{1}{4}$ Sy
                  $\frac{1}{2}$ s × $\frac{1}{2}$ Y = $\frac{1}{4}$ sY
                  $\frac{1}{2}$ s × $\frac{1}{2}$ y = $\frac{1}{4}$ sy

The cross, SsYy × SsYy, then, will produce

	¼ SY	¼ Sy	¼ sY	¼ sy
¼ SY	¹⁄₁₆ SSYY	¹⁄₁₆ SSYy	¹⁄₁₆ SsYY	¹⁄₁₆ SsYy
¼ Sy	¹⁄₁₆ SSYy	¹⁄₁₆ SSyy	¹⁄₁₆ SsYy	¹⁄₁₆ Ssyy
¼ sY	¹⁄₁₆ SsYY	¹⁄₁₆ SsYy	¹⁄₁₆ ssYY	¹⁄₁₆ ssYy
¼ sy	¹⁄₁₆ SsYy	¹⁄₁₆ Ssyy	¹⁄₁₆ ssYy	¹⁄₁₆ ssyy

Phenotypic ratio   Genotypic ratio
S-Y- = ⁹⁄₁₆   SSYY ¹⁄₁₆   SsYY ²⁄₁₆   ssYY ¹⁄₁₆
S-yy = ³⁄₁₆   SSYy ²⁄₁₆   SsYy ⁴⁄₁₆   ssYy ²⁄₁₆
ssY- = ³⁄₁₆   SSyy ¹⁄₁₆   Ssyy ²⁄₁₆   ssyy ¹⁄₁₆
ssyy = ¹⁄₁₆   1:2:1:2:4:2:1:2:1
9:3:3:1

*P = probability

Figure 1.13 A probabilistic approach to phenotypic and genotypic ratios.

penny falls heads about half the time and that the blue penny falls heads about half the time. In other words, the pennies behave the same way independently of whether they are tossed one at a time or both at the same time.

The rules explaining these head-tail observations are called the **laws of chance,** one part of which states that *the probability of two independent events happening together is the product of their separate probabilities.* Thus the probability of one penny landing heads is 1/2; the probability of two pennies at the same time landing heads is $1/2 \times 1/2 = 1/4$. By the same reasoning, if the probability of one monohybrid F_2 plant having yellow seeds is 3/4 and the probability of another monohybrid F_2 plant having smooth seeds is 3/4, then the probability of a dihybrid F_2 plant having yellow/smooth seeds is $3/4 \times 3/4 = 9/16$.

Armed with this statistical logic, the $9:3:3:1$ ratio is easy to understand. In monohybrid F_1 plants, ratios of $3:1$ are observed for both contrasting forms of phenotype, yellow versus green and smooth versus wrinkled. From this we can calculate the probability of all phenotypic classes among F_2 progeny: 3/4 yellow $\times$ 3/4 smooth = 9/16 yellow/smooth; 3/4 yellow $\times$ 1/4 wrinkled = 3/16 yellow/wrinkled; 1/4 green $\times$ 3/4 smooth = 3/16 green/smooth; and 1/4 green $\times$ 1/4 wrinkled = 1/16 green/wrinkled, a phenotypic ratio of $9:3:3:1$ (Figures 1.13 and 1.14).

There is another way to visualize relationships among ratios of F_2 phenotypes, genotypes, and the laws of chance. In the examples above, F_1 plants are heterozygous at two gene loci, S and Y. Their genotype is SsYy. In both the female and the male germ line cells, four types of gamete are produced: SY, Sy, sY, and sy (notice that half the gametes carry S and half carry s; and that half carry Y and half y). The four types of gamete are produced in equal frequency (i.e., 1/4 are SY, 1/4 Sy, 1/4 sY, and 1/4 sy). Now, if 1/4 of the egg cells are SY and 1/4 of the pollen grains are SY, then the probability of an SSYY F_2 plant is $1/4 \times 1/4 = 1/16$. From Figure 1.13 you can calculate the probabilities of all F_2 genotypes and phenotypes.

Mendel observed that each of the seven pairs of alleles segregate independently of all the others, which is to say that every gamete is "packaged" with genes as predicted from the laws of chance. Mendel was correct as far as his experiments took him, but if he had studied an additional seven pairs of alleles he would have discovered exceptions to his "law" of independent segregation. In short, *the segregation of alleles observed by Mendel can be explained by the segregation of homologous chromosomes during meiosis.*

Peas have seven pairs of chromosomes in their germ line cells, but surely there are more than seven pairs of pea genes. In each human cell there are 23 pairs of chromosomes and at least 100,000 pairs of genes. While Mendel couldn't have known it, today it is common knowledge that each chromosome carries hundreds and possibly thousands of genes. This fact suggests that Mendel's "law" of independent segregation, while true for gene pairs located on different chromosomes, is not true for pairs of genes located close together on the same chromosome. Genes located adjacent to one another on the same chromosome will segregate together, that is, dependently, not independently (Figure 1.14). In other words, the segregation of one pair of alleles will depend on the segregation of an adjacent pair but will segregate independently of a pair of alleles located on a different chromosome.

Dependent segregation and chromosome mapping is explained here with hypothetical genes and chromosomes. Imagine one pair of chromosomes carrying two pairs of alleles (Figure 1.14 and 1.11b). The genes are called A and

---

Perspective 1.2

Most geneticists agree that the Mendelian method of studying genes is "old fashioned" and that mapping genes on chromosomes via independent and dependent segregation is hard work. Today there are different ways to accomplish the same tasks.

On each chromosome there is a sequence of genes. One way to determine this sequence is by way of Mendelianlike breeding experiments—difficult to do and difficult to teach. Even so, the information gained thereby is central to genetics, and that information has had a profound influence upon the kinds of scientists who became attracted to genetics. Many geneticists were educated in mathematics, physics, and chemistry, and many of them expressed the feeling that the major problems of heredity will ultimately yield to the exact and precise scientific methods of these disciplines. If true, some of them promised, genetics will be transformed from an art into a science.

Today genetics is a science, but in a much broader way than was predicted from the Mendelian genetics vantage point. As will be discussed in the next chapter, knowledge of the gene was transformed from a Mendelian abstrac-tion to a knowable chemical molecule—a molecule that can now be propagated super fast, without having to wait for cell division.

Further evidence of the importance of locating the positions of genes on chromosomes comes from the fact that today one of the biggest international projects in the history of biology is to map every gene on every human chromosome. This proposed mapping project goes even further, and that is to determine the sequence of chemical molecules within genes that encode the information carried by them (discussed in Chapter 2). There are 3 billion of these encoding molecules along the lengths of the 23 human chromosomes. When this project was first proposed, humorists within the genetics community quipped that the gigantic mapping project would fall flat because geneticists would find the task too boring. One geneticist proposed giving the project to a penal colony, implying that interest in sequencing chromosomes is on a par with interest in breaking rocks. But boring or not, a major fraction of genetics research has been, and will continue to be, that of identifying genes by their locations on chromosomes.

B, and the allelic pairs *A* and *a* and *B* and *b*. If the two dominant alleles are positioned on one homologue and the two recessive alleles on the other, then as the homologous chromosomes segregate toward opposite poles of the cell, one homologue will carry the two dominant alleles to one new cell, and the other homologue will carry the two recessive alleles to the other. In other words, the dominant alleles will appear together in half the gametes, and the recessive alleles will appear together in the other half. This simplistic diagram suggests that all of the genes on a single chromosome are inherited as a block. We know this isn't true because with appropriate breeding experiments it can be shown that blocks of genes are broken up during meiosis.

The classic strategy for observing how blocks of genes break up is called the **test-cross** (Figure 1.15). Test-crosses are designed to reveal *the kinds and the frequencies of gametes produced by an organism whose genotype is unknown.*

To illustrate, recall that SsYy pea plants produce four kinds of gametes—SY, Sy, sY and sy—in a ratio of 1:1:1:1 in both female and male germ lines. Proof of this statement is found in the test-cross SsYy × ssyy (Figure 1.15). You see in the diagram of the cross and from the kinds and ratios of gametes produced by each parent plant that the kinds and ratios of progeny resulting from the cross is a mirror image of the kinds and ratios of gametes produced by the SsYy parent. These test-cross results provide further evidence that the segregation of *S* from *s* is independent of the segregation of *Y* from *y*.

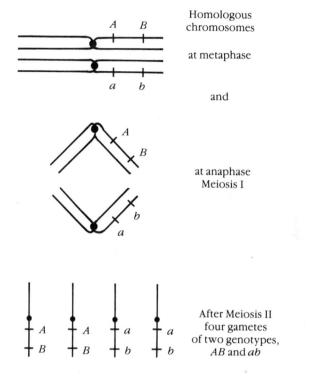

Figure 1.14 Dependent segregation of linked genes.

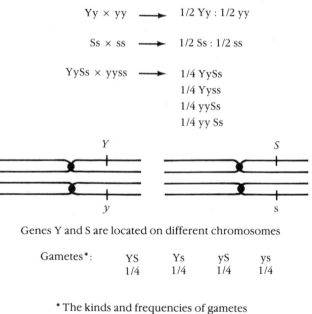

Genes Y and S are located on different chromosomes

Gametes*:
	YS	Ys	yS	ys
	1/4	1/4	1/4	1/4

* The kinds and frequencies of gametes
produced by heterozygous parents
determine the kinds and frequencies
of progeny resulting from test-crosses.

Figure 1.15 A test-cross for genotype identification.

The test-cross: AaBb × aabb

Case 1.   Gametes expected if A and B segregate independently:
¼ AB
¼ Ab ⎤
¼ aB ⎦   50% recombinant types (see Figure 1.12)
¼ ab

Case 2.   Gametes expected if A and B are closely linked:
½ AB
½ ab   no recombinant types (see Figure 1.14)

Case 3.   Gametes expected if chromatid exchanges occur between A and B and A
is 10 map units from B
.45 AB
.05 Ab ⎤
.05 aB ⎦   10% recombinant types (see Figure 1.21)
.45 ab

10% recombinant types = 10 map units

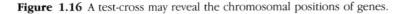

Figure 1.16 A test-cross may reveal the chromosomal positions of genes.

Getting back to the small block of hypothetical genes A and B, the analogous test-cross is AaBb × aabb (Figure 1.16). If the two allelic pairs segregate independently (Case 1), the AaBb parent should produce the following kinds of gamete, AB, Ab, aB, and ab, in a ratio of 1:1:1:1. On the other hand, if the dominant alleles are inherited as one block and the recessive alleles as another block (Case 2), the AaBb parent should produce only two kinds of gamete, AB and ab, in a ratio of 1:1.

Another possibility is that the test-cross progeny are the types and frequencies shown in Case 3 in Figure 1.16, that is, 45% of the plants are phenotypically identical to the AaBb parent, 45% are phenotypically identical to the aabb parent, and 10% of the plants exhibit a recombination of parental phenotypes — neither a 1:1:1:1 nor a 1:1 ratio. The two pairs of alleles do not segregate independently, and they are not inherited as blocks. The chromosomal mechanism that breaks up blocks of genes does not free them to segregate independently. The result of this mechanism is the **recombination of linked** genes. (Genes are said to be linked if they reside on the same chromosome; events that break these linkage arrangements are called recombination events.) The observation of recombination among linked genes is the distinguishing feature of dependent segregation (Figure 1.14).

The chromosomal mechanisms that lead to recombination within blocks of genes were first observed as homologous chromosomes physically exchanging chromatid segments during gamete formation (Figures 1.11b and 1.17). If you follow Figure 1.17 from prior to the first stage of meiosis through the second stage of meiosis, you will see that sometimes both allelic pairs segregate during the first stage (M-I) and that other times one pair will segregate during M-I and the other during M-II.

These observations are explained by the two facts discussed in Perspective 1.2, namely that each homologue is made up of two identical chromatids and that at certain locations along the length of homologous chromatids there are physical exchanges of rather long segments. (Assume that the physical exchanges between homologous chromatids are reciprocal such that each chromatid will carry only one copy of each gene. While reciprocal exchange is the rule, we now know that there are exceptions to the rule, but these are too technical for the discussion here.)

Upon the discovery of these facts and after many naturally appearing mutant forms of genes had been discovered in many plant and animal species, geneticists began to construct **chromosome maps.** In the hypothetical case of the A and B genes, the percentage of progeny showing a chromatid exchange between A and B is extrapolated into a chromosomal map distance between them. For example, 200 of 2,000 progeny (10%) show recombination between A and B. Ten percent recombination is translated literally into a distance on the chromosome of 10 map units (mu). This simple example describes a first step toward constructing a chromosome map (Table 1.4).

The rationale for translating recombination percentages into chromosome distances is based upon evidence that exchanges between homologous chromatids may occur at nearly any point along their length, with nearly equal probability of occurrence at each point. To the extent that this is true, the chance

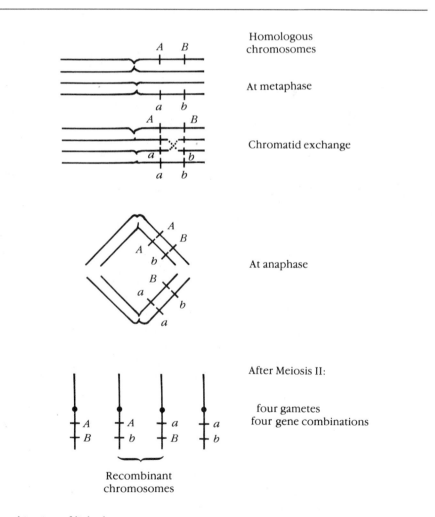

Figure 1.17 The recombination of linked genes.

Table 1.4 Using recombination frequencies to map chromosomes

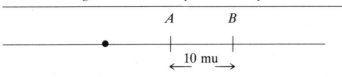

Testcross: AaBb × aabb

Progeny	Number	Percent	Type
AaBb	45	45	parental
Aabb	5	5	recombinant
aaBb	5	5	recombinant
aabb	45	45	parental
Total	100	100	

Summary: 90% of the progeny are parental types, and
           10% are recombinant types; the
           percentage of recombinant types is converted into map units
           on the chromosome.
           Therefore A is located 10 mu from B.

of an exchange occurring between two genes will be a function of the distance between them. It is now known that relationships between recombination percentage and chromosome distance are more complicated than described here; nevertheless, gene loci can be positioned along the lengths of chromosomes, and relative distances between them can be estimated by recombination analyses of the general type described here.

Recombination studies, then, reveal the following: Genetic differences among individuals within populations are maintained by *the segregation of homologous chromosomes during gamete formation* and by *recombining segments of homologous chromatids;* that is, each of us inherits 23 chromosomes from each of our two parents. Thus in each of our own germ line cells there are 23 pairs of chromosomes, and in each of our sex cells there are 23 single chromosomes. Each of the single chromosomes in our sex cells is made up of segments of a pair of parental chromosomes, which is to say that our sex cells not only carry about half of our maternal and half of our paternal chromosomes, but also that each chromosome is made up of maternal and paternal segments. These two kinds of recombination literally create astronomically large numbers of genotypes, so many, in fact, as to preclude the possibility of any two fertilized egg cells ever carrying the same alleles of all their genes.

Human Genes

While the processes of gamete formation and recombination discussed thus far are more alike than different in peas and humans, this book is about human heredity and human societies, not about peas. Today we know that human genes obey the same laws of inheritance, that human genes are located in linear orders on chromosomes, and that human genes influence human phenotypes. In certain basic ways we are very much like peas; it has been a difficult task to demonstrate this resemblance, but in the process of doing so, a number of differences between people and peas have been discovered (e.g., peas do not object to geneticists choosing their mates; people do).

Imagine that you had just read Mendel's paper, in Fall 1900, and that his results had caused you to wonder whether genes exist in humans. How would you have tried to find out? There are no pure lines of humans, as there are of peas. You can't make artificial crosses between different types of humans as you can with peas, at least not without getting into trouble. And if pure lines of humans did exist and if you could persuade members of each, and their children, to reproduce as you would want, to test Mendel's laws of inheritance, you probably wouldn't live long enough to observe the adult phenotypes of the F_2 generation!

It turns out that the first methods used to identify human genes, their effects upon human phenotypes, and their locations on chromosomes, were different from those used by Mendel to study pea genes. As a result, our knowledge of human genetics lagged behind our knowledge of genetics of other species. Nevertheless, during the first decade of this century it was postulated that Mendelian genes can explain what the physician Archibald Garrod called in 1903 **inborn errors of metabolism.** Like Mendel, Garrod was four decades ahead of his scientific colleagues.

Alcaptonuria

Garrod's primary knowledge was in the field of medicine. But he studied genetics and biochemistry at the suggestion of William Bateson, a famous geneticist in England who also coined the name genetics for this new science. Garrod observed many London children with a disease called alcaptonuria. Upon observing the parents and siblings of alcaptonuric children, he concluded first that the disease is initiated by a homozygous recessive genotype, and second that the symptoms of the disease are caused by an accumulation in the urine of a chemical compound that can be **digested** by homozygous dominant and heterozygous people but not by homozygous recessive alcaptonurics. The undigested molecules (of homogentisic acid) first accumulate in body cells, but eventually leave the body via the urine, which turns black as soon as it is excreted and exposed to atmospheric oxygen. (People who expect urine to be pale yellow in color are somewhat disquieted by the sight of black urine, but the black urine itself causes no physical damage.)

About 60% of the alcaptonuric children observed by Garrod were born of parents who were first cousins, and in the families he studied the parents were free of the disease. He suggested that the parents were heterozygous by the fact that their phenotypes were identical to those of the general population of nonalcaptonuric people and that they differed from most nonalcaptonuric adults by having one or more alcaptonuric children. He judged that the high frequency of alcaptonuric children among first-cousin matings is best explained by the fact that rare genes tend to "run in families," thereby increasing the frequency of related heterozygotes, the frequency of matings between heterozygotes, and hence the frequency of homozygous-recessive children. Garrod was correct in all of his suggestions (Figure 1.18).

Yes, the methods for doing human genetics were and still are different from those used by Mendel. But reflect on the similarities between Mendel and

Garrod's conclusions:

Parents of alcaptonuric children are nonalcaptonuric; therefore, their genotype is Aa. Parents of alcaptonuric children are often first cousins; therefore, the mating is Aa × Aa. Inheritance of alcaptonuria can be symbolized as follows (also see Figure 1.19):

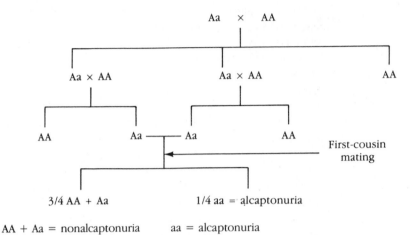

AA + Aa = nonalcaptonuria   aa = alcaptonuria

Since Aa individuals are rare in the general population, most will mate with AA individuals, unless they mate with relatives; the *a* allele may be more frequent within families than in the general population.

Figure 1.18 Garrod's analysis of the inheritance of alcaptonuria, 1900–1909.

Garrod's selection of individuals to be studied. Mendel searched populations of peas for a few plants that differed from the majority in particular ways (e.g., the majority of pea plants were tall, but Mendel found a few short plants). Garrod wasn't looking for rare humans for use in mating experiments; nevertheless he recognized that the vast majority of humans are nonalcaptonuric and that only a few are alcaptonuric. In both cases the key was to follow the inheritance patterns of contrasting forms of the same phenotypic traits and then to infer a causative agent, namely, contrasting forms of a gene—a different form of the gene for each form of the trait.

Garrod introduced what came to be a popular method for studying Mendelian genes within human families, that is, by examining **family pedigrees.** The standard way of summarizing family pedigrees is by **pedigree charts,** from which *it is relatively easy to compare individuals within generations, between generations, and of different families within extended families* (Figure 1.19).

The hypothetical pedigree chart (Figure 1.19) shows an inheritance pattern of the two forms of the gene associated with alcaptonuria, as it was explained by Garrod: two nonalcaptonuric individuals may have an alcaptonuric child *only if alcaptonuria is initiated by a homozygous recessive genotype (aa) and only if both parents are heterozygous (Aa).*

Sickle Cell Anemia

Sickle cell anemia (SCA) is a disease that can be described from many different vantage points. It is a genetic disease; it is one of many kinds of anemia, diseases that result from oxygen deprivation; it is a disease resulting from a mutant form

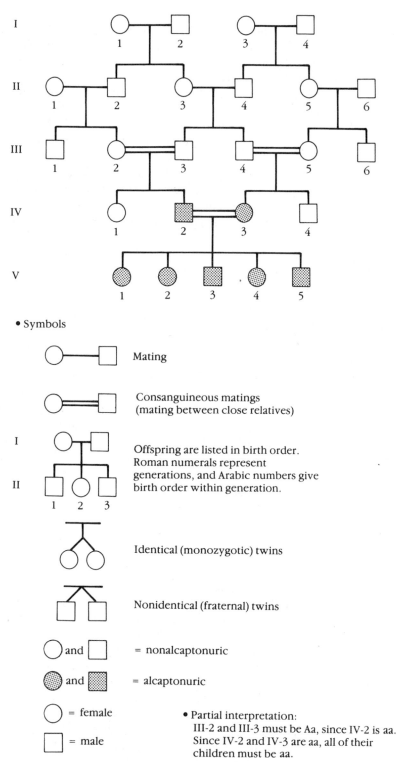

Figure 1.19 A hypothetical pedigree chart illustrating inheritance of alcaptonuria.

of **hemoglobin,** the protein of red blood cells that carries oxygen from the lungs to the cells of the body. The mutant form of hemoglobin gives rise to a mutant shape of the red blood cells, shapes so irregular that they stop the flow of blood through the tiny capillaries between the arteries and the veins. Children with the disease experience nearly every kind of biological disorder imaginable, disorders that shorten the life span by several decades. SCA is notorious in the United States because most of the people with the disease have African ancestry. Very briefly, the allele that causes SCA has had a beneficial effect upon heterozygous individuals with populations of people who have been exposed for centuries to malaria; heterozygous individuals are more resistant to falciparum malaria that are homozygous individuals. The gene is essentially nonexistent in populations that have not been exposed to malaria. We will return to this disease in Chapters 2 and 4, where we will discuss the chemical properties of genes and how genes influence the chemistry of cells and where we will discuss the genetic composition of gene pools. Our focus on SCA in this chapter is restricted to its genetics base.

In addition to pedigree charts, there are other ways of identifying Mendelian genes in human populations. Consider the following real but simplified example: One hundred SCA children are in the care of a medical genetics unit of a large city hospital. The medical geneticist in the group makes a survey of the 100 families of the SCA children, which includes 100 sets of parents and 300 non-SCA **siblings** (sibs) of the SCA children.

The geneticist chooses the symbol S for the SCA gene, and the following genotype-phenotype symbols: SS genotypes are correlated with normal red blood cells and healthy phenotypes; Ss genotypes with normal health and two kinds of red blood cells, normal and sickled, a phenotype called sickle cell trait (SCT); and ss genotypes with sickled red blood cells and very ill health, SCA. She chooses these symbols to match the expectations of Mendel's laws (i.e., none of the parents is SCA, and about 3/4 of their children are non-SCA). If she is correct it must mean that every parent of an SCA child is phenotypically SCT and genotypically Ss, and that about 2/3 of the non-SCA progeny are SCT and Ss. Upon examination of the red blood cells of the 200 parents and their 300 non-SCA children, the results shown in Figure 1.20 were observed.

These results point to a flaw in the Mendelian model of the gene. Recall that Mendel postulated that one of each pair of alleles is dominant over its recessive allele. Recessive alleles were so named because they appeared to be phenotypically silent in heterozygous individuals, which is to say that heterozygous individuals appeared to be phenotypically identical to individuals homozygous for the dominant allele. But as we see in the case of the S gene and its alleles, SS and Ss individuals are not phenotypically identical. SS individuals have normal red blood cells and Ss individuals have both normal and sickled red blood cells. This must mean that both alleles are active within heterozygous individuals, each giving rise to a different kind of red cell. Obviously enough normal red cells are produced to ensure normal health for SCT individuals, but it isn't true that one allele is dominant over the other.

In both the alcaptonuria and sickle cell anemia examples, the **data** were treated in such a way as to conform with the expectations generated by Mendel's model of inheritance, which, incidentally, has served as well for human gene-phenotype relationships as for pea gene-phenotype relationships. But could Garrod have constructed a Mendelian model of alcaptonuria inheritance if he had not known Mendel's model beforehand? Would the medical geneticist

The observation:
    100 SCA children (ss genotype)
    100 sets of parents, all non-SCA (Ss genotype)
    300 siblings of the SCA children, all non-SCA (SS or Ss genotype)

Sample Pedigree:

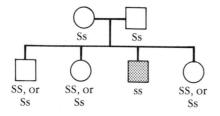

Genotypes of the siblings:
    phenoytpe: non-SCA
    genotype: SS (⅓) or Ss (⅔)

Problem of genetics counseling:
A female from one of the 100 families and a male from another wish to have children. What is the probability that their first child will be SCA?

a)   For the child to be SCA, both must be Ss; the probability of this is
     ⅔ × ⅔ = 4/9

b)   If they are heterozygous, the probability that their first child will be ss is ¼.

c)   4/9 × ¼ = 4/36 = 1/9

Figure 1.20 Sickle cell anemia inheritance.

---

have analyzed the SCA children and their families in the way she did, in the absence of knowledge of Mendelian genes? Probably not, but of course we'll never know.

From Families to Populations

It has been estimated that between 150 and 250 million people worldwide have one or another genetic disease. This is some 3% to 5% of the world population. If this estimate is correct, then hundreds of millions more people are **carriers** of genetic diseases. This point is easily illustrated with the disease SCA.

In the United States about 1 of every 570 babies born to parents of African descent has SCA. The genotypes of SCA babies are ss, and both parents of an SCA baby are genotype Ss. After it became possible to determine by blood examination the genotypes of SS, Ss, and ss individuals, it was discovered that about 1 of every 12 (1/12) Americans of African descent is SCT, genotype Ss (i.e., a carrier). The majority (11/12) of African Americans are genotype SS. Now, consider another aspect of human genetics. Most people do not check the genotypes of their prospective mates prior to having children. Therefore matings are random with respect to most genotypes, including SS and Ss (SCA persons, ss genotypes, rarely live long enough or are healthy enough to have children).

There are, then, within the African American population SS × SS, SS × Ss, Ss × SS, and Ss × Ss matings. Since 11/12 of African Americans are SS and 1/12

An Aside

Family pedigrees have been kept for centuries. Garrod wasn't the first to construct one. As a matter of fact, hereditary diseases have been known at least since the time people began to record their history. In some cases the telltale signs of a "family disease" were interpreted much as Garrod interpreted the alcaptonuric pedigrees. In other cases the telltale signs were not noticed. But people have always tried to explain the observations they make of the world around them. The point for studying history is to understand that contemporary explanations may be only slightly better than the older explanations, but surely they are not absolutely correct. As our predecessors did, we make only tiny contributions to the quality of our explanations, but we aren't more intelligent than they were, and our explanations aren't perfect. However, because of their efforts we have more information with which to fine tune our explanations.

Family pedigrees, however, often record money, positions of political power, royal privileges, and social status. For example, some families appear to be accident prone. Citizens in the United States are reminded every November of the assassination of John F. Kennedy, of his two brothers who died in accidents, and of his disabled sister. Were the parents, Rose and Joseph Kennedy, heterozygous for an accident-prone gene?

While such a question appears silly when viewed in one light, it may serve as a warning if viewed in a different light. It is possible to construct fake pedigree charts of accident proneness; as a matter of fact, dreamers, mystics, cranks and con artists have drawn up fake pedigree charts to prove the genetic causes of wealth, music, art, drug addiction, occupation, and sex perversion (Figure 1.21). To nongeneticists, fake pedigree charts may look like real ones, which, of course, explains why the fake charts were constructed in the first place. The explanatory power of science-based pedigree charts is impressive, and for this reason alone it may be pilfered to explain "phenotypes" that arise from social, not genetic causes.

Ss, about 1 of every 144 matings must be between Ss and Ss individuals (i.e., $1/12 \times 1/12 = 1/144$). Every zygote of an Ss $\times$ Ss mating has a probability of 1/4 of being ss. Therefore the probability that the next child born of African American parents will be SCA is 1 in 576: $1/144 \times 1/4 = 1/576$, close to the observed frequency of 1 in 570.

The above illustration has other ramifications. Within the estimated population of 40 million African Americans, 1 of every 570 babies born is SCA (more than 1,300 each year), and about 1 of every 12 (about 3 and one third million people) African Americans carries one copy of the s allele of the S gene. From these numbers it can be calculated that 1 of every 24 S genes is in the s allelic form. Recall that each individual carries two S genes, thus 11 of every 12 individuals will carry 22 S alleles, and 1 of 12 will carry one S and one s; this adds up to an estimated 23 S alleles and one s allele carried by every 12 African Americans.

If 1 of every 24 S genes is in the s allelic form, the probability that the next baby born to African American parents will inherit two s alleles is $1/24 \times 1/24 = 1/576$, again close to the observed frequency of SCA children within the African American population.

While pedigree charts can tell us whether the phenotypes observed within the populations of people related by descent are explained by Mendelian genes and can tell us the expected frequencies of the corresponding phenotypes in specific families, pedigree charts cannot tell us the expected frequencies of the same phenotypes in the greater population. This is explained in Chapters 3 and 4.

Phenylketonuria

Consider the genetic disease phenylketonuria, or PKU. Unlike SCA, which is more frequent in families from certain parts of Africa, PKU is far more frequent in families of Northern European descent. As yet no one has provided a data-based explanation for this. PKU is like alcaptonuria in that it is an inborn error of metabolism, but in this case the undigested molecule, the amino acid phenylalanine, accumulates in the blood and ultimately affects the developing brains of children such that untreated PKU children are severely mentally retarded.

In the United States about 1 of every 10,000 babies born to parents of Northern European descent is PKU. PKU babies are genotype pp; PP and Pp individuals are phenotypically normal. Of the three genotypes, PP is common and pp is rare (i.e., 1/10,000). If the frequency of pp = 1/10,000, then the

Martin Kallikak

He married a worthy Quakeress.

She bore him seven upright, worthy children,

He dallied with a feeble-minded tavern girl, who bore a son, "Old Horror" who had ten children just like him,

from which hundreds of the "lowest" types were born.

from which hundreds of the "highest" types were born.

Implication: The Quakeress is homozygous worthy; the tavern girl is homozygous unworthy; and Martin is heterozygous

Figure 1.21 A fake pedigree chart that appeared in many college textbooks during the first half of this century, and in one as late as 1965.

Perspective 1.3

In addition to Mendel's use of the laws of chance to explain inheritance patterns, a second use of probability in genetics can be explained by what is called a binomial expansion, that is, an expansion of two numbers. If we let p and q symbolize these two numbers and let $p + q = 1$ and if we expand the two numbers once by squaring them, then $(p + q)^2 = p^2 + 2pq + q^2 = 1$. Now consider the genotypes discussed above for PKU, PP, Pp, and pp. If we let allele $P = p$ and allele $p = q$, then genotype $PP = p^2$, genotype $Pp = 2pq$, and genotype $pp = q^2$; the sum of all genotypes will equal 1.

With this symbolism it is possible to convert known frequencies of alleles into frequencies of genotypes and/or known frequencies of genotypes into frequencies of alleles. For example, if 1 of every 576 African Americans is ss, then about 1 of every 24 S genes is s (24 is the square root of 576). If the frequency of s is 1/24, then the frequency of S is 23/24. From this we can calculate the frequency of Ss genotypes in the population as $2 \times 23/24 \times 1/24 = 0.08$, or 1/12, the number observed in the population.

---

frequency of p = the square root of 1/10,000 = 1/100. See Perspective 1.3. In other words, 1 of every 100 P genes is p, and about 1 of every 50 Euro-American persons is Pp.

Going full circle, if 1/50 = Pp, then the frequency of Pp × Pp matings is predicted to be 1/2,500 (1/50 × 1/50); one fourth of the children of such matings is expected to be PKU, which in turn predicts that one baby in 10,000 (1/4 × 1/2,500 = 1/10,000) will be PKU, that is, 49 of every 50 Euro-Americans are PP, and 1 of every 50 is Pp, or 1 of every 100 P genes is in the p form.

Huntington's Disease

Refer back to the first pages of Chapter 1. Huntington's disease, unlike alcaptonuria and PKU, is initiated by a dominant allele—the disease always develops in the presence of the *Hd* allele of the Hd gene. The *Hd* allele is rare in most population **gene pools,** as evidenced by the fact that the vast majority of the world's population are genotype hdhd. Among the few people who are diseased, the vast majority are heterozygous, Hdhd. It is not known for sure whether anyone is genotype HdHd.

The disease is difficult to diagnose because most individuals with the disease live 40 years before exhibiting any of its symptoms, the earliest of which are emotional changes and cognitive and motor disturbances. Mental impairment may precede the decline of motor skills, but since these often show up as little more than restlessness, lack of concentration, and irritability, they often are overlooked until after the more obvious motor signs of the disease appear. Within a few years of the first symptoms, there is a gradual wasting away of mental, nervous, and muscle control, until death.

As is the case with phenotypes initiated by dominant alleles, people with Huntington's disease have at least one diseased parent, and about half of the children of a diseased parent are diseased (Hdhd × hdhd → 1/2 Hdhd and 1/2 hdhd). In other words, the probability of a child inheriting the *Hd* allele from a diseased parent is 1/2.

While the *Hd* allele is rare (one per 40,000 *hd* alleles), it is not distributed evenly throughout the world population. Within the European population on the island of Mauritius and in the small villages on the shores of Lake Maracaibo, the frequency of the *Hd* allele is much higher than it is in other world populations. On the island of Mauritius the frequency is about 18 to 19 *Hd* alleles per 26,000 *hd* alleles, and in the Lake Maracaibo population there are approximately 650 Hd alleles within the gene pool of 5,200 Hd genes, that is, within the extended family studied by Nancy Wexler.

Interactions Among Genes

Even more complex relationships among alleles and between genes were discovered during the early years of genetics. One, of biological interest and of commercial value, was that of coat color in mice and other mammals. The results of these experiments are somewhat analogous to norms of reaction in that the experiments show clearly that the effects of genes upon phenotypes can be modified both by environmental factors and by other genes. In other words, there are *interactions between genes and environments,* and there are *interactions among genes.*

Variations of coat colors of many mammals are initiated by interactions among at least five major genes (five pairs of alleles). In mice the genes have been named A, B, C, D, and S. The *A* allele gives rise to a color called agouti, the color most of us know as "mousy." The mousy color appears gray because each single hair of an agouti mouse is mostly black but with a small band of yellow near the outer end (the ends are black; Table 1.5. The *a* allele leads to the absence of the yellow band, that is, aa mice are solid in color.

At least two alleles of the B gene are known. The AABB mice are agouti, aaBB mice are solid black, AAbb mice are cinnamon (mousy brown), and aabb mice

Table 1.5 The inheritance of coat color in mice, showing gene-gene interaction

Genotype	Phenotype	Hair type
A _ B _ C_	agouti	Black Yellow Black
A _ bbC _	cinnamon	Brown Yellow Brown
aabbC _	brown	Solid brown
aaB _ C _	black	Solid black
A _ B _ cc A _ bbcc aabbcc aaB _ cc	albino	Colorless
A _ B _ C _ dd A _ bbC _ dd aabbC _ dd aaB _ C _ dd	dilute agouti dilute cinnamon dilute brown dilute black	
A _ B _ C _ D _ ss aaB _ C _ D _ ss aabbC _ D _ ss	agouti piebald black piebald brown piebald	

are solid brown; that is, AA and Aa genotypes produce a yellow band on each hair, BB and Bb genotypes produce the black color, and the bb genotype produces a brown color. Brown hairs with yellow bands are cinnamon in appearance. Crosses between cinnamon (AAbb) and black (aaBB) mice and between agouti (AABB) and brown (aabb) mice both give rise to agouti (AaBb) progeny. Brother-sister matings among these agouti progeny give rise to agouti (A-B-), cinnamon (A-bb), black (aaB-) and brown (aabb) progeny in the ratio 9:3:3:1. (Where a dash, -, appears as a substitute for an allele in a genotype, it means that the phenotype will be the same whether the allele is in the dominant or the recessive form. Thus A-B- is phenotypically equivalent to AABB, AABb, AaBB, and AaBb.)

The C gene expresses itself almost like a switch, color on (CC and Cc) and color off (cc). Mice of genotype cc are colorless, or albino. In the presence of the C allele the A and B genes are expressed as described above, but these mice are colorless in the presence of the cc genotype. An interesting and different F_2 phenotypic ratio is observed as well, for example, a cross between black mice (BBCC) and albino mice (bbcc) gives rise to black (BbCc) progeny, which in turn gives rise to an F_2 population of black (B-C-), brown (bbC-), and albino (B-cc and bbcc) mice, in the phenotypic ratio 9:3:4.

The two alleles of the D gene differ in that D permits the full expression of the A and B genes, but dd genotypes dilute the other colors (agouti, black, cinnamon, and brown), making them appear "milky." It is possible to distinguish among the four milky colors, but each color is a pale counterpart of its expression in the presence of the D allele. In other words, the dd genotype modifies the expression of the A and B genes.

The S gene is "silent" in the S allelic form, but ss mice of all colors have white spots, called piebald. There are in fact interesting and beautiful color combinations that appear in response to interactions among these (and other) genes. These color combinations have been observed in many mammalian species, some of which have been exploited for their esthetic value (Figure 1.22). As we move toward the chemical composition of genes (Chapter 2), we will see other examples of gene-gene and gene-environment interactions that will change many of the views of genes that we have derived from classical, Mendelian experiments.

Summary

The main points of Chapter 1 are as follows: Genes influence phenotypes of individuals but not independently of environmental factors (see norms of reaction) or of interactions with other genes. Mendel's discovery of inheritance patterns that obey the rules of chance suggested to him that each trait is determined by one gene and that the contrasting forms of each trait are determined by contrasting forms of each gene. Contrasting forms of genes are called alleles. The presence of contrasting forms of phenotypic traits is not prima facie evidence for the existence of genes; it is a specific inheritance pattern of contrasting phenotypic traits that signals the presence of genes.

The Mendelian method for identifying genes does not include direct observations of genes but rather the observation of specific kinds of inheritance patterns. That is, contrasting phenotypic traits are visible; Mendelian genes are invisible. Indeed, Mendelian genes are but abstractions of unobserved but real

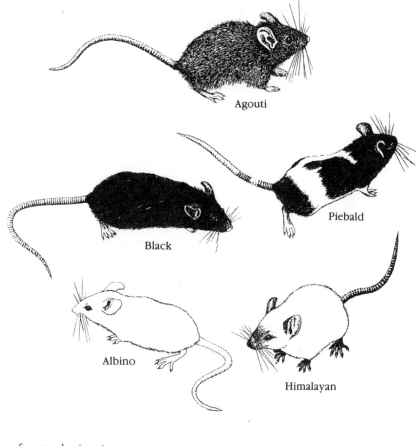

Figure 1.22 Variation of coat color in mice.

internal properties of cells. The abstract alleles of Mendelian genes were credited with initiating the development of contrasting forms of phenotype, and they were postulated to appear in gametes and to unite to form zygotes, according to the law of chance.

Genes located on different chromosomes segregate into gametes independently of one another. After the discovery of meiosis, the segregation patterns of alleles were understood in terms of segregation patterns of homologous chromosomes during meiosis. Since Mendel, it has been discovered that hundreds, even thousands of gene pairs reside on each chromosome pair. These blocks of linked genes tend to be inherited as units, but physical exchanges between homologous chromosomes break up these blocks of genes. To the extent that exchanges between homologues occur at random along their length, the distance between genes can be estimated from the frequencies of recombination between them. The test-cross is the classical method for estimating the frequency of recombination between linked genes.

Based upon the introductory remarks made here about human genes, there seems every reason to believe that pea genes and human genes obey the same "laws" of inheritance and exert their influences upon phenotype by similar biological processes. However, the study of human genes by way of family pedigrees is cumbersome and fragmentary. After we study the chemical and functional properties of genes in Chapter 2, it will become even more evident that human genes are more like than different from genes of other species of plant and animal. What we do not know much about is the way that genes interact in living systems, in particular, the ways in which specific gene activity is modified — modulated — by the activities of other genes. But we do know enough to understand that no gene is an island, or has a mind of its own, or an agenda that it carries out behind the backs of the other genes. The human genome is more like a 100,000-piece orchestra than like 100,000 solo performances.

Study Problems

1. Using the language and logic of genetics, describe one contrasting phenotypic trait within your family that is best explained as resulting from (a) contrasting forms of one or more genes, and (b) contrasting experiences. Do members of your family exhibit any gene-influenced phenotype that has been modified by experience? By medical intervention?

2. Explain why it is impossible today to determine norms of reaction of human genotypes.

3. Does knowledge of the fact that the "genes" discovered by Mendel were abstract symbols, not real biological chemicals or structures, and were not observed but only inferred from inheritance patterns change your previous view of science in any way? Explain.

4. Refer to Figure 1.10. Explain how parents with mid-digital hair can have a child without mid-digital hair. Of all the possible matings among the three genotypes, which ones will never give rise to a child without m-d hair? Which will give rise only to children without m-d hair?

5. Would you ever expect two persons without mid-digital hair to have a child with mid-digital hair? Explain. (As you proceed with these kinds of questions, it will speed your understanding of genetics if you *write out the kinds and ratios of gametes produced by each prospective parent.*)

6. Paula and Paul want to have a child, but they are afraid. Paula's brother and Paul's sister have PKU. They come to you with the following questions:
    (a) What are the genotypes of Paul's parents? Of Paula's parents?
    (b) What are the genotypes of Paula's PKU brother? Of Paul's PKU sister?
    (c) What is the probability that Paula is heterozygous, Pp?
    (d) What is the probability that both Paula and Paul are heterozygous?
    (e) What is the probability that the first child of Paula and Paul will have PKU? Explain.
    (f) What is the probability that their first child will be a PKU boy?

7. Below are partial pedigrees of three families. The hatched circles and squares in families 1 and 2 denote PKU; in family 3 they denote Huntington's disease.
    (a) What are the genotypes of I-1, I-2, I-3, and I-4?
    (b) What is the probability that II-3 carries the p allele? The Hd allele?
    (c) What is the probability that III-6 carries the Hd allele?
    (d) If III-2 and III-4 have children, what genotypes are possible? In what frequencies would you expect to find these genotypes among their children?

Generation

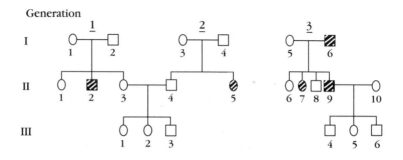

8. Given the state of medical knowledge (1991), which of the two genotypes, ss (sickle cell anemia) or pp (PKU), has the widest norm of reaction, that is, which disease can be modified to the greatest extent by environmental intervention?

9. A woman and a man "fall in love," make plans for a wedding, and hope to have children. In your opinion, is it important for them to learn as much as possible about one another's genotype before they start having a family? Explain.

10. You are conducting genetics studies of garden peas; mice got into your field seed house and ate the labels off the seed packets. You can't remember any of the genotypes of any of the seeds. You are dismayed, but you have the presence of mind to sort the seeds by phenotype. (So far, we are enacting a true story!) You find about 200 seeds that are smooth/yellow. Answer the following:

    (a) What are the possible genotypes of these seeds?

    (b) What crosses will you make to determine the genotypes of these seeds? Show the results of these crosses.

    (c) How would the results in (b) above, have been different if genes S and Y were linked, i.e., located on the same chromosome, about 10 map units apart?

    (d) For practice answer these same questions for the smooth/green, wrinkled/yellow, and wrinkled/green seeds scattered about by those agouti pests.

11. Explain the biological basis of MZ twin formation after the fertilization of one egg cell by one sperm cell using mitosis as a basis for the explanation.

12. Explain the biological basis for genetic variation among non-MZ siblings using meiosis as a basis for the explanation.

13. Recall that the frequencies of phenotypes and genes in populations cannot be determined by their frequencies in families; assume a population of 1 million Americans of African descent, in which 1 of every 10 is heterozygous at the S gene locus. Answer the following:

    (a) If matings are random with respect to genotype SS and Ss, what fraction of matings will be Ss × Ss? Explain.

    (b) What fraction of babies born in this population is expected to be SCA? SCT? SS? Explain.

    (c) What fraction of S genes is in the s form? The S form? Explain.

    (d) On the basis of what you know about the reproductive potential of ss individuals, would you predict that the ratio of S to s in the population gene pool is changing? If so, in what direction?

14. Based on the information given about Huntington's disease, calculate the ratios of Hd/hd in the gene pools of the Europeans living on the island of

Mauritius and of the extended family studied by Nancy Wexler and living on the shores of Lake Maracaibo.

15. Give at least one example of environmental influence upon phenotypic expression and of the influence of gene-gene interaction upon phenotypic expression.

16. In the discussion of the influence of gene-gene interactions upon phenotypic outcomes (coat color of mice), it was stated that the F_2 progeny of BbCc black mice are found in the ratio of 9/16 black, 3/16 brown, and 4/16 albino. Illustrate the cross, the genotypes, and the phenotypes of all the progeny and the genotypic and phenotypic ratios.

Suggested Reading

1. Bishop, J. E. and Waldholz, M. 1990. *Genome*. New York: Simon and Schuster. Both authors have been staff reporters for the *Wall Street Journal;* for this book they directed their investigative talents toward understanding the scientific events leading up to the massive, international project designed to locate every gene on every chromosome within the human genome.

2. Cummings, M. R. 1991. *Human Heredity: Principles and Issues.* 2nd ed. St. Paul, MN: West Publishing Co. Cummings' book is a bit more advanced and comprehensive than this text, but the sections dealing with Mendelian genetics will be helpful to you as you strive to solve the problems presented here. Cummings' book includes many more examples of human genetic diseases and of genetics methodology than does this text.

3. Herrnstein, R. 1971. IQ. *Atlantic*. September, Vol. 228 pp. 63-64.

4. Kevles, D. J. 1986. *In the Name of Eugenics: Genetics and the Uses of Human Heredity.* Harmondsworth, Middlesex, England: Penguin Books. This is a readable and exceptionally well-documented history of ideas about relations between our biological and our social "natures." The questions about justifying social policies and ethical values as outcomes of our biological natures are as much alive today as they were in Nazi Germany, or during the eugenics period of U.S. history, or during the colonial period of British history. Kevles' book presents the facts of this history; it doesn't tell its readers how to interpret the facts.

5. Wexler, A. 1991. *Huntington's Disease*. Times Books/Random House. Submitted.

2

A Second Method for Sighting Genes

Chapter Outline

A Genetic Response to Social Disease

Between 1976 and 1983 the South American country of Argentina was ruled by a fascist, military government. During that government's reign of terror, at least 12,000 so-called political prisoners were murdered, many of whom were women, many of whom had children. As far as can be determined, most of the children of the women who were murdered (at least 210) survived. The children of the murdered mothers were scattered throughout Argentina; some were adopted by military personnel; some were sold in the underground adoption trade; and some were used as barter by military personnel for a variety of services, including prostitution.

One year after the military government acquired power, a human rights organization like no other in history began watch dogging the military and its actions toward political prisoners. It was called "The Grandmothers of the Plaza of May," and was made up of grandmothers of the missing children. Not only did these grandmothers, the mothers of the murdered women and men, rally every week on the plaza outside the military offices, they formed an intelligence network that traced the kidnappings, the murders of the parents, and the deployment of the children. By the time the military relinquished power in 1983, the grandmothers had gathered and filed an enormous cache of information about the missing children. They also had discovered the burial sites and remains of some of the murdered parents of the children. The quality of their work has impressed human rights organizations all over the globe, especially organizations formed to combat fascism, a virulent social disease.

A primary goal of the grandmothers was to find the children and to unite them with their biological families. About all the grandmothers knew of genetics, at the time, was that many cases of disputed parentage were settled in the courts using blood compatibility tests to exclude one of two or more men suspected of being the father of a child. They also were aware that it is not possible, by use of conventional blood tests, to prove that any suspect *is the father*. But the grandmothers were motivated to use whatever evidence that could be admitted into the courts to identify their grandchildren.

So motivated, they sent representatives of their organization to the United States in 1984 to confer with the leadership of the largest scientific organization in the Americas, the American Association for the Advancement of Science (AAAS). After lengthy discussions, phone calls, and further discussions, the Argentine organization was routed to Mary-Claire King, a geneticist at the University of California, Berkeley.

King was about the age of the murdered mothers, and she had a daughter about the age of the missing children. But she was more. King had an impeccable scientific reputation at the time, and she believed and had believed since she was an undergraduate student at Carleton College in Minnesota that scientists ought to do the kinds of science that help people, particularly people who support science with tax monies and who, for whatever reason, do not have the economic base with which to finance the health care, education, food, and shelter needed to compete in today's upwardly mobile and technological society. While many scientists agree with King on many of these issues, very few accompany their words with actions. She is one of the few.

In June 1984, King made her first trip to Argentina. Her task was to find a genetic test that could unequivocally match each missing child with one or more of its biological grandparents. The first criterion of genetic relatedness used was that of the *proteins* located on the membranes of white blood cells. One large family of the many kinds of blood protein is called the **human leukocyte antigen** (HLA) family. The leukocyte antigens had been of great interest to physicians and scientists who study the biology of blood transfusions (each combination of donor and recipient must share the same or very nearly the same blood proteins, otherwise the recipient's blood will coagulate and death will often follow). These HLA proteins were of interest to geneticists because they are **encoded** by genes and because HLAs are so important within the branch of biology called **immunology.** There are many kinds of human blood proteins, and each is unique because of a unique gene that encodes it. The uniqueness of each gene and hence of each protein has led immunologists and geneticists to discover the processes that make it possible for immune systems to distinguish self-antigens from nonself-antigens and for them to attack foreign (nonself) proteins and not destroy self-proteins.

Human leukocyte antigens are part of the immune system. Within the entire human population there are more than a hundred different kinds of HLAs, but each person will carry only 4 to 8 of them. The more than 100 different kinds of antigen are encoded by as many alleles of four major HLA genes. Each person will inherit 4 to 8 of these alleles — four from each parent. The four pairs of genes may be homozygous, in which case that person will possess 4 HLAs, or heterozygous, in which case the person will possess 8 HLAs (Figure 2.1). If an

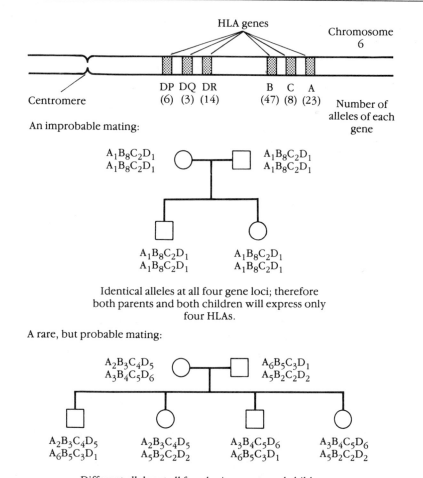

Figure 2.1 An outline of the HLA complex, which includes four pairs of genes and from four to eight antigens per individual.

analogy is made between the HLAs and a deck of cards, from which each zygote is dealt 4 to 8 at the time of fertilization, it helps to understand that it is highly unlikely that any two people will carry the same few antigens, since there are in fact thousands of combinations of more than 100 alleles. But, unlike dealing from a deck of cards, closely related people are more likely to carry the same antigens than are nonrelated people—not true for hands of poker or bridge. With HLAs, King was able to match 49 children with one or more grandparent.

There is a problem with the HLA compatibility approach to matching grandchildren with grandparents. The test is not foolproof. King estimated the probability of a correct match in each case, and the highest degree of certainty was 99.8%—good, of course, but not perfect. Recall that most court cases have two sides; that each case is heard by a jury and a judge in a courtroom. The grandmothers were represented by lawyers and King's evidence, and each missing child was represented by foster parents, friends of the foster parents, lawyers, and, in many cases, the military. In this emotional setting, margins of error in any of the evidence delayed the process of uniting the children with their grandparents. Both parties in each trial wanted to win, but the grandmothers won most of the initial cases. However, as the trials continued the opposition gained experience, determination, and momentum. In fact, the opposition began to demand that some of the early cases be retried. From King's point of view, she and the grandmothers needed a more accurate (foolproof) genetic criterion of relatedness. King came up with one.

You will understand some of King's logic if you recall that the vast majority of the 100,000 or so genes each of us inherited from our parents arrived in 46 packages, called chromosomes. These 46 packages were in turn packaged in two groups of 23, one an egg cell, the other a sperm cell. These 100,000 chromosomal, or nuclear, genes have captured most of the attention of most geneticists. Nuclear genes, however, are not the only genes that appear in zygotes. One of the several types of cellular organelle, the mitochondria, also carries about 37 genes, all of which are located on 1 mitochondrial (mt) chromosome (Figure 2.2).

Mitochondria are vitally important organelles; they are the sites of the chemical reactions that convert sugarlike molecules into energy and the energy into a kind of chemical bond that releases energy as it is needed during the performance of biological work. Mitochondria are sometimes referred to as cell powerhouses because within them the breakdown products of the food we eat are converted to energy, and the energy is converted to a form that can be used for biological work.

Some of the mitochondrial proteins involved in the transformation of small biological molecules into high-energy molecules are encoded by nuclear genes; others of the necessary proteins are encoded by genes (mt-genes) located on the mitochondrial chromosome (mt-chromosome). A few biologists have specu-lated that mitochondria originated from a bacteriumlike organism that, over long periods of time, developed a *symbiotic* relationship with cells, the invader providing the energy needed by the cell and the cell providing a congenial environment for the invader. This is a long and fascinating side-track of basic genetics; this short book hasn't the space to accommodate it. (See *The Search for Eve* by M. H. Brown in Suggested Reading.)

There are a few significant differences between the inheritance patterns of mt- and nuclear genes. Mitochondrial genes are transferred from parents to

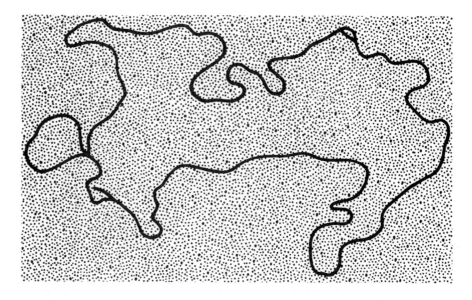

Figure 2.2 The mitochondrial chromosome carries 37 genes. Mitochondrial DNA, the only DNA found outside the cell nucleus in humans, appears as a convoluted loop.

progeny through egg cells only, not through sperm cells. Children, male and female, inherit mt-genes only from their mothers, but only the daughters transmit mt-genes to the next generation of children. Mitochondria, mt-chromosomes, and mt-genes are passed from generation to generation only by females. Males carry maternal mt-genes, but they do not transmit them to their children.

This should explain a second difference between mt-genes and nuclear **genomes.** Recombination never joins parts of one mt-chromosome with parts of another. Mitochondrial chromosomes are transferred from mother to daughter to granddaughter and so on in the exact same configuration. The only changes that occur within mt-genes are mutations, which occur randomly and infrequently but in measurable frequencies through time.

The missing children, King understood, could be expected to carry the mitochondrial chromosomes of their grandmothers, not of their fathers or grandfathers. King understood all of this beforehand because she had studied mt-genes in a different context (see Chapter 3); but the information was a surprise to the grandmothers and a stimulus for their renewed hope.

Without going into detail, the vast proportion of the mt-chromosome is comprised of a sequence of some 37 genes that encode 37 different kinds of mt-protein. Since these proteins are essential to the processing of energy from foodstuffs, it is to be expected that most people living in Buenos Aries carry similar genes along the length of their mt-chromosomes — not always identical but generally similar. This means that with respect to mt-genes, many people could be found to match any given mt-chromosome. However, a feature of mt-chromosomes that made it possible for King to develop a foolproof criterion of matching is that each chromosome possesses a short segment of DNA that *does not encode* for proteins.

Now, to reiterate, during the passage of time mt-mutations occur, and some of these are retained within the mt-chromosomes and are passed to future generations. Since mutations are rare, many generations are needed for the accumulation of significant changes within the mt-chromosome (it is estimated that about 3% to 7% of the chromosome will change through mutation over a time span of 1 million years). However, changes will accumulate more rapidly within the short, noncoding segment of the chromosome because mutations in that region do not interfere with the normal functions of the mitochondria, that is, the mutations are not harmful, and therefore they will not be eliminated from the population through the deaths of the mitochondria that carry them.

If we now put these facts into the logic used by King, it should be obvious that grandmothers and their grandchildren will carry identical or nearly identical, noncoding segments of their mt-chromosomes and that unrelated persons will not. Therefore King set out to isolate mt-chromosomes from white blood cells, the same cells that were used for HLA typing, collected from all of the people involved in the investigation. In those cases in which the short segments of mt-chromosome matching were perfect between missing child and putative grandmother, and imperfect between child and foster parents, the cases should have been closed. The evidence is about as close to certainty as it is possible to get in the otherwise fuzzy arena of human biology.

But the procedures for matching mt-chromosomes were new to the courts and to the judges, so there was a reluctance to admit mt-chromosome matching as evidence. In addition, the opposition obtained "expert" witnesses to disclaim King's genetic technology. This fact seems unbelievable to geneticists, who clearly are impressed by the simplicity and the accuracy of the genetic procedures used by King and her colleagues; nevertheless the societal uses of the technology developed for matching the missing children with their grandparents is likely to be delayed by politics and economics, especially in light of the fact that since 1983 the military leadership in Argentina has begun a vigorous program to return to political power, and not last on their agenda is the plan to destabilize the organization called the Grandmothers of the Plaza of May. Soon, of course, the grandmothers will have grown old and died. (See "Genes of War" by David Noonan in Suggested Reading.)

After a more detailed discussion of the chemical composition of genes and chromosomes, the procedures used by King to help the grandmothers locate their grandchildren will take on new meaning. In addition, it will become easier to understand the kinds of evidence, called DNA fingerprinting, that is being used more and more in the courts to identify victims of murder, rape, and bodily injury and to identify those guilty of such crimes. In other words, once we are able to answer the question "What are genes?" we should be able to understand how the resolving power of genetic analysis has increased so greatly over that provided by Mendelian genetics.

What Are Genes?

Even after Mendel's papers were discovered and his experiments confirmed by three biologists in 1900, the press did not jump on the story; only a few biologists were impressed. Partly this was because Mendel had discovered a new branch of biology based not on cataloging facts but on contingent rules of inheritance and partly because between 1860 and 1910 biology was a diverse subject, many subdivisions of which were controlled as the "personal property" of one or a few opinionated scientists. There was little interaction among many of the branches of biology; little momentum was generated from interactions among ideas.

First Step . . . Genes are Located on Chromosomes

The first signs of unity between the branches of biology that would later become known as genetics came out of the observation that chromosome behavior during meiosis is congruent with Mendel's "law" of segregation. In 1902 an American and a European biologist independently proposed that the (abstract) genes described by Mendel are located on (real) chromosomes. One thread of genetic logic took the view that to know the gene it is necessary first to know the chromosome. The first proof of the physical relationship between segregating genes and the homologous chromosomes upon which they reside was provided in 1931 by Barbara McClintock and Harriet Creighton, working with corn, and Curt Stern, working with *Drosophila* (see Perspective 2.1).

The unity between Mendelian genetics and **cytogenetics** was a leap forward, but during the first decade of the 20th century, fewer than 100 biologists paid any attention to it. Archibald Garrod explained "inborn errors of metabolism" as arising from mutant forms of Mendelian genes, but it wasn't until about 1940 that Garrod's ideas became integrated within mainstream genetics. Another important discovery was of the A, B, AB, and O blood types by Karl Landsteiner in 1900. But again, many years passed before it was shown that blood proteins are encoded by Mendelian genes.

The purpose for calling upon history here is that during the 20th century genetics became a centerpiece of scientific achievement and ultimately newsworthy as a result of interactions among what would have been, in the

---

Perspective 2.1

Creighton and McClintock studied two genes located on chromosome 9 of corn. They studied the segregation of these two genes in conjunction with two unusual chromosomal structural formations, one on each end of chromosome 9 (these structural formations are visible by use of the light microscope). Creighton and McClintock created a hybrid plant, one chromosome 9 of which carried the two visible formations (a "knob" at one end and a long, additional segment of a chromosome at the other end) and the dominant alleles of two genes; the homologous chromosome 9 carried neither structural formation, only the recessive alleles of the two genes, as shown:

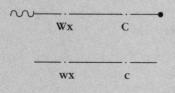

Among the progeny plants, Creighton and McClintock observed nonrecombinant chromosomes like those of the parent plants above and recombinant chromosomes as shown below:

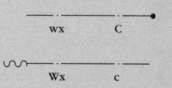

The cross-over event that gave rise to these recombinant chromosomes occurred between Wx and C. While it had been suggested earlier than 1931 that genetic recombination is accomplished by physical exchanges of segments between homologous chromosomes, this was the first reported evidence that the suggestion was in fact correct. Even so, this experiment did not reveal what genes are, only that they reside on chromosomes. Many years later McClintock received the Nobel Prize for a different discovery, but many geneticists have expressed the opinion that the work outlined here also warranted Nobel recognition.

"good old days," separate and jealously guarded scientific fiefdoms. Ultimately discoveries in biochemistry, chemistry, physics, cytology, mathematics, microbiology, immunology, etc. jelled to become the modern science of genetics. Interactions among these scientific disciplines led to an understanding of what *genes are.* As will be described later, these interactions, aimed at blood proteins and inborn errors of metabolism, led to a better understanding of *what genes do,* and with help from chemistry and physics it was discovered *how genes do what they do.*

Today we know that genes are long segments of even longer molecules of **deoxyribonucleic acid,** or **DNA.** The chemical structure of each gene specifies the biological **information** it contains, and the information carried by genes is translated from the language of DNA into the language of proteins (protein chemical structure). In general, each gene encodes one protein, and each protein (usually) performs one specific biological function. The well-being of a cell or an organism is the result of the integrative actions of dozens, hundreds, even thousands of gene-encoded proteins.

Genes Are Segments of DNA Molecules

After it had been demonstrated that genes are located on chromosomes it appeared obvious (to some geneticists) that if it could be demonstrated what chromosomes are, we might know what genes are. But the early, popular models of genes on chromosomes looked like a string of beads, the beads symbolizing the genes and the strings symbolizing the structural components that hold the genes in a linear alignment. This model encouraged geneticists to postulate that genes are different from chromosomes, which "simply evolved as carriers of the genes."

The beads-on-a-string model of chromosomes clearly is suggested by the early light microscope views of the large salivary gland chromosomes of *Drosophila melanogaster,* the tiny fruit fly made famous by geneticists. If these very large chromosomes are stained in a certain way they exhibit alternating dark and light bands (Figure 2.3). Many geneticists, upon observing these strange chromosomes, must have imagined seeing Mendel's "factors."

By 1933, the year *Drosophila* salivary gland chromosomes were discovered, it was known that the main chemical ingredients of chromosomes are proteins and nucleic acids, but it wasn't until many years later that it was discovered how these ingredients are organized into the structures of chromosomes (Figure 2.4). Even as late as 1952, most geneticists felt that the structural backbones of chromosomes were nucleic acids and that genes were proteins. This view persisted for two decades during which time evidence to the contrary surfaced. Three examples of such evidence are discussed here.

Ultraviolet Light Causes Gene Mutations

Between 1925 and 1933, Louis J. Stadler demonstrated that ultraviolet light (UV) causes mutations in corn pollen. (H. J. Muller won a Nobel Prize for demonstrating that x-rays cause mutations in *Drosophila;* his paper was published in 1927. Stadler, working with corn, required a longer time to get the necessary evidence, partly because he could observe only two generations of

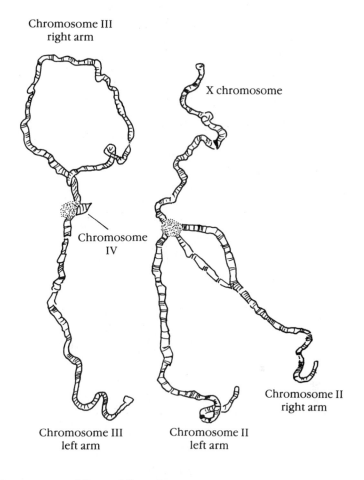

Chromosome III
right arm

X chromosome

Chromosome
IV

Chromosome II
right arm

Chromosome III
left arm

Chromosome II
left arm

Figure 2.3 Salivary gland chromosomes of *Drosophila melanogaster*.

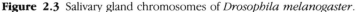

corn per year, whereas in the same year Muller might observe 20 generations of *Drosophila*. Both discoveries were momentous, but in a paper written by Stadler just before his death, he credited Muller with being the first to discover the induction of mutations.)

UV, like visible light, is a spectrum of different wavelengths of light, but in the UV spectrum, light waves are shorter and of higher energy than are wavelengths in the visible spectrum (Figure 2.5). If you look closely at a lower rainbow you will observe that the upper band is red and that the inner band is violet. (Outer rainbows show the reverse order of colors.) We can't see light waves longer than red (infrared) or shorter than violet (UV). What you see in a rainbow is visible light separated into discrete wavelengths by a prism, in this case sunlight passing through a "prism" of rain drops.

Wavelengths of UV light can be separated in much the same way, that is, by passing UV light through a prism (Figure 2.5). This is what Stadler did to test each wavelength for its efficacy as a mutagen. The rationale for testing each wavelength stems from the fact that different chemical substances absorb different wavelengths of light to a different degree. For example, proteins absorb maximally at wavelength 280 **nanometers** (nm) and nucleic acids absorb maximally at 260 nm.

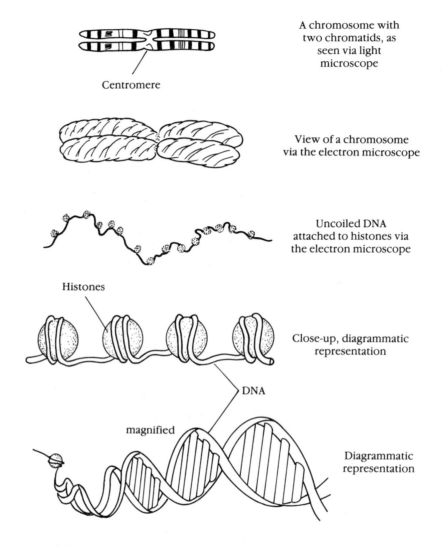

A chromosome with two chromatids, as seen via light microscope

Centromere

View of a chromosome via the electron microscope

Uncoiled DNA attached to histones via the electron microscope

Histones

Close-up, diagrammatic representation

DNA

magnified

Diagrammatic representation

Figure 2.4 The structure of chromosomes.

Stadler discovered that the most effective wavelength for causing gene mutations is 260 nm, the wavelength absorbed maximally by nucleic acids (Figure 2.6). This discovery did not prove that genes are made of nucleic acids, but it did show that it is possible to "bombard" cellular proteins with electromagnetic energy and not induce genetic mutations and that the absorption of similar energy by nucleic acids is accompanied by an increase of genetic mutations.

Transformation of Bacterial Genes

A more formal proof that genes are composed of nucleic acids was published by Oswald Avery and his associates in 1944. Avery used bacteria in his experiments, specifically *Pneumococcus* bacteria. Pneumococci causes pneumonia in many species of mammal.

The contrasting phenotypic traits of two Pneumococcus strains studied by Avery were **virulent** and **avirulent.** Virulent bacteria caused pneumonia in mice, and avirulent bacteria did not; otherwise the two strains were genetically similar. Mice artificially infected with virulent bacteria quickly became ill with pneumonia and died within a few days. If the infected mice were sacrificed before they died, it was possible to isolate virulent bacteria from their dead carcasses and with these to infect healthy mice. Mice infected with avirulent bacteria never became ill with pneumonia, but from such mice avirulent bacteria could be isolated.

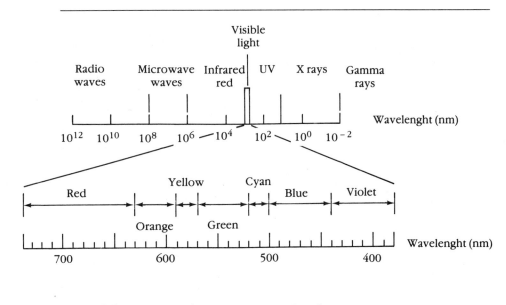

Colors corresponding to various wavelengths

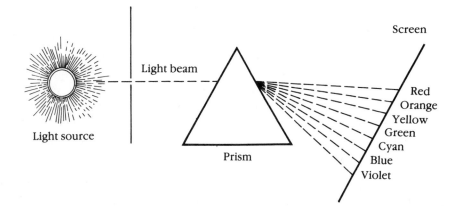

Figure 2.5 The electromagnetic spectrum, showing the relationship between ultraviolet and visible light waves.

Redrawn from: Waldman, Gary. "Introduction to Light: The Physics of Light, Vision & Color" Prentice-Hall, Inc., Englewood Cliffs, N.J. Fig. 2.1., 2.3., 2.10

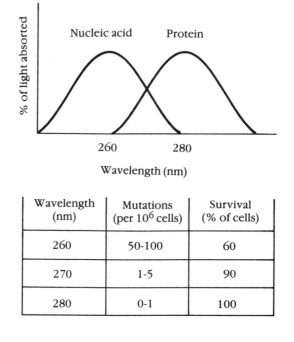

Wavelength (nm)	Mutations (per 10^6 cells)	Survival (% of cells)
260	50-100	60
270	1-5	90
280	0-1	100

Figure 2.6 The wavelengths absorbed maximally by nucleic acids induce mutations.

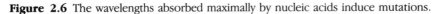

If mice were coinfected with heat-killed, virulent bacteria and with live, avirulent bacteria, they often developed pneumonia. From such mice both avirulent and virulent bacteria were isolated; the live, virulent bacteria were phenotypically identical to the heat-killed bacteria (Figure 2.7). Avery predicted that specific chemical(s) found in the dead bacteria program the phenotype called virulent and that these chemicals enter the avirulent bacteria and transform them into virulent bacteria. If correct, Avery had only to isolate the transforming chemical(s) to discover the chemical composition of the hereditary material.

In other words, which chemical leaves the genomes of the dead, virulent bacteria and enters the genomes of the live, avirulent bacteria? Or which chemical is the gene? What Avery did to answer these questions was to fractionate the heat-killed, virulent bacteria into their major chemical components—carbohydrates, lipids, proteins, and nucleic acids—and then to coinfect mice with each component plus live, avirulent bacteria. When this was done it was found that the only active component (transforming substance) of the dead bacteria was nucleic acid.

It turns out that there are two kinds of nucleic acid (Figure 2.8). One is **ribonucleic acid, RNA,** and the other is deoxyribonucleic acid, DNA. By use of specific enzymes that digest one or the other of these two kinds of nucleic acid, it was discovered that if RNA is destroyed the activity of the transforming substance is not changed, but if DNA is destroyed the activity of the transforming substance is reduced to zero. By 1950 the chemical components of the virulent bacteria had been sufficiently refined and purified as to prove beyond doubt that a segment of DNA from the dead, virulent bacteria enters the avirulent, live bacteria, integrates into their genomes, and changes their phenotypes from avirulent to virulent. (If Avery had lived longer he would have won a Nobel Prize for his work, clearly one of the benchmark experiments in genetics and also a benchmark in uniting bacteriology with mainstream genetics.)

Yet even in 1950 the majority of geneticists would have bet on proteins as being the stuff of which genes are made. Old ideas die slowly, but luckily new ideas grow up around them. The idea that genes are DNA came into its own, relatively quickly, between 1952 and 1953. The clincher experiment, using bacterial viruses, was done by Martha Chase and Alfred Hershey, 1952.

Bacteriophage Chromosomes are DNA

Bacterial viruses, called **bacteriophage,** or phage, are composed of proteins and DNA, nothing more. It is now known that proteins make up the outer coat and that DNA resides inside the protein "heads" of the bacteriophage. The key to the Hershey-Chase experiment involved **labeling** the protein and the DNA such that the fates of each molecular species could be followed during the time course of viral infection of bacterial cells. Proteins contain sulfur and DNA does not; DNA contains phosphorus and the proteins of bacteriophage do not. Prior to virus multiplication, bacterial cells were made **radioactive** with the heavy **isotope** of sulfur, ^{35}S. Viruses grown on these "hot" bacteria incorporated

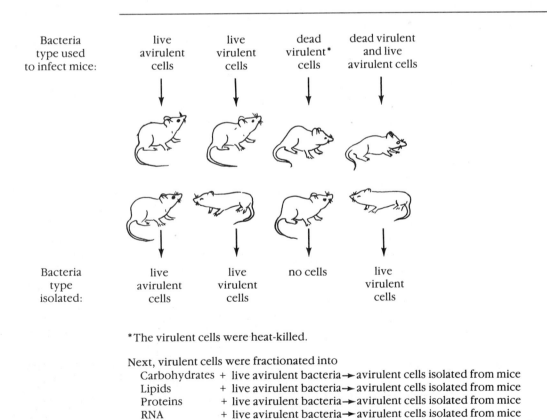

*The virulent cells were heat-killed.

Next, virulent cells were fractionated into
Carbohydrates + live avirulent bacteria → avirulent cells isolated from mice
Lipids   + live avirulent bacteria → avirulent cells isolated from mice
Proteins   + live avirulent bacteria → avirulent cells isolated from mice
RNA   + live avirulent bacteria → avirulent cells isolated from mice
DNA   + live avirulent bacteria → VIRULENT CELLS isolated from mice

Figure 2.7 Bacterial transformation via deoxyribonucleic acid.

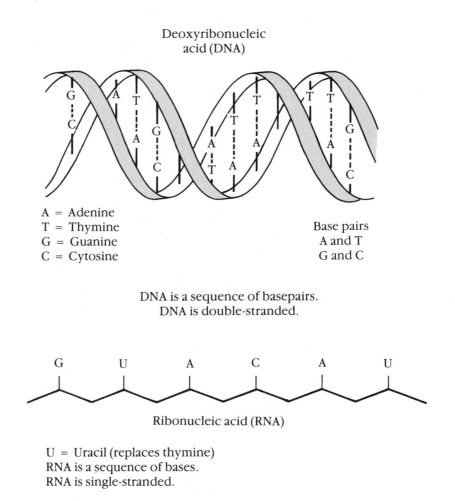

Deoxyribonucleic
acid (DNA)

A = Adenine
T = Thymine   Base pairs
G = Guanine   A and T
C = Cytosine   G and C

DNA is a sequence of basepairs.
DNA is double-stranded.

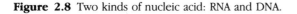

G   U   A   C   A   U

Ribonucleic acid (RNA)

U = Uracil (replaces thymine)
RNA is a sequence of bases.
RNA is single-stranded.

Figure 2.8 Two kinds of nucleic acid: RNA and DNA.

radioactive sulfur into their protein coats. Bacterial cells were made radio-active with the heavy isotope of phosphorus, ^{32}P, and the viruses grown on them incorporated radioactive phosphorus into their DNA genomes (Figure 2.9).

Later the "hot" viruses were allowed to infect "cold" bacteria, upon which the fates of their protein coats and DNA genomes could be followed by their radioactivity. It was discovered that whole **virus particles** do not enter bacterial cells. The radioactive sulfur remained outside the bacterial cell wall, and the radioactive phosphorus was found inside the bacterial cells. Since it is inside the bacterial cells that new virus particles emerge, the results of these experiments were interpreted to mean that viral DNA carries the instructions for making new viral particles. (Hershey received a Nobel Prize for this work, which convinced most geneticists that genes are DNA. In addition, this work further broadened the scope of genetics. Viruses became popular and useful targets of genetics investigation, and chemistry became an important strategy in the design of genetics experiments.)

By the end of 1952 the majority of geneticists had become convinced that DNA is the substance of which genes are composed. Previously, most geneticists had studied phenotypes, not genes, and they could have gone on forever in that mode without ever knowing what genes are and how genes influence phenotypes. But with a new and developing technology permitting direct analyses of DNA, RNA, and proteins, new kinds of genetics experiments became possible.

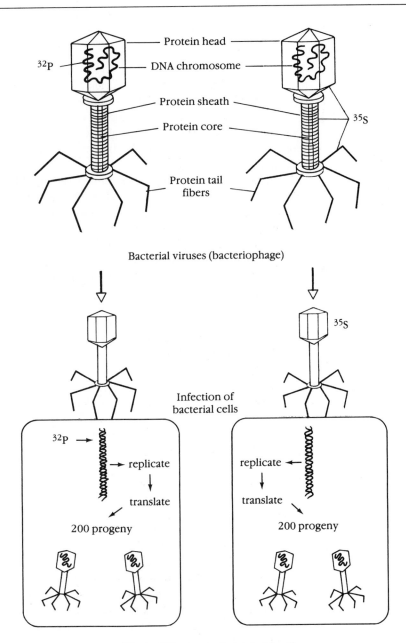

Figure 2.9 Evidence that bacteriophage chromosomes are DNA.

If you consider the *subject of genetics* to include all of the biological processes that link genes with phenotypes, the significance of the discovery that genes are segments of DNA molecules becomes apparent. This discovery opened the door to the use of chemistry as a means of studying the genetic phenomena discussed in Chapter 1—meiosis, chromosome duplication, mutation, recombination of linked genes, and so on—and, significantly, for studying genes directly by studying DNA (Figure 2.10).

What is DNA?

In 1953 James Watson and Francis Crick published a model of the structure of DNA in one of the most celebrated scientific papers of all time (Figure 2.11). The proposed structure of DNA suggested a way to explain all of the properties of genes known at that time, and it suggested experiments needed to back up the explanations, none of which were obvious from the vantage point of Mendelian breeding experiments.

Look back on Figure 2.4, a diagrammatic sketch of the molecular structure of a chromosome. One feature of this structure suggests that a single molecule of DNA extends from one end of a chromosome to the other. As far as is known, each giant molecule of DNA is unbroken and unmarked by other kinds of chemical formations from end to end.

To understand the impact upon Mendelian genetics of DNA chemistry we must pause to consider what biologists refer to as the relationships between structure and function. The view of DNA described in the last paragraph is that of a chemist, and it is a structural, not a functional view. The geneticist, guided by the Mendelian dictum that different forms of phenotypic traits are caused by different forms of genes, seeks as much to discover what genes do (function) as what they are (structure).

The knowledge that a single DNA molecule spans the full length of a chromosome does not tell us what genes are or what genes do; at most we can conclude that genes are portions of large DNA molecules. But the structure of DNA does not reveal that a molecule of DNA is a linear sequence of (hundreds to thousands) genes. What marks the beginning and ending of a gene? The answer cannot be derived from DNA structure, but only with help from function, as we shall see.

The structure of DNA does not tell us how such large molecules are packaged into chromosomes (or phage heads, or bacterial circular chromosomes). Packaging arrangements had to be discovered, one of which is described in Figure 2.4. Here we find that special kinds of proteins serve to keep long DNA molecules "untangled" during their transitions between condensed and relaxed states during the different phases of the cell division cycle. In the diagram you see the proteins (called **histones**) that serve as "spools" around which DNA is looped while in the condensed state.

Structure-function relationships are more predictable in the context of information transfer. If the chemical composition (structure) of DNA is in fact stored biological information, and if information transfer (function) from DNA to RNA to protein is facilitated by knowable chemical processes, then the relations between structure and function can be expected to be precise and predictable. See Perspective 2.2.

The structure of DNA proposed by Watson and Crick can be illustrated from several vantage points. Going from the sketch of a chromosome (Figure 2.4) to

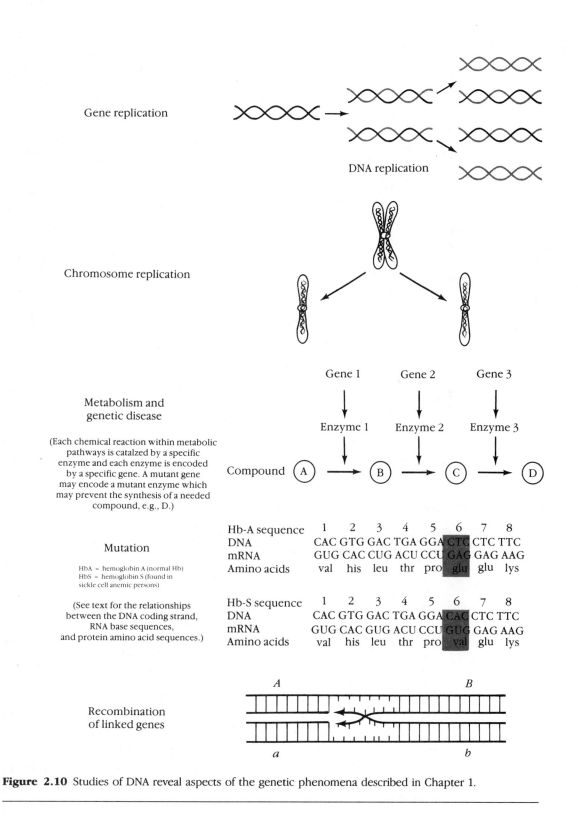

Figure 2.10 Studies of DNA reveal aspects of the genetic phenomena described in Chapter 1.

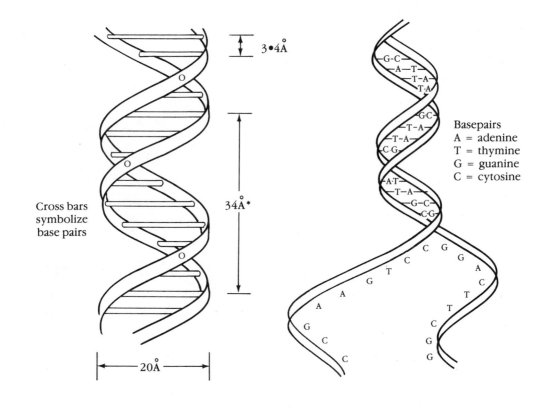

3•4Å

Basepairs
A = adenine
T = thymine
G = guanine
C = cytosine

Cross bars
symbolize
base pairs

34Å *

20Å

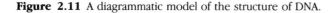

* Å symbolizes a unit of measure called an angstrom,
which is 10^{-8} (0.00000001) cm., or 3.937×10^{-9} (0.000000003937) inch.

Figure 2.11 A diagrammatic model of the structure of DNA.

three representational views of DNA (Figure 2.12), all representations show each molecule to be a double helix, symmetrical in shape. The two strands of each molecule appear to be wound 'round one another as you would wind two shoe laces by holding two ends in place while twisting the opposite ends. However, the two strands are held together, at a constant distance apart, such that they appear to form a "ribbon," which in turn is helical.

From one end of a DNA molecule to the other, each strand is composed of alternating phosphate and sugar subunits (Figure 2.13). The sugar is deoxyribose, hence the name, deoxyribonucleic acid. A single molecule of DNA is made of two strands, each of which possesses millions of alternating phosphate and deoxyribose subunits. But that is not all. Each deoxyribose sugar subunit is attached to one of four kinds of nitrogen-bases, two of which are **purines** (**adenine** and **guanine**) and two of which are **pyrimidines** (**thymine** and **cytosine**).

The two strands are held together by attachments of purines from one strand to pyrimidines of the other (each attachment is called a **base pair**), such that, unwound, DNA looks like a ladder, the supports being the two phosphate-sugar strands and the rungs being base pairs (Figure 2.13). In its helical form, DNA looks more like a spiral staircase than like a ladder.

To be more specific about base pairs, adenine always pairs with thymine, and guanine always pairs with cytostine. In shorthand these are called A-T and G-C base pairs (Figure 2.13). What this means is, if an adenine is found at a specified position on one strand, we can expect to find thymine in the corresponding position on the **complementary** strand. In other words, from one end of a DNA molecule to the other, A-T and G-C base pairs connect the two strands, one base pair connecting each pair of complementary deoxyribose sugar subunits.

This discussion of molecular genetics does not require a detailed knowledge of DNA chemistry, only enough to explain the outstanding properties of genes. However, many readers will recognize, as we go along, that the two strands of a DNA molecule are not identical (homologous) in the sense that a photocopy is identical to its **template.** So in anticipation of the discussion of what genes do, study the diagrams of DNA molecules (Figures 2.13 and 2.14) showing the two strands "running" in *opposite directions* and as being *complementary*. More technical discussions of DNA chemistry symbolize direction as 5' to 3' or as 3' to 5'; in Figures 2.13 and 2.14 these symbols are used only to illustrate that each strand has a shape, just as people do, and that each strand can be thought of as "right side up," or "upside down." When the two strands of one molecule are fitted together, as shown in Figure 2.14, the meaning of complementarity augments the meaning of direction; one strand is a *complement* of the other in that if you know the base sequence and the direction of one strand, the base sequence and the direction of the other strand is specified. *These two characteristics of DNA molecules are key to the link between Mendelian genetics/meiosis and molecular genetics.*

Perspective 2.2

As we begin to discover relationships between genes and phenotypes, we will be tempted to describe them by use of metaphor. The popular metaphor, adopted by many geneticists, projects DNA as orchestrating all living processes. In many ways this metaphor is misleading. In reality DNA is static, but it does store information. The information contained within DNA molecules is stored as a sequence of chemical units, called alphabet letters, but information does not flow from DNA until it is acted upon by proteins. In other words, the metaphor used in this text presents the gene as an inactive participant in the processes we call life; rather, it is an ensemble of cellular proteins that orchestrate living processes— growth, metabolism, reproduction, and so on. With DNA providing information and gene-informed proteins providing the action, it

becomes possible to distinguish between information and phenotype, in particular between information storage (gene structure) and growth and development (biological functions), processes that facilitate the emergence of phenotypes. Simply, genetic instructions are written in the language of DNA and the language of DNA is translated into the language of proteins. Ultimately an ensemble of gene-informed proteins orchestrate living processes. *Proteins bring genetic information to life.*

This metaphor avoids the silly argument of whether DNA or proteins are the most important biological molecules. DNA would be hopelessly static without proteins, and proteins would be hopelessly uninformed without instruction from DNA; therefore it should be obvious that both kinds of molecule are absolutely necessary to life. More on the metaphor as we go along.

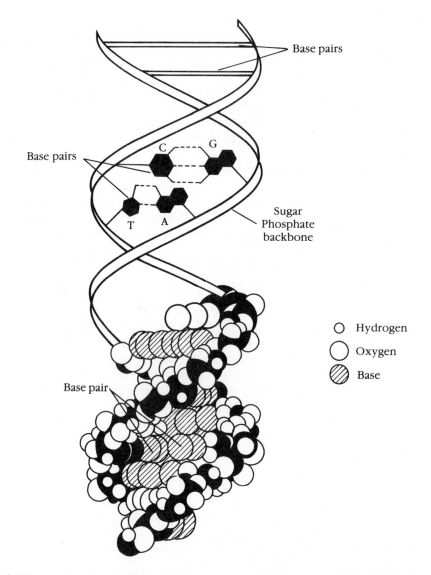

Figure 2.12 A view of DNA structure that shows a relatively accurate chemical structure below, two base pair silhouettes, and cross-bar symbols of base pairs.

What Does DNA Do?

What are genes? What do genes do? How do genes do what they do? These questions were posed before it was known that DNA is the answer to the first question. But even though we now know this, we still have not explained the important physical features of genes. In other words, the experimental evidence "that genes are DNA," discussed above, is slightly misleading; the experimental evidence presented showed that bacteriophage genomes are DNA, that DNA is able to transform phenotypes in bacteria, and that wavelengths of UV light absorbed by DNA increases the frequency of gene mutations. Now, DNA molecules extend the full length of chromosomes, and each chromosome carries hundreds, even thousands of genes; therefore each gene might be a small segment of a DNA molecule. But what is it that defines a gene segment? What marks the boundaries of a gene? What separates one gene from another? How do genes make exact copies of themselves prior to every cell division? How do genes influence phenotypes?

In making the transition from Mendelian to molecular genes, once it is known that *genes are segments of DNA molecules,* the wording of the second question can be changed from "What do genes do?" to "What does DNA do?" (Recall the metaphor: DNA might not *do* anything. Rather it may have something *done to it.*) At any rate, the change of emphasis from gene to DNA permits us to put aside preconceived notions we may have had about genes, and it encourages us to investigate the properties of DNA. Now we can ask about the *chemical structure of DNA and the biological consequences of that structure* without losing historical contact with genes. In hindsight we see that knowledge of DNA became a launching pad from which much was learned about genes, proteins, and even more complex aspects of phenotype—indeed, this knowledge revolutionized the science of genetics.

Replication of DNA

The replication of a parent molecule of DNA results in the formation of two progeny DNA molecules, each precisely identical to the parent molecule. DNA replication precedes cell division, both mitotic and meiotic, but the relationship between DNA replication and mitotic cell divisions is easier to illustrate

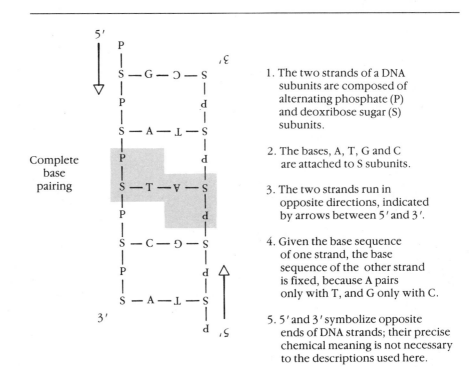

1. The two strands of a DNA subunits are composed of alternating phosphate (P) and deoxribose sugar (S) subunits.

2. The bases, A, T, G and C are attached to S subunits.

3. The two strands run in opposite directions, indicated by arrows between 5' and 3'.

4. Given the base sequence of one strand, the base sequence of the other strand is fixed, because A pairs only with T, and G only with C.

5. 5' and 3' symbolize opposite ends of DNA strands; their precise chemical meaning is not necessary to the descriptions used here.

Figure 2.13 The "backbones" of DNA molecules are alternating sugar and phosphate subunits that run in opposite directions.

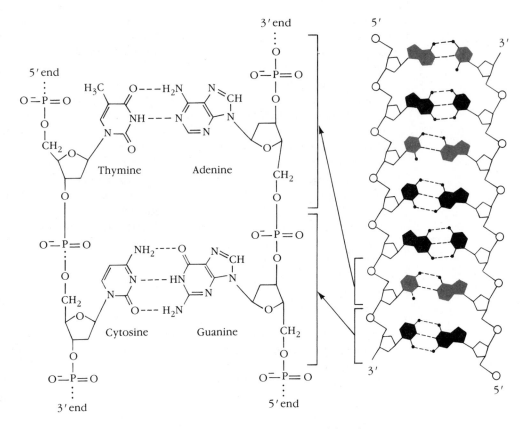

The two strands are nonidentical, but the
structure of one can be determined from the other.

Figure 2.14 The two strands of a DNA molecule are complementary to one another.

(Figure 2.15) than it is prior to meiosis and during gamete formation. The generalities are the same, however.

Since the two strands of a DNA molecule are complementary, the base sequence of each parent strand is expected to determine the base sequence of its progeny strand (i.e., an A in a parent strand will attract a T to its progeny strand; a G in a parent strand will attract a C to its progeny strand, and so on). Because the rules of complementarity specify structural and chemical relationships between complements, each strand of a parent molecule will act as a *template* upon which the progeny, complementary strand is synthesized. While the model of DNA structure proposed by Watson and Crick suggests such a scheme for DNA replication, the experimental evidence favoring it did not surface until 1956, in two very ingenious experiments.

1. The Meselson-Stahl experiment: Prior to the acquisition of experimental evidence, three models of DNA replication were considered: **conservative, semiconservative,** and **dispersive** replication (Figure 2.16). The most promising model was that of semiconservative replication, since if each of the two strands of a DNA molecule serves as a template upon which a new, complementary strand is synthesized, the two progeny molecules arising from the parent molecule would each be "half new" and

"half old." One experiment that led to the conviction that this was the correct model is shown in Figure 2.17. This work was done with bacterial DNA.

Bacteria were grown for many generations in a food supply containing isotopically labeled ^{15}N, a heavy isotope of ^{14}N, common nitrogen. ^{15}N becomes incorporated into the DNA of the bacteria. ^{15}N-labeled bacterial chromosomes are expected to be heavier than ^{14}N chromosomes. This was demonstrated to be so by the fact that labeled DNA moves further from the starting point in a centrifugal field (Figure 2.17).

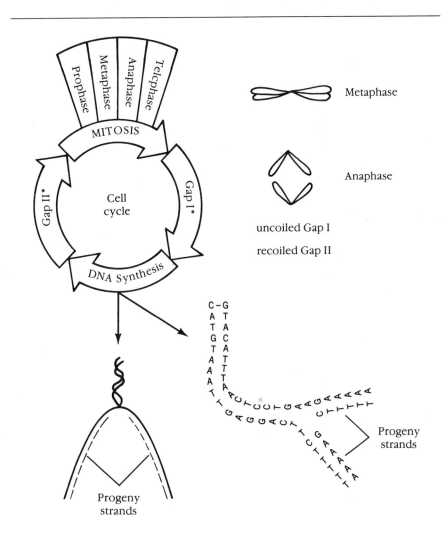

*Gap I symbolizes a period during the cell cycle between telophase and the start of DNA synthesis. The chromosomes change from a condensed state to an uncondensed state; i.e., the DNA molecules become relaxed prior to DNA synthesis.

Gap II the period during which the newly synthesized DNA molecules become condensed into chromosomes prior to mitosis.

Figure 2.15 DNA replication in the context of cell division.

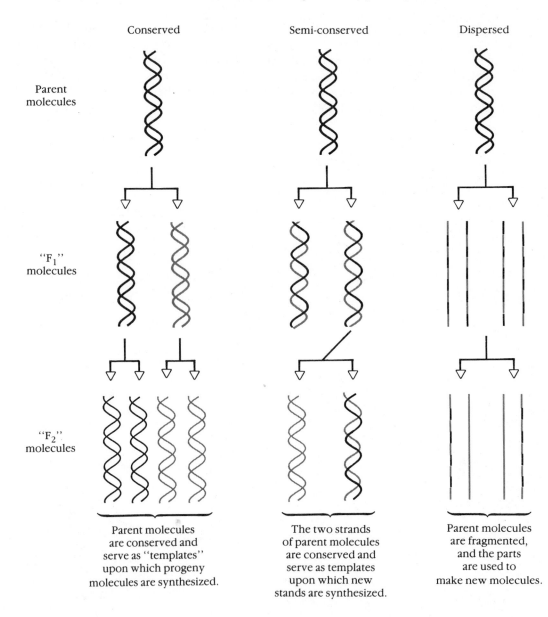

Three models showing that parent molecules are conserved,
semi-conserved, or dispersed during DNA synthesis.

| Conserved | Semi-conserved | Dispersed |

Parent
molecules

"F_1"
molecules

"F_2"
molecules

| Parent molecules are conserved and serve as "templates" upon which progeny molecules are synthesized. | The two strands of parent molecules are conserved and serve as templates upon which new stands are synthesized. | Parent molecules are fragmented, and the parts are used to make new molecules. |

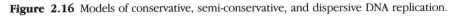

Figure 2.16 Models of conservative, semi-conservative, and dispersive DNA replication.

If heavy chromosomes are allowed to undergo one round of replication in ^{14}N
media, their progeny chromosomes are exactly half as heavy, as evidenced by the
fact that they travel midway between heavy and normal chromosomes in a
centrifugal field. A second generation of bacterial chromosomes is made up of
equal numbers of normal and half-heavy chromosomes. Follow this progression
in Figure 2.17, to verify that the data support the semiconservative model of
replication.

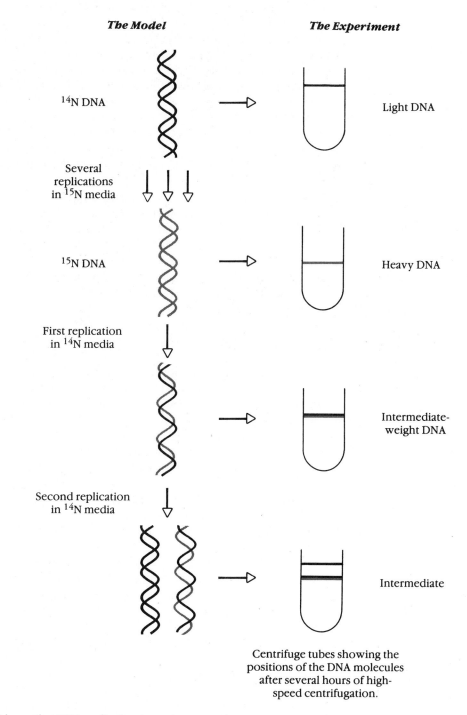

¹⁴N DNA   Light DNA

Several
replications
in ¹⁵N media

¹⁵N DNA   Heavy DNA

First replication
in ¹⁴N media

                                                Intermediate-
                                                weight DNA

Second replication
in ¹⁴N media

                                                Intermediate

Centrifuge tubes showing the
positions of the DNA molecules
after several hours of high-
speed centrifugation.

Figure 2.17 Evidence that DNA replication is semi-conservative.

2. Taylor's experiment: The second experiment showing that DNA replication is semiconservative, performed by Herbert Taylor, required "round-the-clock" observations of the replication of whole chromosomes of broad beans. Taylor's experiment also required the use of isotopic substances, this time, radioactive hydrogen, or tritium. By the use of chemical methods we will not discuss in this text, thymine was made radioactive with tritium, then bean plants were grown in a medium containing the "hot" thymine. The hot thymine was incorporated into the DNA of the bean chromosomes.

If bean plants with "hot" DNA are transferred to "cold" medium long enough for one cell division, it is observed that the chromosomes are "half hot"; after two rounds of cell division, half the chromosomes are cold and half are half hot (Figure 2.18).

Both kinds of experiments are interpreted to mean that during DNA replication, each of the two strands of a parent DNA molecule serves as a template upon which a new strand is synthesized; both progeny molecules of DNA are half old and half new. A diagrammatic sketch of DNA synthesis (Figure 2.19) shows the relationship between templates (the two strands of the parent DNA molecule), which are not changed during synthesis, and the progeny molecules of DNA. The sketch also shows that DNA synthesis proceeds in opposite directions along the two template strands.

Transcription of DNA

Think of transcription as a form of communication as well as a chemical process. In genetics the word transcription is used interchangeably with the phrase RNA synthesis in a context not unlike that of using a tape recorder to record a rock concert. Even though the word communication is used as metaphor, the actual meaning of the words transcription and synthesis is not compromised; the chemical information contained within DNA molecules is transcribed into the chemical structures of RNA molecules as they are synthesized.

RNA synthesis resembles DNA synthesis, but the two processes are not identical. For example, RNA molecules are synthesized on short segments of DNA molecules, not whole chromosome DNA, and only one of the two DNA strands within this segment serves as a template during RNA synthesis. Indeed, these two facts mark the first step of our learning to define genes biochemically, that is, *the template upon which RNA is synthesized is the gene* (this definition will be modified as we go along; Perspective 2.3; Figure 2.20).

That only one of the two DNA strands acts as a template is evidenced by the fact that the base sequences of new RNA molecules are complementary to the template strands, also called **sense** strands, not to nontemplate, or nonsense strands. In other words, RNA molecules are similar to nonsense DNA strands but complementary only to sense, or gene strands. (Similar to nonsense strands but not identical; the four bases of RNA are A, **U,** G, C, not A, T, G, C. The base, **uracil,** is normally found in RNA molecules, and thymine is normally found in DNA molecules.)

After it became known that genes are DNA and after the Watson-Crick model of DNA structure suggested that DNA can be replicated and transcribed, as outlined above, it was predicted that the genetic information carried by DNA is

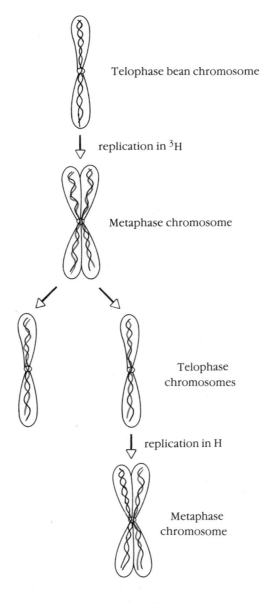

Telophase bean chromosome

replication in ^{3}H

Metaphase chromosome

Telophase chromosomes

replication in H

Metaphase chromosome

Figure 2.18 DNA replication is semi-conservative in higher plant chromosomes.

transmitted, via **messenger RNA** molecules, to proteins and that this information informs proteins of their chemical structures. *This predicted relationship between genes and proteins was a crucial step in understanding how genes influence phenotypes.*

In summary, all of the DNA in a genome is replicated prior to cell division, but the information contained within genes is transcribed into mRNA transcripts, one at a time or several in concert, at nearly any time in the life of a cell. Then from mRNA transcripts the information is **translated** into the **primary chemical structures** of proteins. The overall scheme can be symbolized in two ways: (1) One gene → one transcript → one functional

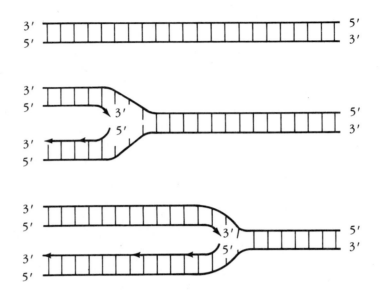

Figure 2.19 A diagram of DNA synthesis. New bases are added to the 3′ end of the sugar attached to the last base. The parent strands are copied in opposite directions. The new 5′ to 3′ strand is synthesized first, and the other new strand is synthesized in the opposite direction, in segments of about 150 bases. The symbols 3′ and 5′ are explained in Fig 2.13.

protein; or (2) DNA → RNA → Protein. (The two languages, of nucleic acids and of proteins, and the processes of translating the messages encoded within sequences of RNA bases into sequences of amino acids of proteins are described below.)

An Example of DNA → RNA → Protein

As mentioned earlier, sickle cell anemia is a disease that always accompanies $\beta^S\beta^S$ genotypes, where β^S is a mutant allele of β^A, the gene that specifies the primary structure of β-hemoglobin chains. There are two kinds of protein in hemoglobin molecules, α and β chains; the α proteins are encoded by the α gene located on chromosome 16, and β proteins are encoded by the β gene located on chromosome 11. Normal hemoglobin is abbreviated Hb-A, and sickle cell hemoglobin Hb-S.

This example is somewhat confusing in that the two levels of phenotype, blood proteins and health, represent two very different levels of biological complexity. The least complex level is that of hemoglobin, Hb-A versus Hb-S. The more complex level is that of individual health, normal versus sickle cell anemia (Figure 2.21).

The β gene is a small segment of the DNA molecule that runs the length of chromosome 11; it has a unique base pair sequence; therefore the difference between the two alleles β^A and β^S must exist as a chemical difference within the base pair sequence of the β gene segment. We further expect that the chemical difference between β^A and β^S must show up in the RNA transcripts copied from them. Finally we expect to see the chemical difference between the two alleles and their RNA transcripts to show up in the amino acid sequences of the two β proteins encoded by them.

A diagram of the previous paragraph is shown in Figure 2.22. Here you see the first 18 base pairs of the β^A gene. The bottom strand of this DNA segment is C A C G T A A A T T G A G G A C T C, and the complementary strand is G T G C A T T T A A C T C C T G A G. When these two strands are held together by chemical bonds between matching bases, we see

C A C G T A . . .

| | | | | |

G T G C A T . . . base pairs.

So far it has not been said how we know which of the two strands acts as a template upon which RNA transcripts are synthesized, nor has it been discussed how RNA transcripts inform β proteins. If only one of the two

---

Perspective 2.3

The β gene that encodes the primary structure of the β protein of hemoglobin is more complicated than has been described so far. In Chapter 1 the β gene was called S, and was described as having 2 alleles, S and s. From now on we call this the β gene with alleles β^A and β^S. A few years ago it was discovered that the template region of the β gene (the transcriptional segment) does not correspond, base pair to base pair, to the mature mRNA. Indeed, the "gene" is longer than the mature mRNA. The gene is a sequence of alternating coding (called **exons**) and non-coding regions (called **introns**). The following cartoon of the β gene, the pre—mRNA, the modification of the pre-mRNA, and the mature mRNA, is typical of many genes, except for the fact that the numbers (1 to 18) and sizes (a few to thousands of base pairs) of introns vary from gene to gene. It is not known why genes evolved in this way, what it is that determines the numbers of introns within a transcriptional unit, or what functions introns serve or may have served in the past. The presence of introns does, however, pose a problem with the definition of a gene: Is a gene the entire transcriptional segment or merely the sum of the exons, or coding sequences, within the transcriptional segment?

The human β globin gene

5' ncr   E-1   I   E-2   I   E-3   ncr 3'

5' ___ 3' See Figure 2.13.
ncr = non-coding region
E-1 = Exon-1, which encodes amino acids 1–30
E-2 = Exon-2, which encodes amino acids 31–104
E-3 = Exon-3, which encodes amino acids 105–146
 I = introns

The pre-mRNA, which is complementary to the transcribing segment, is modified by **enzymes** that "snip" out the bases encoded by introns and that "splice" together the three segments encoded by exons. This cutting and splicing process succeeds in fashioning an RNA transcript with a sequence of bases that are complementary to the three exons and needed to encode functional β chains. The mature mRNA moves out of the nucleus, where it is synthesized and modified, into the cytoplasm, where it will translate its genetic message into **amino acid** sequences of proteins.

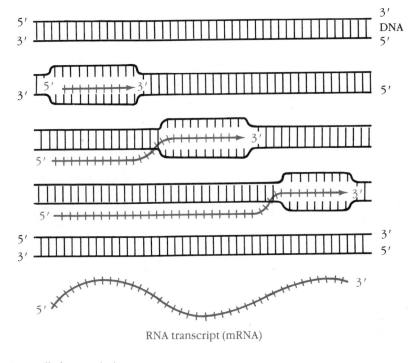

RNA transcript (mRNA)

Figure 2.20 RNA synthesis, called transcription.

strands of the β^A allele serves as a template for RNA synthesis, we should expect that the template strands of the two alleles will differ from one another in one or more bases and that this difference will show up as a difference of amino acid sequences between the two β chains of Hb-A and Hb-S. Consider the following explanation of the statement that only one of the two DNA strands is the sense strand. If C A C G T A . . . , is the sense strand, the RNA transcript will have the base sequence, G U G C A U . . . ; if G T G C A T . . . , is the sense strand, the RNA transcript will have the base sequence, C A C G U A . . . (Figure 2.23). If the sequence of bases in RNA transcripts is genetic information (i.e., a message from the gene), then these two RNA transcripts are as different from one another as the English word, L o n d o n is from D n o l n o. One sequence of letters makes sense in the English language; the other sequence is nonsense.

If we assume that C A C G T A A A T T G A G G A C T C . . . is the sense strand, the RNA transcript will be G U G C A U U U A A C U C C U G A G . . . . Earlier it was suggested that *the transcribing strand of DNA is the gene;* a few years ago, when this appeared to be the best chemical definition of the gene, the transcribing, sense strand of DNA was called the *gene strand.*)

The "language of life" has a simple alphabet of four letters, A, T, G, and C; simple "words" (defined later); and precise expressions of word meanings (first the amino acid sequences, the **primary structures** of proteins, and second, the protein functions). The example of sickle cell anemia shows both the meaning of the language and how it was discovered.

To further explore the language metaphor, consider a few of the differences between cultural languages like Spanish, and chemical languages like the genetic language. The alphabets and words of cultural languages are arbitrary and agreed upon by convention; they change through time, sometimes rapidly; and they are different from culture to culture—compare Spanish with Chinese. The alphabets and words of the genetic language, on the other hand, were determined billions of years ago by the chemical properties of the molecules called alphabet letters and the combinations of those molecules called words. These letters and words were **selected** by the environments within which life had its origin; they have not changed; and they are the same for all living forms, past and present. The differences are so many, in fact, that metaphor is the only means available to us for uniting them with English words.

Having covered some of the evidence that genes are DNA, a few models of how DNA is replicated and transcribed, and having inferred that RNA messages inform proteins which do biological work, it is time to take a closer look at proteins.

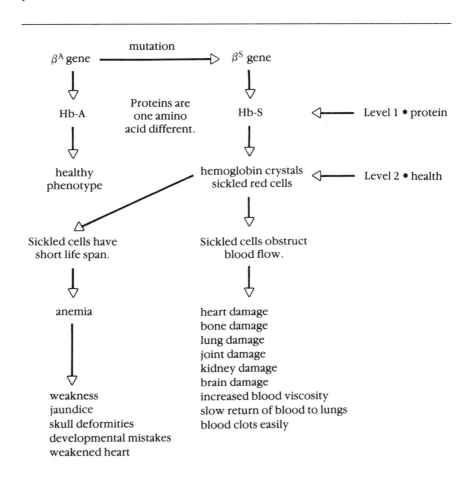

(Not all SCA persons show all of the symptoms listed above;
there is great variation.)

Figure 2.21 The phenotypes of SCA children are expressed at two levels of biological complexity.

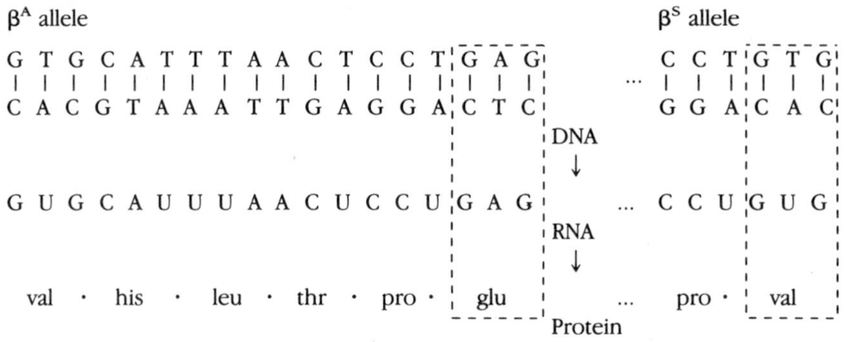

β^A allele   β^S allele

```
G T G C A T T T A A C T C C T│G A G│        C C T│G T G│
│ │ │ │ │ │ │ │ │ │ │ │ │ │ │ │ │ │ │    ...  │ │ │ │ │ │
C A C G T A A A T T G A G G A│C T C│        G G A│C A C│
                             │      │ DNA
                             │      ↓
G U G C A U U U A A C U C C U│G A G│    ... C C U│G U G│
                             │      │ RNA
                             │      ↓
val · his · leu · thr · pro ·│ glu │    ... pro ·│ val │
                              ─────── Protein
```

- Each gene is a packet of information.
- The information carried by each gene is transcribed into mRNA molecules.
- The information carried by mRNA molecules is translated into amino acid sequences of proteins.
- Alleles carry similar, but not identical, information.
- Mutant information may initiate the development of mutant phenotypes, as does the information carried by the β^S allele.

Figure 2.22 The β^A and β^S alleles, their transcripts, and the β proteins encoded by them.

Proteins

Hemoglobin (heme means iron, and globin means protein; therefore hemoglobin means iron-containing protein) is a protein with a specific biological function which is to carry oxygen from the lungs to all cells in the body. Persons with Hb-A are phenotypically normal, in contrast to sickle cell anemia, a disease of persons with only Hb-S. Not only do we want to know how the two kinds of β protein are encoded within DNA molecules, we also want to know how these two kinds of hemoglobin lead to states of health as different as normal health is from sickle cell anemia (SCA).

The α proteins (two per hemoglobin molecule) are identical in Hb-A and Hb-S. Therefore we assume that the α gene is identical in SCA and non-SCA persons. The β proteins are different; Hb-A has normal β protein, and Hb-S has abnormal β protein. Normal β protein is found in persons of β^Aβ^A and β^Aβ^S genotype, and abnormal β protein is found in persons of β^Aβ^S and β^Sβ^S genotype. We assume therefrom that β^A encodes normal β and that β^S encodes abnormal β protein (Figures 2.21 and 2.22).

Linus Pauling and his coworkers discovered in 1949 that Hb-A and Hb-S differ by electric charge (the two hemoglobins migrate at different speeds toward the positive pole [anode] within an electric field). Hb-A migrates toward the anode faster than Hb-S, which means that Hb-A carries a higher negative charge than Hb-S (Figure 2.24). This was the first evidence of a chemical difference between the two kinds of hemoglobin.

Seven years later Vernon Ingram discovered why the two hemoglobins differ in electric charge. Ingram discovered that Hb-A carries an electronegative amino acid in the same position that Hb-S carries an electroneutral amino acid (Figure 2.25).

Each β protein is a chain of 146 amino acids (Figure 2.25). Ingram discovered that the sixth amino acid of the β-A protein is glutamic acid (electronegative) and that the sixth amino acid of the β-S protein is valine (electroneutral; Figure 2.25). The remaining 145 amino acids in the two kinds of protein are identical in kind, and in position within the amino acid "chains" (See Figure 2.26 for the kinds and structures of the 20 amino acids that appear in the proteins of all living creatures, plants, animals, and microorganisms).

Ingram was the first to discover the kind of relationship between contrasting alleles (e.g., β^A and β^S) and contrasting phenotypes (β-A and β-S proteins) that can be explained as a *substitution of one amino acid for another*. His discovery preceded the discovery of mRNA, and the discovery of the base-pair sequence of any gene, that is, Ingram's discovery preceded all knowledge of how genes determine the sequence of bases in RNA transcripts and how RNA transcripts (mRNA) determine the amino acid sequences of proteins. Indeed, the discovery that inherited differences between normal and abnormal protein can be so simple a change as *an amino acid substitution* set off an avalanche of experiments designed to "decode" the language of life.

As an introduction to the language of life, follow the diagram in Figure 2.22. Working upward from the partial Hb-A protein chain, notice that the RNA transcript immediately above it has three base molecules for each amino acid in the protein, and notice that the DNA molecule above the transcript has the same number of base pairs as there are bases in the transcript. In short, the nucleic acid language (employing a four-letter alphabet) appears to have only three-letter "words." Furthermore, each three-letter word *expresses itself* in the form of one amino acid within the primary structures of proteins. Notice, also, that the protein language employs a 20-letter alphabet!

Tracing the right portion of the diagram in Figure 2.22, you see that the mRNA transcript above the Hb-S protein differs from its counterpart above the Hb-A protein. The three letters above glutamic acid in the Hb-A protein are G A G;

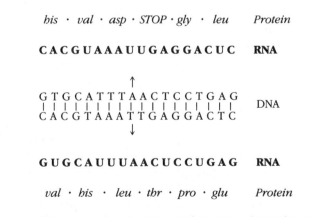

his · val · asp · STOP · gly · leu   *Protein*

C A C G U A A A U U G A G G A C U C   **RNA**

↑
G T G C A T T T A A C T C C T G A G
| | | | | | | | | | | | | | | | | | | DNA
C A C G T A A A T T G A G G A C T C
↓

G U G C A U U U A A C U C C U G A G   **RNA**

val · his · leu · thr · pro · glu   *Protein*

This protein has an amino acid sequence that makes "sense."
The top amino acid sequence is "nonsense."

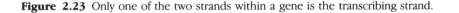

Figure 2.23 Only one of the two strands within a gene is the transcribing strand.

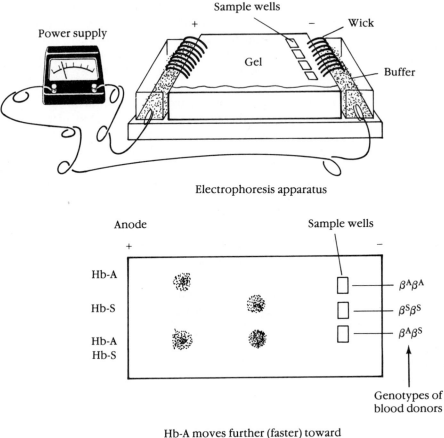

Electrophoresis apparatus

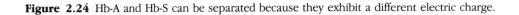

Hb-A moves further (faster) toward
the anode in an electric field.

Figure 2.24 Hb-A and Hb-S can be separated because they exhibit a different electric charge.

above valine in the Hb-S protein, the RNA letters are G U G. This difference is reflected in the base-pair sequences of the β^A and β^S alleles. The sixth base-pair triplet of the β^A allele is G A G
$\quad\quad\quad\quad\quad$ | | |
$\quad\quad\quad\quad\quad$ C T C, and of the β^S allele, G T G
$\quad\quad\quad\quad\quad\quad\quad\quad\quad\quad\quad\quad\quad$ | | |
$\quad\quad\quad\quad\quad\quad\quad\quad\quad\quad\quad\quad\quad$ C A C

The relationships that exist between the base-pair sequences of DNA molecules, the base sequences of messenger RNA molecules, and the amino acid sequences of proteins are collectively referred to as **genetic coding** relationships; specifically the sequences of bases found in mRNA molecules are referred to as a **genetic code,** while each set of three bases is called a **codon.**

In the case of the two β alleles, it was discovered that the inversion of one base pair in the middle of the sixth base-pair triplet of the gene gives rise to one base difference in the middle of the sixth triplet of RNA bases; this change subsequently results in valine (found in Hb-S) replacing glutamic acid (found in Hb-A). This amino acid substitution alters the shape and function of hemoglobin molecules sufficiently to alter their oxygen-carrying function, which in turn alters the development of human beings such that their phenotypes are as strikingly different from normal health as are those of persons with sickle cell anemia.

Hundreds of human diseases can be explained by analogous genetic changes and analogous abnormal development. Experiments show, in precise detail, how information encoded by DNA is translated into protein primary structures. Yet we still do not understand how the information contained within the amino acid sequences of proteins is "transmitted" into their **tertiary structures.** But it is known that changes in protein primary structure may (but not always) lead to changes of protein function.

In other words, it is possible to trace the information encoded within genes through RNA into the primary structures of protein, but after that point we lose the tracing, that is, it is as yet impossible to follow the DNA base pair, RNA base, protein amino acid sequences to the next step, which is the folding of a protein chain into a functional, tertiary structure. Indeed it might be said that the folded structures of proteins are the *first phenotypic outcomes of gene activity,* and that protein functions are the second stage of the phenotypes to emerge, followed by metabolism, physiology, growth, and development. Phenotypes, spanning processes as diverse as protein folding and those that lead to the shapes of noses, are as dependent upon nongenetic as upon genetic inputs—a fact that in no way detracts from the exquisite beauty of the primary sequences.

Or of Mendel's tall and short peas. Mendelian genetics still works, even after DNA. Genetic diseases can be traced to specific genes, and *many genetic diseases can be explained from the vantage point of inborn errors of*

Hb-A β chain

val · his · leu · thr · pro · glu · glu · lys · ser · ala · val · thr · ala · leu · tyr · gly · lys · val · asn · val · asp · glu · val · gly · gly
1   6   10   20   glu

pro · thr · ser · leu · asp · gly · phe · ser · glu · phe · phe · arg · gln · thr · try · pro · tyr · val · val · leu · leu · arg · gly · leu · ala
asp 50   40   30

ala · val · met · gly · asn · pro · lys · val · lys · ala · his · gly · lys · lys · val · leu · gly · ala · phe · ser · asp · gly · leu · ala
             60   70   his

asn · glu · pro · asp · val · his · leu · lys · asp · cys · his · leu · gln · ser · leu · thr · ala · phe · thr · gly · lys · leu · asp · asp · leu
phe 100   90   80

arg · leu · leu · gly · asn · val · leu · val · cys · val · leu · ala · his · his · phe · gly · lys · glu · phe · thr · pro · pro · val · gln · ala
           110   120   ala

his · tyr · lys · his · ala · leu · ala · asp · ala · val · gly · ala · val · val · lys · gln · tyr
146   140   130

HbS β chain

val · his · leu · thr · pro · val · glu · lys · ser · ala . . . . . (Hb-S is identical to Hb-A from here to amino acid 146.)
1   6   10

Figure 2.25 Hb-A and Hb-S have a different electric charge because of a difference in one amino acid at the sixth position from the starting end.

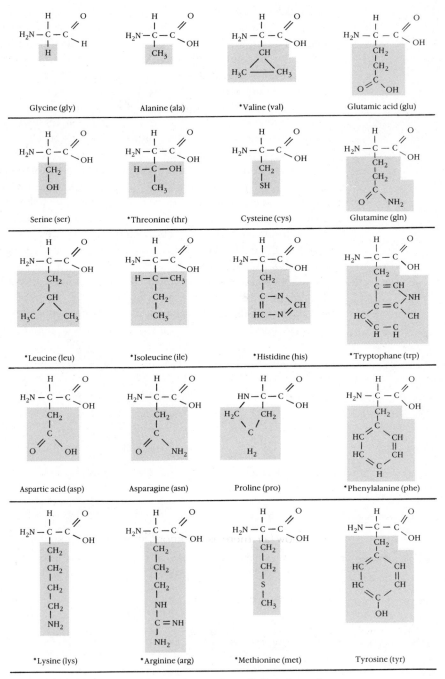

Glycine (gly)

Alanine (ala)

*Valine (val)

Glutamic acid (glu)

Serine (ser)

*Threonine (thr)

Cysteine (cys)

Glutamine (gln)

*Leucine (leu)

*Isoleucine (ile)

*Histidine (his)

*Tryptophane (trp)

Aspartic acid (asp)

Asparagine (asn)

Proline (pro)

*Phenylalanine (phe)

*Lysine (lys)

*Arginine (arg)

*Methionine (met)

Tyrosine (tyr)

*These amino acids are "essential" amino acids that must be provided in the human diet.

1. All amino acids have in common one amino (NH_2) and one carboxyl (COOH) group.

2. Each amino acid is unique at its R (marked by the shaded box).

Figure 2.26 The 20 amino acids that make up the proteins of plants, animals, and microorganisms.

metabolism. And most inborn errors of metabolism can be traced to an altered enzyme. As one protein chemist has said, "If you use a verb in a discussion of biology, you are talking about the actions of proteins; proteins are the verbs within the context of biological growth, development, reproduction, hunting and gathering, etc." Proteins "breathe life" into the information encoded within DNA molecules.

But again, there are different levels of biological complexity from which to view this fact. The simplest level is that of a single protein (e.g., hemoglobin). It is possible to determine the primary structures of hemoglobin molecules, the folded structures (in particular, their different shapes as they become attached and unattached to oxygen), and so forth. In short, we can strive to discover "everything there is to know about hemoglobin."

A higher level of biological complexity includes an entire ensemble of proteins whose coordinated actions sustain the life of cells, tissues, organs, and individuals. Every healthy human being carries millions of red blood cells within their blood streams, and every healthy red blood cell is packed with millions of hemoglobin molecules. The vast majority of humans possess the same muscle proteins, actin and myosin, the same battery of **digestive** enzymes (carbohydrates are digested into simple sugars; proteins into amino acids; lipids into glycerol and fatty acids; nucleic acids into A, T, G, C, and U bases, etc.), and the same battery of **synthetic** enzymes, enzymes that catalyze the reactions necessary for everyone to make their own proteins, carbohydrates, nucleic acids, and lipids. Incidentally, everyone makes their own proteins, etc., using the breakdown products of the foods they eat (e.g., the amino acids derived from digesting chicken and bean proteins). Now, it is far more difficult to discover how the thousands of species of protein are synthesized in the proper place, at the proper time, in the proper concentrations, while maintaining the proper cellular and organismal environments for the performance of their individual functions in concert, than it is to "discover everything there is to know" about any one of them. However, a philosophical question that has not been answered to the satisfaction of all biologists and that complicates the statements made above is whether it is possible to discover everything there is to know about any one protein before we know all there is to know about the ensemble.

In this short introductory book about heredity, it is impossible to give proteins a "fair hearing;" nevertheless, any description of heredity is obliged to emphasize that *the development of phenotypes is a coordination of processes mediated by an incomparably complex ensemble of gene-informed proteins.* At least one gene in each of our genomes informs each species of protein, and in this way genes influence phenotype; but phenotypes develop in concert with protein action within rather wide ranges of environmental inputs (review norms of reaction, Chapter 1).

A review of what has been said illustrates this point. The sum of the biological processes that include the base-pair sequences of the β^A and β^S alleles, the base sequences of their transcripts, the amino acid sequences of the β proteins encoded therefrom, and the health of the individuals of all three genotypes is easily abbreviated in ways that obscure the whole. The statement that "the β^S allele causes SCA," while not absolutely wrong, emphatically blurs our understanding of what genes are and do and how genes influence phenotype. Each of us is far more complex than any slogan that may appear, superficially, to describe us.

How Does DNA Inform Proteins?

There are two sets of processes that make up the whole: (1) the sense strand of DNA informs an mRNA transcript by serving as a template upon which the transcript is synthesized, as described above; and (2) the information contained within mRNA transcripts is translated into amino acid sequences of proteins during protein synthesis. A discussion of this second set of processes is next.

Protein Synthesis

The vast majority of proteins are synthesized in the cell cytoplasm, not in the cell nuclei where chromosomes and genes reside and where RNA transcripts are synthesized (Figure 2.27). Many kinds of protein and a few kinds of cellular organelle participate in protein synthesis, but only a few of these are included in this brief summary. First, of course, is an mRNA transcript. Next, protein synthesis occurs on tiny organelles called **ribosomes** (Figure 2.28). These tiny organelles are composed of RNA and protein (**ribosomal RNA, or rRNA,** and ribosomal proteins), all of which are informed by genes. A third species of RNA (**transfer RNA, or tRNA**) transports amino acids from the cell cytoplasm to the

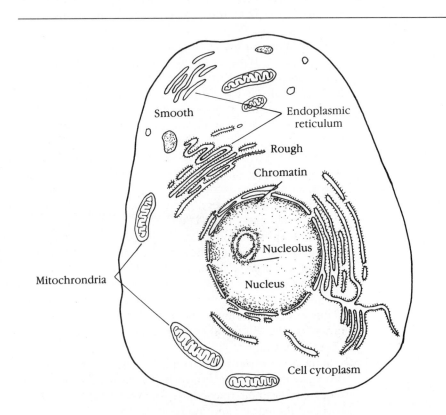

Smooth

Endoplasmic reticulum

Rough

Chromatin

Nucleolus

Mitochrondria

Nucleus

Cell cytoplasm

The endoplasmic reticulum is internal membrane,
some smooth, some rough. The "roughness" is due to the attachment of
thousands of ribosomes. Cellullar proteins are synthesized on the ribosomes.

Figure 2.27 A generalized human cell. Proteins are synthesized in the cell cytoplasm.

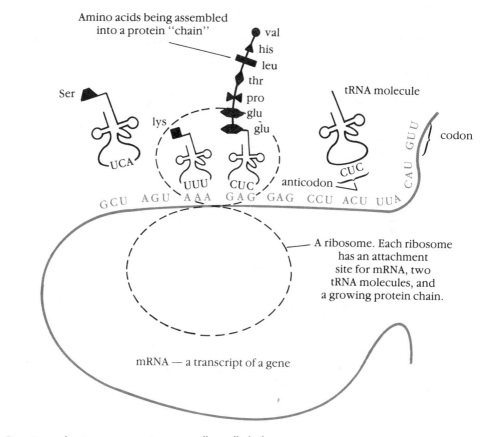

Figure 2.28 Protein synthesis occurs on tiny organelles called ribosomes.

sites of protein synthesis on the ribosomes. There are some 61 different kinds of tRNA molecule, each with a sequence of bases encoded by a specific gene (Figure 2.29).

If a movie could be made of protein synthesis over a time span of a tiny fraction of a second, we would see frames of structures and process that are shown diagrammatically in Figure 2.30. The long mRNA molecule is attached to many ribosomes, shown at the top of the figure. In the "magnified" view in Figure 2.30 the bases to the right of the ribosome have not been translated; the six bases on the ribosome are being translated, and the bases to the left of the ribosome have been translated. Now to the coding relationships between mRNA and protein primary structures.

The Genetic Code

Keep an eye on Figure 2.30. The bases are translated in sets of three at a time. For each set of three bases on or to the left of the ribosome, there is one amino acid at the top of the ribosome or on the growing protein chain. Specifically, the amino acid located at the top end of the growing protein is met (methionine), and the first set of three bases located at the extreme left end of the mRNA is A U G. Whenever A U G appears in the mRNA, met appears in a corresponding position on the growing protein molecule. The next set of three bases on the

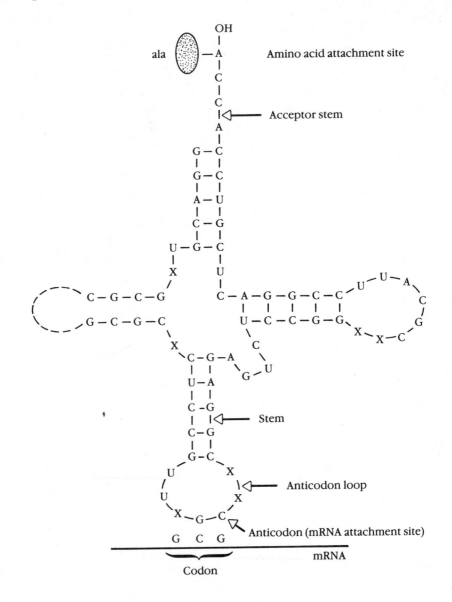

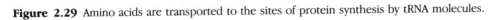

Figure 2.29 Amino acids are transported to the sites of protein synthesis by tRNA molecules.

mRNA is U U U. Attached to U U U is a tRNA molecule with an A A A sequence at the point of attachment, as predicted from the base-pairing rules. But what is important here is that the tRNA molecule that base pairs with a set of three mRNA bases always carries a specific amino acid at its other end. In this case the AAA tRNA molecule carries phe (phenylalanine). At the time the movie was taken, this tRNA molecule was, in one frame, caught in the act of transferring phe to met at the growing end of the new protein molecule.

There are, in fact, about the same number of tRNA molecules as there are kinds of mRNA triplets. Given an alphabet of 4 letters and a vocabulary consisting only of 3-letter words, we calculate a "dictionary" of 64 words: $(4)^3 = 64$ (Figure 2.31). In Figure 2.31 you see not only the vocabulary used by all living

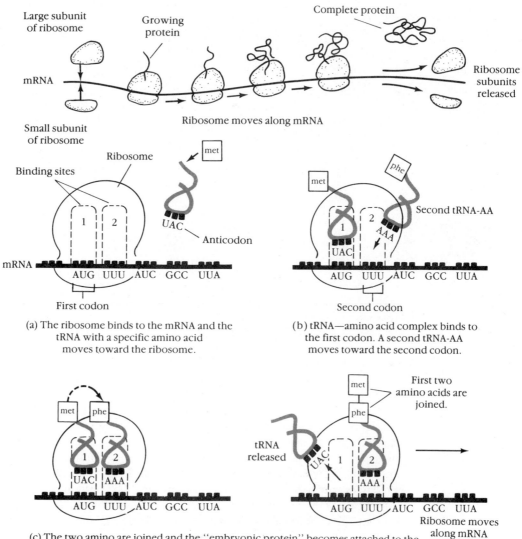

(a) The ribosome binds to the mRNA and the tRNA with a specific amino acid moves toward the ribosome.

(b) tRNA—amino acid complex binds to the first codon. A second tRNA-AA moves toward the second codon.

(c) The two amino are joined and the "embryonic protein" becomes attached to the second tRNA. The first tRNA then vacates the first binding site.

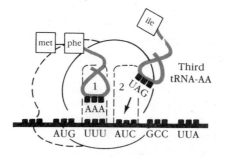

(d) The ribosome moves down the mRNA, transferring the tRNA-two amino acids to the first binding site and opening up the second binding site to a third tRNA.

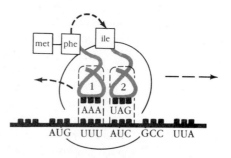

(e) The two amino acids are linked to the third amino acid, forming a three amino acid complex. This process repeats itself until the last codon has been transcribed.

Figure 2.30 A brief moment during protein synthesis.

1	2 U		C		A		G		3
U	UUU	*phe*	UCU	*ser*	UAU	*tyr*	UGU	*cys*	U
	UUC	*phe*	UCC	*ser*	UAC	*tyr*	UGC	*cys*	C
	UUA	*leu*	UCA	*ser*	UAA	*STOP*	UGA	*STOP*	A
	UUG	*leu*	UCG	*ser*	UAG	*STOP*	UGG	*trp*	G
C	CUU	*leu*	CCU	*pro*	CAU	*his*	CGU	*arg*	U
	CUC	*leu*	CCC	*pro*	CAC	*his*	CGC	*arg*	C
	CUA	*leu*	CCA	*pro*	CAA	*gln*	CGA	*arg*	A
	CUG	*leu*	CCG	*pro*	CAG	*gln*	CGG	*arg*	G
A	AUU	*ile*	ACU	*thr*	AAU	*asp*	AGU	*ser*	U
	AUC	*ile*	ACC	*thr*	AAC	*asp*	AGC	*ser*	C
	AUA	*ile*	ACA	*thr*	AAA	*lys*	AGA	*arg*	A
	AUG	*met*	ACG	*thr*	AAG	*lys*	AGG	*arg*	G
G	GUU	*val*	GCU	*ala*	GAU	*asp*	GGU	*gly*	U
	GUC	*val*	GCC	*ala*	GAC	*asp*	GGC	*gly*	C
	GUA	*val*	GCA	*ala*	GAA	*glu*	GGA	*gly*	A
	GUG	*val*	GCG	*ala*	GAG	*glu*	GGG	*gly*	G

Capital letters are mRNA codons
Letters in italics are abbreviations of the 20 amino acids

Figure 2.31 The genetic "dictionary."

creatures but also the meaning of each word. Each word of the RNA vocabulary translates into a specific amino acid. Using the communication metaphor, you may analogize an mRNA molecule with a sentence, a sentence written in the language of nucleic acids. In the same context, protein synthesis is a translation process during which words (codons) of the nucleic acid language are translated into letters (amino acids) of the protein language. The protein language has an alphabet of 20 letters (the 20 amino acids), and protein words range in length between 7 to 8 amino acids to several hundred. (The β-protein of hemoglobin is 146 amino acids in length.) The meaning of protein words is found in their tertiary structures and in their biological functions.

The three-letter words of mRNA are called codons. The three-letter recognition sites on tRNA molecules are called anticodons. While this is not a good name, we are stuck with it. What it means is that codons attach, by the base-pairing rules, to the three-letter complements of specific tRNA molecules; through this attraction process, each mRNA codon "invites" a specific amino acid to the growing end of a new protein molecule. That each anticodon is accompanied by a specific amino acid is explained by the fact that in the cytoplasm of the cell there are enzymes that recognize both the tRNA molecule and the amino acid and that **catalyze** their chemical union. There is at least one such enzyme for each tRNA-amino acid combination.

The first experiment to reveal that codons attract specific amino acids to the sites of protein synthesis, done by Marshall Nirenberg and for which he received a Nobel Prize, included all of the cellular components needed for protein synthesis except mRNA. Nirenberg added artificial mRNA to this system and observed the chemical composition of the new proteins synthesized therein. If the artificial mRNA was U U U U U U U U U . . . (poly U), the new protein was phe-

phe-phe . . . (poly phe). If the artificial mRNA was poly C, the new protein was poly pro. Poly A gave rise to poly lys, and poly G gave rise to poly gly.

Nirenberg thereby discovered the meaning of four codons. Eventually he was able to make artificial mRNA molecules, 3 letters in length, in all of the 64 combinations using a 4-letter alphabet. He then allowed each to become attached to its complementary anti-codon on tRNA molecules "loaded" with their appropriate amino acids. From each tRNA–3-letter mRNA pairing, he was able to identify all of the amino acid-codon combinations shown in the genetic dictionary (Figure 2.31).

As the language of life was being decoded, genetics was more and more in the news. The question whether the genetic dictionary is a *universal language* of life created excitement not only within the community of biologists but in all of science as well as in the public sector. So simple a language: 4-letter alphabet, 61 words, 3 punctuation marks (all "periods," which signal the end of synthesis of a protein). Can so simple a language explain bacterial reproduction and variation as well as redwood tree, python, and human reproduction and variation? As we know today, with a few exceptions that will not be included here, the answer is a resounding yes!

A review here will provide direction in anticipating what is ahead. Every species of animal, plant, and microorganism possesses a "library" of information. This information is stored and replicated as base-pair sequences within DNA molecules. No individual possesses all of its species information, nor do individuals "mail to the next generation" all of the information carried by their own genomes. So not only do individuals inherit but a sample of their parent's genes, each new generation inherits only a sample of the genes carried by its parent generation. However, individuals within species are far more alike genetically than different, that is, the vast majority of individuals within a species inherits the genes necessary for encoding the thousands upon thousands of proteins necessary for normal growth and development (Figure 2.32).

As far as is known, all genes are subject to being copied incorrectly or to being altered by any one of thousands of mutagens. The β^A allele that encodes the β

---

decrease in genetic variation

Species (Homo sapiens)
$\cong$ 100,000 genes
$\cong$ 30,000 genes with two or more alleles
an unknown number of genes with three or more alleles
Examples: I gene, I^A, I^B, I^O alleles (chromosome 9)
100^+ alleles of the four HLA genes

Individual genotypes
$\cong$ 100,000 genes
maximum of two alleles of any one gene
Examples: I^AI^A, I^AI^B, I^AI^O, I^BI^B, I^BI^O, I^OI^O
A_1A_3, B_2B_4, C_3C_5, D_4D_6

Gametes
$\cong$ 100,000 genes
only one allele of each
Examples: I^A, I^B, or I^O

Figure 2.32 Similarities and differences, from the perspective of DNA.

Table 2.1 Some functions of proteins

Protein	Function
Actin and myosin	Contractile proteins of muscle cells
Hemoglobin	Transport protein; carries oxygen to and CO_2 away from metabolizing body cells
Collagen and keratin	Connective tissues and hair
Antibodies and others proteins	Act within the immune system
Histones	Structural proteins that complex with DNA in chromosomes
Enzymes	Catalyze single biochemical reactions that break complex molecules into simpler ones or create complex molecules from simple ones

protein of HbA, when changed by one base pair out of 438, gives rise to a β protein changed by one amino acid, and this change may derail development from normal to severe ill health. If mistakes can occur within DNA molecules, why shouldn't we be able to correct those mistakes by changing the base pair sequence of the gene back to its normal state or by inserting normal genes into the chromosomes of persons with mistaken ones (Figure 2.33)? These questions are being addressed by "genetic engineers," but to more fully understand the task of the genetic engineers it is helpful to advance the discussion of proteins.

Protein Functions

To begin, see Table 2.1 for a short list of protein functions, and Figure 2.34 for an introduction to protein structure. Hemoglobin carries oxygen to all cells of the body, an important function considering the fact that cells die quickly without oxygen. Muscle cells expand and contract, a fact that makes it possible for us to behave as animals as opposed to plants (Table 2.1). Enzymes make it possible for us to break down complex foodstuffs into small molecules, to transport these small molecules through gut membranes into the blood stream, to transport them out of the blood into cells, and there to digest them further into carbon dioxide and water and therefrom to extract energy, or to use them as building blocks during synthesis of our own complex molecules (Figure 2.35).

The subject of enzyme functions brings us back to Garrod's inborn errors of metabolism. But before getting to specifics, examine the general concept of synthesis shown in Figure 2.35. The letters A, B, C, etc. symbolize a family of chemical molecules; the numbers, 1, 2, 3, etc. symbolize a family of enzymes, each number a different enzyme. The arrow from A to B symbolizes the chemical transformation of molecule A into molecule B, and the number 1 above the arrow symbolizes the specific enzyme that catalyzes this transformation. (See Perspective 2.4, page 95)

When the muscle on a T-bone steak is completely digested, 20 amino acids remain, the same 20 amino acids required to make β-A and β-S proteins (Figures 2.25 and 2.26). These 20 amino acids are recycled among plants, animals, and bacteria, (all animals can synthesize some of their own amino acids) and they function in all organisms as building blocks of proteins and as a source of energy. Consider the relationships among human beings, cows, and grass. Grass makes its proteins from CO_2, H_2O, and energy from the sun; cows ingest the grass,

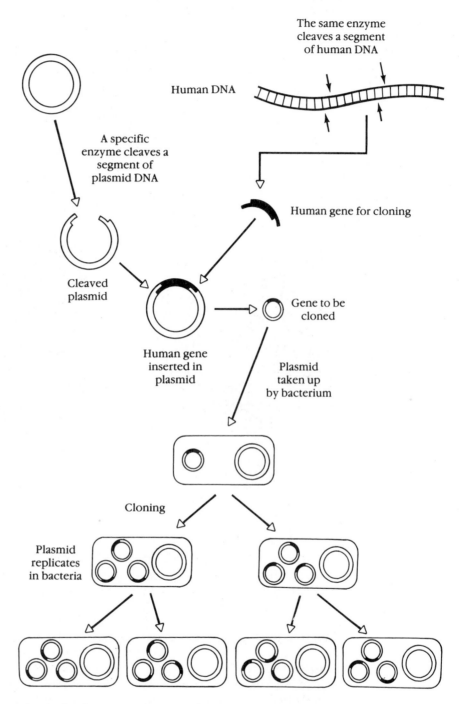

Figure 2.33 Can genetic mistakes be corrected? A normal human gene can be isolated and inserted into a bacterial plasmid—a small, circular DNA molecule found in many species of bacteria. The recombinant plasmid can then be returned to a bacterial cell where, as the bacterium divides, the plasmid replicates, thereby amplifying the human gene (gene cloning).

Once cloned, the human gene can be used to (a) make RNA transcripts, which in turn may be used to make normal proteins, (b) correct the action of its mutant allele by inserting it into a human cell, after which the cell carrying the normal allele may be inserted into the human tissue needing the normal protein, or (c) at some time in the future the normal allele may be inserted into a fertilized egg, thus permitting the individual that emerges from the recombinant egg to pass the normal allele to her or his children.

The primary structure of a protein
is its amino acid sequence, determined
by the base pair sequence of a gene
(see also Figure 2.25)

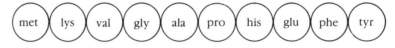

The folded structure is no
doubt influenced by the primary
structure, but we do not know how
it is assumed.

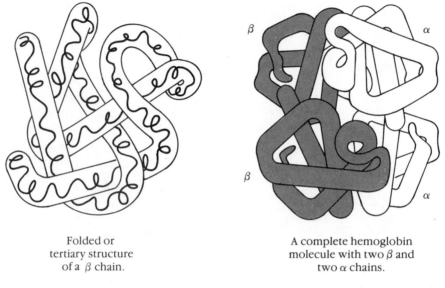

Folded or
tertiary structure
of a β chain.

A complete hemoglobin
molecule with two β and
two α chains.

Figure 2.34 Protein structure.

digest grass proteins, and synthesize their own proteins from amino acids retrieved from grass proteins; humans ingest cows, digest cow proteins, and synthesize their own proteins using the amino acids retrieved from cow proteins.

Genes, Enzymes, Metabolism, and Disease

Consider two of the 20 amino acids, phenylalanine and tyrosine. Almost every protein we eat, whether from fish or fowl, beans or Big Macs, contains phenylalanine and tyrosine. Upon digestion of protein, these amino acids enter the blood stream where they are carried to all cells, tissues, and organs of the body. These two amino acids are acted upon by a number of enzymes, and they may enter into a number of metabolic pathways (Figure 2.35). One pathway is that of further digestion into CO_2, H_2O, and energy; another pathway leads toward melanin pigment, the pigment that gives color to skin and hair; another leads toward the hormone, thyroxin. In addition, both amino acids are used as building blocks in the synthesis of body proteins.

These pathways illustrate how gene mutations lead to genetic diseases (inborn errors of metabolism). Consider the disease, alcaptonuria, described by

Garrod. Observe the metabolic pathway by which tyrosine is digested into CO_2 and H_2O (Figure 2.35). Part way down the pathway homogentisic acid is transformed into maleylacetoacetic acid by the enzyme homogentisic acid oxidase. If the gene that encodes homogentisic acid oxidase is altered (into a mutant allele), the enzyme may become altered and as a consequence fail to transform homogentisic acid into maleylacetoacetic acid. Individuals who inherit two copies of such a mutant allele will not change tyrosine into CO_2 and H_2O but into homogentisic acid, which accumulates in cells, in the blood stream, and in the urine. Upon excretion of urine with relatively high concentrations of homogentisic acid, the urine turns black, the first diagnostic criterion of alcaptonuria.

If a mutation occurs in any one of the genes that encodes an enzyme in the metabolic pathway from tyrosine to melanin, the phenotypic consequence (of having two such mutant alleles) is albinism. Persons who lack melanin are albino. As you can predict upon examining this metabolic pathway, several genes, in mutant form, lead to albinism.

Probably the most studied of the inborn errors of phenylalanine and tyrosine metabolism is **phenylketonuria** (PKU). As shown in Figure 2.35, phenylalanine is transformed into tyrosine by the enzyme *phenylalanine hydroxylase*. Functional phenylalanine hydroxylase is encoded by the *P* allele of the P gene. Persons homozygous for the *p* allele do not synthesize phenylalanine hydroxylase, and in the absence of this enzyme phenylalanine accumulates, first in certain liver cells, then in the blood stream and urine. Untreated PKU children, who a few days after birth begin to accumulate relatively high concentrations of phenylalanine in their blood, are characterized by retarded mental development, eczema, pungent urine odor, reduced pigmentation (i.e., reduced melanin pigment), and specific neurological symptoms. The clinical diagnostic feature of PKU is elevated concentrations of phenylalanine in the blood of babies about a week after birth.

It is not known how high concentrations of blood phenylalanine lead to mental retardation, but there is no doubt that it does. Untreated PKU children develop exceedingly few mental skills (96% score below 60 on IQ tests). However, the reduced pigmentation is explained by the inhibition of the enzyme tyrosine hydroxylase by high concentrations of phenylalanine. Tyrosine hydroxylase transforms tyrosine into a chemical called DOPA, a step in the metabolic pathway toward melanin. It is postulated that mental retardation and the

---

Perspective 2.4

Almost all chemical reactions that take place in our body cells and tissues can occur outside our bodies but at much slower rates. It takes years for the iron in an old car to become **oxidized** (rust), but it takes only a tiny fraction of a second for the iron in hemoglobin in the blood stream to become oxidized. It is possible to speed up chemical reactions outside the body by supplying heat, but temperatures much above 37° C are not healthy for human cells. As it turns out, the tasks of making chemical reactions happen quickly at relatively low temperatures is accomplished by synthesizing catalysts, chemicals that are not changed by the chemical reactions they speed up. Biological catalysts are enzymes, and all enzymes are gene-informed proteins. Enzyme-catalyzed reactions that occur in our bodies occur at rates thousands of times faster than they would at the same temperature without a catalyst. Think of it this way. You digest the proteins of the beans you eat into amino acids in about two hours. In the absence of enzymes the process would require hundreds of years at 37° C.

I. Synthesis

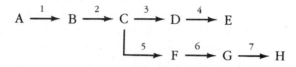

- A, B, C, etc. are biochemical molecules.
- 1, 2, 3, etc. are enzymes.
- E and H are essential for life.

II. Digestion

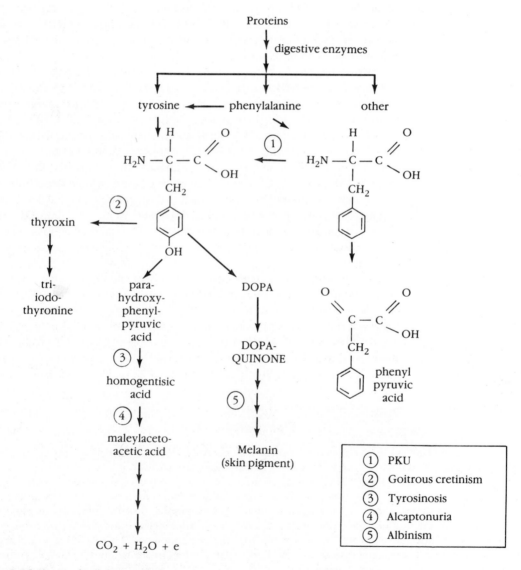

Figure 2.35 Types of cellular metabolism.

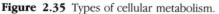

neurological symptoms (irritability, tremors, hyperactivity, and seizures) also result from inhibition by phenylalanine of enzymes that facilitate metabolism within the nervous system.

PKU children will develop normally if they are given a phenylalanine-free diet after about one month of age until the age of 10 to 12 years. This is costly, and the food prepared without phenylalanine is anything but tasty, but the results are striking; PKU children on phenylalanine-free diets exhibit intelligence levels indistinguishable from those of their non-PKU siblings. This is an excellent example showing that fantastically different phenotypes may emerge from the same genotype if that genotype initiates development in different environments, in this case, different intake levels of phenylalanine.

It is interesting to find that phenotypes of the kind studied by Mendel also can be explained as "inborn errors of metabolism." For example, the wrinkled seeds of peas are explained by the fact that they possess a different kind of starch than is found in round seeds. Starch molecules are long chains of glucose subunits, and glucose is the simple sugar used to sweeten coffee, ice cream, and breakfast cereals. Long starch chains are synthesized in cells by two different enzymes, one of which merely adds glucose molecules to the growing ends of starch molecules and the other of which causes branching of starch molecules (Figure 2.36).

The S gene that encodes the "branching enzyme" is mutant *(s)* in wrinkled seed plants; the enzyme is missing, and the starch molecules are unbranched. Unbranched starch takes up less space within the seeds; hence the seeds are wrinkled. The discovery of the metabolic consequences of the *s* allele of the S

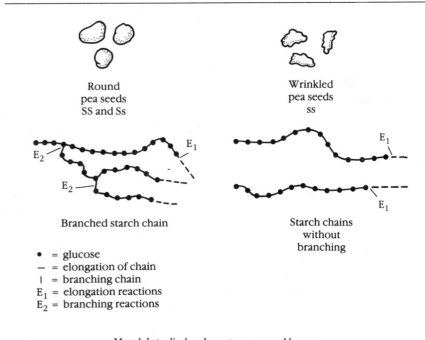

Mendel studied a phenotype caused by an
"inborn error of metabolism". (Branching enzyme is mutant.)

Figure 2.36 A new understanding of the wrinkled peas studied by Mendel.

gene studied by Mendel was made nearly 125 years after Mendel published the results of his studies of inheritance patterns. In short, normal phenotypes emerge as outcomes of metabolic processes catalyzed by normal enzymes; mutant phenotypes emerge as outcomes of metabolic processes disrupted by mutant enzymes; mutant enzymes are encoded by mutant genes.

Genetic Engineering

In Chapter 1 it was mentioned that about 1 of 12 Americans of African descent carries one β^S allele and that about 1 of every 570 babies born to African Americans carries two β^S alleles. Let us suppose that it will become possible to insert β^A alleles into bone marrow cells of $\beta^S\beta^S$ babies and that the transplanted β^A alleles will initiate the synthesis of enough Hb-A proteins to ensure good health for these babies. When these babies mature and have children, they will pass to each of their children one copy of the β^S allele because genotypes of germ line cells are not changed by gene transplants to somatic, in this case, bone marrow cells.

However, the health of these babies will be bettered by the insertion of β^A alleles into their bone marrow cells, and this can't help but be of staggering importance to those individuals and their families. But such a procedure fails to improve species health; in fact, to the contrary, such a procedure increases the chances that additional β^S alleles will be returned into the species *gene pool*. This conflict of interest—between individual and species good—is seldom addressed in the classroom.

Throughout most of human history people have been ignorant of reproduction, growth and development, metabolism, physiology, health, and disease. Most of what is known about these phenomena has been discovered during the 20th century. But even today hundreds of millions of people are ignorant of the human life cycle, the causes of disease and malnutrition.

Ignorance of human biology has not deterred speculation about it. Human beings have created all kinds of ideas about human biology and relationships between humans and other forms of life. Some of these old and incorrect ideas have influenced modern attitudes toward genetic engineering, for example, the complementary myths of human monsters and *the ideal* human being.

A common thread of the human monster myth has taken the form of imagining hybrids between human and nonhuman forms—the head of a man and the body of a horse; the head and torso of a woman and the lower torso and tail of a fish; unicorns; people with dragon tails and donkey ears; devils, angels, and so on. The better the artist the more convincing the myth.

The second myth rests on the premise that perfection is possible. While many people may boast of having reached perfection, the state of perfection is antithetical to material reality (in the real world everything is in flux). The idea of perfection is as much of a myth as are mermaids.

Many of the expressed fears of genetic engineering are fears of creating monsters, while one of the most advertised promises of genetic engineering is that of transforming human beings—and cows, horses, wheat, dogs, corn, and so on—into "perfect" forms. Even seasoned and popular television news analysts titilate us with visions of parents shopping for just the right parts for their future children—voice of Mahalia Jackson, body type of Jose Conseco or Farah Fawcett, brain of Richard Feynman, musical ability of Vladimer Horowitz,

An Aside

There are traditions within scientific research communities as there are within communities of artists, architects, and angels (Hell's). Many community traditions have unknown origins, and many are arbitrary, but those who believe in them often give the impression of believing that traditions are equivalent to the laws of nature—they are real, they are uncompromising, and they are there to be obeyed.

We hear it said that **objectivity** is a tradition within science communities, and that scientists are in fact objective in their research and research publications. To a large extent, this is correct; but it is not, by any reckoning, absolutely correct. What may appear on the surface to be objective behavior may be behavior spurred by very different motives. For example, every scientist knows that another scientist, somewhere, sometime, may repeat his or her experiments, and that if anything is wrong or incomplete about them, the community of scientists will hear about it. The fear of being found wrong may provoke some scientists to stay on their objective toes. Other scientists try their best to be objective for the simple reason that they believe that objectivity is an inviolate rule that scientists must follow, analogous to one of the 10 commandments. (If you have traveled by automobile from state to state, you may have noticed that the signs reminding you to buckle your seat belt vary, as scientists vary with respect to objectivity. To wit, one state will tell you to "Buckle up: It's the Law." Another state will tell you to "Buckle up: For Your Life." The first warning appeals to our fears—a fear of being stopped and fined by a highway patrol person; the second warning appeals to our moral sense—we are being told what is good for us).

The tradition of objectivity is a unifying theme among scientists, but at the same time the word means different things to different scientists. Some scientists feel that it is absolutely wrong to make public statements about societal issues, saying, in effect, that the scientist's expertise fizzles at the border between science and society. Others feel that the process of acquiring expertise in science gives one an advantage in acquiring expertise in other areas of human activity and that therefore scientists have a societal obligation to share their skills and their expertise with members of the larger community. In short, some scientists feel that science and scientists *should* share with others the responsibility of improving the quality of life for all.

These two views of the scientist's social responsibilities are as different from one another as are the views of conservatives and liberals, and the resolution of the differences seems as far from reality today as it did during the Edward Teller-Linus Pauling debates over the future of developing and using atom and hydrogen bombs during the 1950s. Both Teller and Pauling are Nobelists (Pauling won two Nobel Prizes!); both are devastatingly smart, by my standards; yet they presented for all the world to hear opposite points of view on nearly every societal question of war and peace. Were their differences rooted within a misunderstanding of the atom or of how to make a bomb? *Should* they have kept quiet, leaving the questions of war to the generals, and the questions of peace to the supreme court judges? Would the debates have been better if scientists had stayed out of them? *Should* nonscientists in society try to understand some of the things that scientists understand and vice versa?

What are the correct societal responses to progress in science? There are no absolutely right and wrong answers just as there are no absolutely right or wrong answers to the questions of how to teach science-society relationships. If we ignore these issues we can expect continuous warfare among competing special interest groups. On the other hand, if we begin to make scientific assessments of both the dangers and the causes of pollution, disease, species extinctions, malnutrition, sexism, and racism, we may be able to reverse the damage before reaching the point of no return. It is my view that scientists and teachers have a social responsibility to contribute to the discourse as well as to the discovery process.

running style of Florence Joyner Griffith, and the moral fortitude of Carrie Nation coupled with Donald Trump's privilege of operating outside the rules.

Knowledge of genetics has expanded and improved our understanding of human health and disease, and that is something to shout about. But our knowledge of genetics does not explain Chopin's musical creations or

Horowitz's rendition of them. To combine Mahalia Jackson's singing voice with Feynman's genius in physics will require more than a knowledge of DNA base-pair sequences.

Early Genetic Engineering Programs

Long before genetics became a science, social programs had been set in motion to improve the quality of the members of tribes and nation states. Matings were forbidden across territorial boundaries; parents selected spouses for their children; weak and handicapped people were forbidden from having children. The list is long, but at root in most cases were the complementary myths of monsters and the ideal type.

During the past 150 years the most notorious form of human genetic engineering has been **eugenics,** a pseudoscientific approach to "improving the quality" of human beings. The founder of eugenics, Sir Francis Galton of England, proposed to "improve the human stock" by encouraging matings among genetically superior people and by discouraging childbearing among the genetically inferior, the latter by forced sterilization.

To date no laws have been enacted to increase the numbers of "genetically superior" children. But laws have been enacted to decrease the numbers of "genetically inferior" children. The first eugenics law was enacted in the state of Indiana in 1907, and by 1930 some 30 states in the United States had enacted forced sterilization laws. Among the traits used to diagnose genetically inferior people were drunkenness, sexual perversion, venereal disease, poverty, homelessness, blindness, deafness, idiocy, imbecility, moronism (this name was invented in the 1920s), epilepsy, and the state of being orphaned. Even avid genetic determinists today disagree that genes direct children to become orphans and that syphilis is caused by a particular base-pair sequence.

It turns out that the major proponents of eugenics were not so much concerned with the health of unborn babies as with the issue of human races (discussed in Chapter 4). Sterilization programs were and are now aimed at ethnic minorities, women, non-nationals, and poor people. However, before we move on to modern techniques for genetic engineering, the reader must be warned not to equate the motives for using eugenics to improve the "quality of the human stock" with modern genetic engineering motives to ameliorate or cure human diseases. They are different indeed. At the same time, it would be equally wrong to assume that the kinds of motives that breathed life into the eugenics movement have died and are now "resting in peace." These motives are not dead, and they have not been buried.

Modern Proposals for Genetic Engineering

Consider that "inborn errors of metabolism" initiated by mutant genes often can be treated without correcting the genes. In the case of elevated levels of phenylalanine, the solution is to lower the intake of phenylalanine. But treating genetic diseases in this way actually increases the numbers of deleterious genes that are returned to the gene pool.

Now there are different ways to treat genetic diseases. But treatment is different from cure. Unless the gene itself in germ line cells is corrected or replaced, that mutant gene will increase in population gene pools as the lives of diseased individuals are prolonged by medical intervention.

Contemporary genetic engineering programs include the diagnosis, treatment, and elimination of diseases that can be traced to single gene mutations, chromosome abnormalities, and abnormal numbers of chromosomes. So far the only programs that have been proposed for improving the genetic health of populations have been programs to control human exposure to mutagens (e.g., reducing exposure to x-rays and reducing the concentrations of mutagens and carcinogens in food, medicine, air, and water).

A few attempts have been made to screen certain populations for known deleterious mutations and to implement genetic counseling centers for prospective parents, but more than 99% of babies born are conceived by parents who have not been screened or counseled. The only widespread attempts to use counseling to prevent births of genetically diseased children have been in populations with high frequencies of thalassemia and SCA, and these attempts have been judged to have failed.

Diagnosis of Genetic Diseases

Whether the goal is to eliminate or simply ameliorate the symptoms of genetic diseases, it is necessary to diagnose them early. Today it is possible to diagnose a few genetic diseases during prenatal development, and in a few cases causative genes can be identified in eggs and sperm cells and in embryos.

The most useful technique for diagnosing genetic diseases during the first trimester (of embryonic and fetal development) is **chorion villus sampling** (Figure 2.37). As early as the eighth week of gestation it is possible to aspirate villi cells (a form of biopsy) from the chorion of a fetus by use of a catheter inserted into the uterus. The villi cells can be cultured, and from a large population of cultured cells DNA can be extracted, after which it can be compared with normal DNA.

During the second trimester it is possible to inspect the fetus directly and to extract fetal cells from the amniotic fluid by means of **amniocentesis** (Figure 2.38). It is possible to remove samples of fetal blood and to biopsy fetal skin and liver cells. Again the cells extracted can be cultured, and from the cultured cells DNA can be isolated. The chromosomes of the cultured cells can be examined microscopically and compared to normal chromosomes.

If a first trimester fetus is discovered to carry a deleterious gene or chromosome, its parents may choose to have it aborted. But if such a discovery is made during the second trimester, abortion becomes more problematic. The later the discovery is made, the more likely it becomes that the attending physicians will treat the disease symptomatically (i.e., treat the symptoms of the disease as opposed to trying to cure the disease). Indeed, to date it is impossible *to cure* genetic diseases. A few of the genetic diseases that can be diagnosed in utero are shown in Table 2.2.

Fetal DNA Analyses

The technology used to detect single gene mutations in fetal DNA are complicated. Only a brief description is presented here. The technology is based on the discovery of enzymes that "cut" DNA molecules at specific sites (i.e., specific short sequences of very long DNA molecules). Enzymes that "cut" DNA

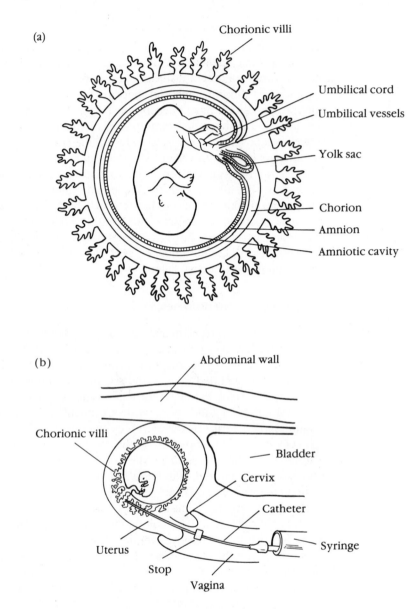

Figure 2.37 Chorion villus sampling. (a) Fetus at eighth week of pregnancy. (b) The villi cells are removed by means of a small tube inserted through the vagina and cervix, into the uterus.

molecules are called **endonucleases,** and enzymes that "cut" DNA molecules at specific sites are called **restriction endonucleases.**

A specific restriction endonuclease will cut identical DNA molecules into identical numbers and sizes of smaller DNA fragments. Therefore it is possible to determine whether an unknown DNA molecule is identical to or different from a known molecule simply by comparing the numbers and sizes of fragments produced following digestion with a specific restriction endonuclease (Figure 2.39). If the unknown and the control samples yield identical sets of fragments, the recognition sites of the restriction endonuclease in the two molecules are identical. If the two sets of fragments are different, the recognition sites in the two molecules are correspondingly different.

It must be remembered that one restriction enzyme recognizes only a tiny fraction of the base-pair sequences within large DNA molecules. For this reason many restriction enzymes are needed to determine similarities and differences among related DNA molecules. As you see in Figure 2.39, each enzyme explores different segments of a DNA molecule's primary structure.

To illustrate with a familiar example, recall the β gene and the β protein chains of hemoglobin. Glutamic acid is found in the sixth position of Hb-A β chains, and valine is found in the corresponding position of Hb-S β chains. The β^A allele differs from the β^S allele at the middle base pair of the sixth set of base pair triplets; it turns out that an inversion of this base pair masks a restriction site of the restriction enzyme called MstII (Figure 2.40).

You see in the diagram three sites at which MstII cuts the DNA in the region of the β^A allele. The site on the far left is 1,100 base pairs (bp) from the sixth bp triplet of the gene. The middle site includes the sixth bp triplet, and the site on the far right is 200 bps in the other direction from the sixth bp triplet. In other words, the DNA of $\beta^A\beta^A$ persons is cleaved into two fragments in the region of the β^A allele, a 1,100-bp and a 200-bp fragment. If the exact same analysis is made of the DNA of $\beta^S\beta^S$ persons, only one fragment is found, a 1,300-bp fragment that includes the sixth bp triplet of the β^S allele. This must mean that the mutation that changes the β^A allele into a β^S allele masks the middle MstII recognition site, a fact that makes is possible to identify the base pair difference between the two

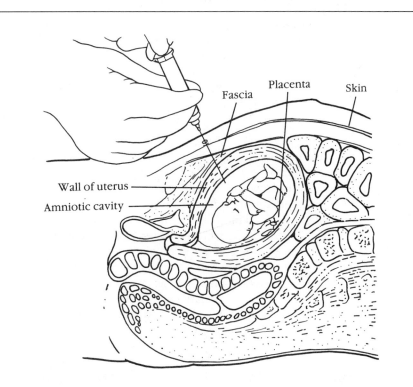

Fascia   Placenta   Skin

Wall of uterus

Amniotic cavity

Figure 2.38 Amniocentesis. Amniocentesis is usually performed during the 12th to 16th weeks of pregnancy, and involves the withdrawal of amniotic fluid.

Table 2.2 A few genetic diseases that can be diagnosed in utero.

Disease	Missing Enzyme
Lipid metabolism	
Fabry's disease	α-galactosidase A
Gaucher's disease	β-glucosidase
Gangliosidoses	β-galactosidases A, B or C
	Hexosaminidases A, or A and B
Krabbe's disease	Galactocerebroside β galactosidase
Metachromic leukodystrophy	Arylsulphatase A
Mucolipidosis	Multiple lysosomal hydrolases
Neimann-Pick disease	Sphingomyelinase
Mucopolysaccharidoses	
Hurler syndrome	α-L-iduronidase
Hunter syndrome	Iduronic acid sulphatase
Sañfillipo A	Heparin sulphamidase
Maroteaux-Lamy	Arylsulphatase B
Carbohydrate metabolism	
Galactosemia	Galactose-1-phosphate uridyl transferase
Glycogen storage II	α-1,4-glucosidase
Amino acid metabolism	
Arginosuccinic aciduria	Arginosuccinase
Homocystinuria	Cystathionine synthetase
Maple syrup urine disease	Branched chain ketoacid decarboxylase
Methylmalonic acidemia I	Methylmalonic-CoA mutase
Others	
Adenosine deaminase deficiency	Adenosine deaminase
Congenital erythropoietic porphyria	Uroporphyrinogen III co-synthetase
Lesch-Nyhan syndrome	Hypoxanthine-guanine-phosphoribosyl transferase
Xeroderma pigmentosa	Defective DNA repair
Hypophosphatasia	Alkaline phosphatases

alleles simply by identifying the fragments of each after digestion by a restriction enzyme.

Consider what has been learned about the β^S allele: it was first recognized (1) by its effects upon health (SCA), (2) then by its effects upon the shape and function of red blood cells, (3) then by its effect upon the amino acid sequence of hemoglobin β chains (valine in place of glutamic acid), (4) then by its base sequence in the coding strand of the gene (CAC in place of CTC), and (5) now by its *restriction enzyme fragment pattern* (Figure 2.40).

Several forms of thalassemia are easily recognized by substitution, deletion, and or duplication of base pairs within a gene. It is anticipated that many other gene mutations will yield to this much improved resolving power of genetic analysis. See An Aside, page 107.

Gene Exchanges Between Species Occur Naturally

When it first became apparent that genes can be isolated from one genome and spliced into another, many biologists became worried that unforeseeable mischief might be inflicted upon contemporary species of plants, animals, and microorganisms — if the DNA of the different species is artificially mixed. This worry was supported in part by the claim that every species is reproductively isolated from all other species and that reproduction within species and the absence of reproduction between species are delicately balanced biological processes that have been selected, throughout evolutionary time (see Chapter 4), to become efficient processes for ensuring the continuance of species. We now know that DNA of different species is being mixed, to some extent, all the time.

Earlier we discussed bacterial viruses (Figure 2.9) but only to show how they were used to prove that DNA is the material of genes. Bacterial viruses are but one of many types of virus. All organisms are prey to viral parasites — even some viruses are infected by simpler viruses. Humans are plagued by many different kinds of viruses.

Viruses are not living cells. Viruses are much smaller than cells; they are composed only of proteins and either DNA or RNA; they exist in a variety of shapes but always with proteins encapsulating the DNA or RNA; viruses have no capacity for metabolism, and they cannot reproduce outside living cells. Viruses infect host cells either by injecting their genomes or by entering cells as intact particles.

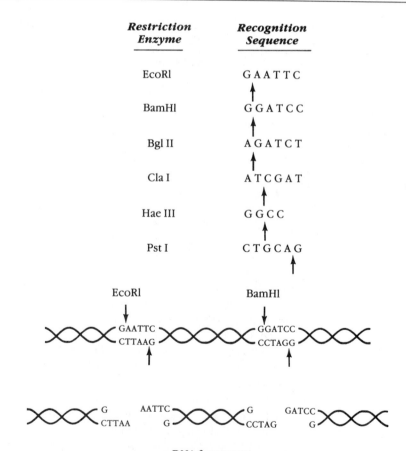

Restriction Enzyme	Recognition Sequence
EcoRl	G A A T T C
BamHl	G G A T C C
Bgl II	A G A T C T
Cla I	A T C G A T
Hae III	G G C C
Pst I	C T G C A G

DNA fragments

Figure 2.39 Digestion of DNA with restriction endonucleases. Restriction enzymes are so named because the action of each is restricted to specific, short sequences of DNA.

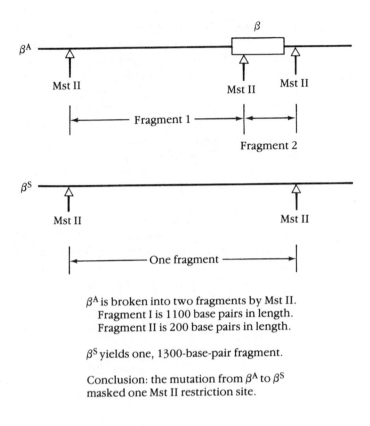

β^A is broken into two fragments by Mst II.
Fragment I is 1100 base pairs in length.
Fragment II is 200 base pairs in length.

β^S yields one, 1300-base-pair fragment.

Conclusion: the mutation from β^A to β^S masked one Mst II restriction site.

Figure 2.40 The use of restriction endonucleases to identify alleles of specific genes.

In the case of viruses that infect humans, the general pathway of infection is as follows. First, the virus particles bind to cell membranes and most often enter as intact particles; second, inside the cell the protein coats open up, releasing the viral genomes into the cell cytoplasm; third, the genomes take one of several pathways leading toward programming new virus particles, depending upon whether the genomes are double-stranded DNA, single-stranded DNA, or RNA. There are in fact three general pathways of viral infection:

1. **Lytic infection:** This is the simplest mode of viral reproduction, and it is best characterized by the viruses that cause common colds and polio. The virus genomes multiply within human cells with the help of human enzymes and human ribosomes (i.e., the host cells provide the medium within which viruses reproduce). After a viral genome has been released from its protein coat, the genome is replicated many times (100 to 200 new genomes may be synthesized). Then the genomes transcribe RNA, which, in turn, translates the genetic information to viral proteins. Once the genomes and coat proteins are synthesized, new virus particles are assembled. Following the assembly of 100 to 200 new viral particles, the host cell bursts open (lysis), releasing the progeny particles, which then are free to infect adjacent cells, thus preparing for a cascade of viral replication, and, of course, destruction of host cells. (Bacterial viruses behave this way; see Figure 2.9.)

An Aside

Restriction sites permit both a more detailed mapping of chromosomes and a diagnostic tool for identifying suspected gene mutations. There is a procedure called **restriction fragment length polymorphism** (RFLP) and linkage analysis that has proved very helpful for identifying unknown gene positions along the lengths of chromosomes. If, for example, within a large extended family many individuals of genotypes AA, Aa, and aa can be identified, and if the A gene's position on a chromosome is known, it may be possible to distinguish *A* from *a* chromosomes by restriction site differences of the types described for the β gene. In addition, the length of fragments carrying the gene in question may vary in length from one chromosome to the next. Thus the gene itself may be identical in all members of a kinship, but the fragment of DNA on which the gene resides may vary in length. This is a complicated procedure, an example of which is seen in Figure 2.41.

As base-pair sequences of more and larger segments of human chromosomes become known, it will become correspondingly easier to identify mutations that initiate genetic diseases. After the genes have been located it will be routine to isolate them and to identify the proteins encoded by them and from there to prescribe treatments for the symptoms of the disease. But the goal of the genetic engineer is to replace mutant alleles with normal alleles within the germ line cells, that is, to cure genetic diseases.

Today it is possible to move genes from one cell type to another, even from one species to another. The technology of "genetic surgery" is based on the properties of restriction endonucleases. Look at Figure 2.41 and notice that after digestion with the restriction enzyme EcoR1, all of the DNA fragments, except for the two original ends, have "sticky ends," that is, ends that will base pair with any other fragment produced by EcoR1. If one wishes to insert a specific fragment (gene) of human DNA into bacterial DNA, all that is necessary is to digest both genomes with EcoR1, isolate from the many fragments of human DNA the specific one to be inserted, and add it to the bacterial DNA with an enzyme that joins the fragment ends that come together naturally because of the base-pairing rules.

With the dozens of restriction enzymes available, it is possible to produce dozens of sets of fragments from any genome. What this means is that it is theoretically possible to insert any one of hundreds of thousands of fragments from any species into the genomes of any other species. While the technology is not well enough developed to achieve all of the theoretical possibilities, there is little doubt but that at least a few genetic diseases will be ameliorated by gene transplants before the end of the 20th century.

To illustrate that "gene surgery" and restriction fragment mapping of chromosomes are not just esoteric pastimes of eccentric people dressed in laboratory coats, the U.S. government has agreed to sponsor what is called The Genome Project, a project aimed at mapping every gene on every human chromosome. This signifies that gene mapping is important to people in high political places, especially in light of the fact that the estimated cost of the project exceeds $3,000,000,000. There still is debate about whether the project will eventually "map" every one of the 3 billion base pairs on the 46 human chromosomes, but given the present interest and financial support, there is little question but that eventually the bp sequences of the 46 chromosomes will become a database within the computer networks that unite the thousands of scientists who will work on the Genome Project.

Three billion dollars is a lot of money, though not in comparison with the money allocated for Star Wars and the space programs, but the uses of tax monies for giant engineering projects raises even more societal questions. The money belongs to the tax payers, but it is allocated and dispensed by a small number of bureaucrats, most of whom are not elected. How will we know how many of us agree with the ways our money is being allocated and dispensed? *Should* scientists speak their views? *Should* students learn about the greater society within which science is sponsored and have a voice in deciding what science *should* and *should not* sponsor? How would you vote on the Genome Project? Is there better use of tax monies?

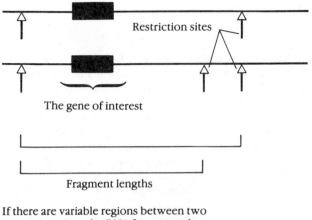

Restriction sites

The gene of interest

Fragment lengths

If there are variable regions between two restriction sites, the DNA fragments that accompany the gene will vary in length.

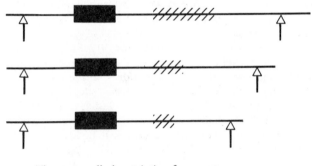

These are called restriction fragment length polymorphisms.

Figure 2.41 Restriction fragment length polymorphisms.

2. **Persistent infection:** Persistent infection differs from lytic infection in that not all of the infected host cells die. In one type of persistent infection all of the cells are infected, but they all survive. They do not, however, behave as normal cells. Such infected cells continually release new viral particles at a slow and regular pace. Their own metabolic functions are correspondingly slowed down since the infected cells are "reading from two genomes" at the same time. It is an energy drain for them to make viral and host proteins at the same time, and as a result neither function is efficient.

Another form of persistent infection is characterized by the maintenance of both uninfected cells and lytic infections within the viral target area. Thus the target area is not destroyed, as it is with lytic infections, but the infection persists, sometimes for years, and sometimes without leading to serious disease. Hepatitis B, human leukemia, and **human immunodeficiency viruses (HIV)** behave in this manner.

3. **Latent infection:** Latent infections do not lead right away to the reproduction of new viral particles. Infected host cells appear healthy, and the virus particles are nowhere to be found. We wouldn't know about latent

infections if the viral genomes remained hidden, but they don't; so we do know about them. The viral genomes hide by integrating themselves into host chromosomes where they behave as if they were host DNA. This is, in every way, a natural analog of splicing a human gene into a bacterial chromosome (i.e., the genetic information of one species is spliced into the genome of another).

Integrated viral genomes every now and then become activated, whereupon their effects upon their host cells are like those described for lytic infections. New viral genomes are synthesized, viral proteins are synthesized, viral particles are assembled, and finally new viral particles are liberated as the host cell is lysed. This explains why latent infections are characterized by periodic viral replication and why the symptoms of the disease come and go. Herpes viruses are of this type (viruses associated with herpes simplex, cold sores, and genital herpes), as is HIV.

HIV illustrates that genes and genomes are transferred from one species to another in nature without help from geneticists. To follow how this happens, as we followed the β gene from its chemical and physical structure to its influence upon individualized and population phenotypes, it is necessary to describe the virus, its genome, how the virus invades human cells, how its genomes become integrated into human chromosomes, and how the viral infection affects the phenotypes of individuals and populations. The description here is divided into three segments: the virus, the kinds of cells it infects, and **acquired immune deficiency syndrome (AIDS),** the disease that follows HIV infection.

HIV

A cross-section of HIV is shown in Figure 2.42. The simplest interpretation of this diagrammatic representation is that there are two RNA genomes encapsulated within two protein coats and an outer lipid membrane. The two RNA genomes are identical, and each carries at least eight genes. In addition, each RNA genome is accompanied by an enzyme called **reverse transcriptase.**

Reverse transcriptase has provoked a great deal of interest within the community of geneticists. At the time the enzyme was discovered, more than a decade before HIV was discovered, a major theme, which came to be known as **the central dogma,** within genetics was symbolized by the slogan DNA → RNA → Protein. The dogma stated that genetic information flows in one direction, from DNA to protein. The activity of reverse transcriptase appeared to turn this theme on its head; reverse transcriptase catalyzed the flow of genetic information from RNA to DNA. The RNA viral genomes serve as templates upon which DNA strands are synthesized — genetic information "flows backwards!" Reverse transcription is no longer startling, but it is rare, being confined to a group of viruses with tiny RNA genomes that encode reverse transcriptase and seven other proteins. These viruses are called **retroviruses,** so named because they "send their genetic messages backwards" from RNA to DNA.

The DNA strands copied from HIV genomes in turn serve as templates upon which complementary DNA strands are synthesized, producing double-stranded DNA molecules. These DNA molecules are called **proviruses.** One strand of a provirus DNA molecule, as with other double-helical DNA molecules, is a template upon which RNA molecules can be synthesized; these RNA molecules

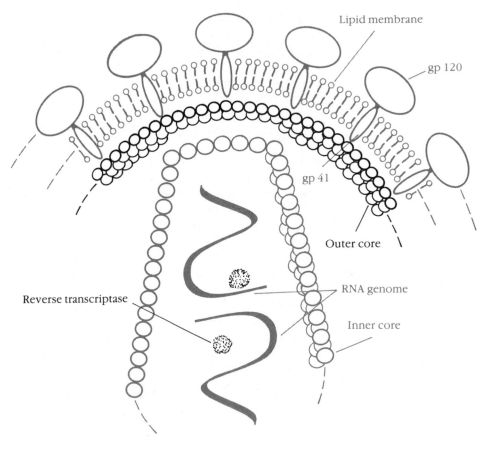

RNA genome = 9,749 bases, 9 genes
Diameter of HIV = 100nm (1/10,000 mm)

Figure 2.42 A cross-sectional view of the human immunodeficiency virus (HIV).

are simultaneously HIV transcripts and HIV genomes. After the transcript-genomes have been translated into proteins, the proteins encapsulate the genome-transcripts, producing new HIV particles. As the HIV particles escape from and destroy infected cells, they "wrap" themselves in the lipid membrane of the host cells, following which they are free to infect other cells (Figure 2.43).

However, before a provirus is translated into RNA genomes, it may become integrated into any of the 46 human chromosomes of the infected cell. Apparently the provirus can integrate at many sites on any chromosome, but the details of how and where integration occurs are as yet unknown. Once integrated, the provirus does not "look like" a virus but only as a "normal" segment of the human chromosome—normal as judged by its synchronous replication with human DNA, not by its transcription of nonhuman RNA.

Proviruses remain integrated within host chromosomes for indefinite and varying time periods. Eventually they become activated, that is, they are provoked to transcribe new HIV genomes. Soon thereafter new HIV particles are formed and the infected cells are destroyed as they release new virus particles that in all likelihood will infect other cells.

HIV's Target Cells

As their name implies, HIV infects cells of the immune system. There are many kinds of cells within immune systems, only a few of which are described here. One type, called T4 cells, stimulates the immune system to respond to invasions of viruses, bacteria, and other kinds of foreign proteins. Another type, called

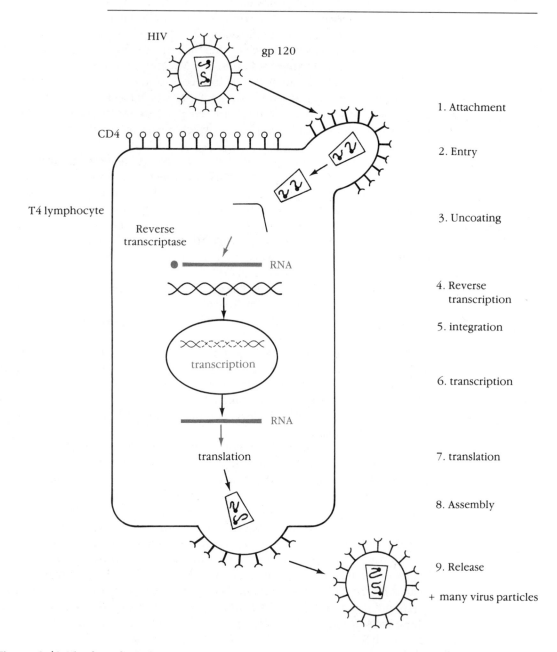

Figure 2.43 The fate of HIV from the time it enters a T4 cell until many copies of it destroy the cell. The gp 120 protein of the virus recognizes the CD4 protein of the T4 cell, and the two proteins become attached, allowing the virus to enter the cell.

macrophages, engulf these foreign objects and present the engulfed proteins (**antigens**) to the T4 cells. In addition to destroying these cells of the immune system, HIV particles also infect brain cells, to which they are thought to be carried by macrophages.

The immune system is far more complex than outlined in Figure 2.44. All cells found in the blood arise from **stem cells** found in bone marrow. The formation of specialized blood cells from unspecialized stem cells is another aspect of development and mimics in some ways the development of embryos from fertilized eggs. The first step in the differentiation process leads to two kinds of secondary stem cell, each of which further differentiates into mature blood cells. One cell type is called a **myeloid stem** cell. Myeloid stem cells give rise to cells that will further differentiate into red blood cells, platelet cells, and several types of white blood cell. A full discussion of immunity would include all of these blood cells, but here only those white cells called **neutrophils,** which attack all bacterial infections, and macrophages, which engulf bacteria and viruses, are mentioned.

The second type of partially differentiated stem cell is the **lymphoid stem** cell, a cell that will become a **T** or **B lymphocyte.** B lymphocytes produce **antibodies** upon being provoked by T4 cells, which in turn are provoked by the antigens presented to them by macrophages. T lymphocytes are of two types, T4 and T8. It is the T4 cells that first respond to "foreign" entities (i.e., to nonself proteins). Upon recognition of nonself proteins, T4 cells send out signals to other cells in the immune system. A key response to these signals is the stimulation of B cells to make antibodies.

T8 cells are called killer cells because they bind directly to the invading organisms, thus inactivating them. T8 cells have other functions as well, one of which is to send an "all-clear signal" after the invading organisms have been inactivated and another to send a "back-to-normal" signal, which has the effect of keeping the immune system in a state of readiness but not active. Certain B and T cells "remember" the antigens to which they have been exposed, such that following a second invasion they will not have to be warned by T4 cells. This is why vaccines work. Vaccines leave "memories" of themselves within some B and T cells, ensuring that should the invader return there will be antibodies prepared and waiting for them.

There is an explanation for how HIV specifically targets T4 cells and macrophages (Figure 2.43). On the membranes of macrophages and T4 cells there are proteins integrated within the other molecules that make up the membranes of cells. One of these proteins is called CD4; it is this protein that is recognized by the virus. Indeed, the virus recognizes the CD4 protein of the host cell with a protein located on its own membrane, a protein called gp120. These two proteins recognize one another almost as vigorously as do opposite poles of a magnet. Once the white cell protein and the viral protein become attached, the virus particle enters the cell. Cells that are not coated by the CD4 protein do not become infected with HIV, and of course if HIV did not make gp120 it would not infect cells that are coated with CD4. These facts are at the base of many research strategies designed to protect against HIV.

HIV infections, then, eliminate macrophages and T4 cells from the blood and from the immune system. In other words, HIV compromises the immune system. There is a suspicion that HIV kills stem cells as well as mature white cells, and this comes from the observation that

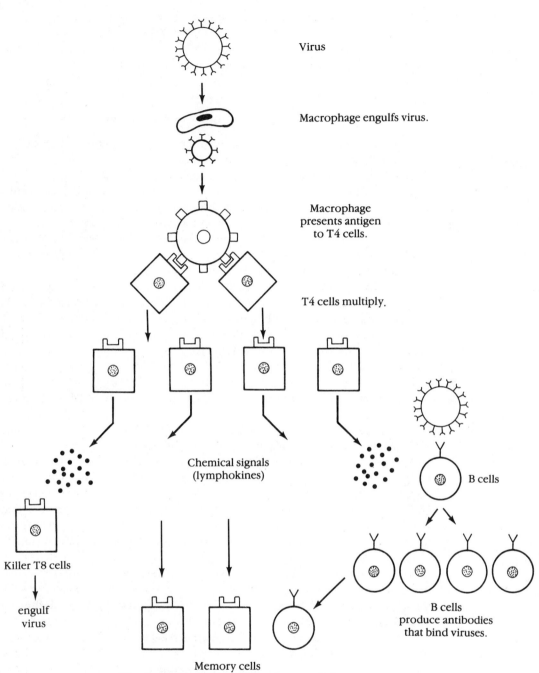

Virus

Macrophage engulfs virus.

Macrophage
presents antigen
to T4 cells.

T4 cells multiply.

Chemical signals
(lymphokines)

B cells

Killer T8 cells

engulf
virus

Memory cells
(Vaccines induce memory in memory cells.)

B cells
produce antibodies
that bind viruses.

Figure 2.44 A brief outline of the immune system. Invader viruses are engulfed by macrophages (scavenger blood cells), which presents antigens of the virus to the T4 cells. This is the key step in an effective immune system.

When HIV invades T4 cells, it (1) reduces production of chemical signals (lymphokines), (2) reduces activation of killer T8 cells, (3) reduces activation of antibody production by B cells, and (4) reduces activation of memory cells. Thus the chances of infection by other invaders such as viruses, bacteria and fungi are greatly increased.

white cells are eliminated from the blood faster than can be accounted for by the deaths of T4 cells and macrophages. If stem cells are killed, fewer T4 cells and macrophages will develop. A great deal remains to be learned about the immune system and in particular about how HIV compromises it.

The Effects of Compromised Immune Systems Upon Phenotypes

Acquired immune deficiency syndrome (AIDS) is a multitude of diseases. To fathom how deficient immune systems lead to a wide variety of illnesses, recall that individuals with any one of a number of kinds of anemia (oxygen deprivation) may have suffered from any one of a number of developmental abnormalities. A mistake within a developmental process may have consequences in all subsequent developmental processes; in short, an initial insult to a biological process can be amplified until, as in the case of all persons with AIDS, death. A deficient immune system is a potential insult to all of the processes needed to maintain a healthy body.

AIDS was first reported in the medical literature in 1981. But in retrospect it seems clear that AIDS has been around much longer. Antibodies to the gp120 proteins located on the membranes of HIV have been found in blood that was collected by blood banks prior to 1960. Doctors who now understand some of the symptoms of AIDS have testified that they had treated patients with similar symptoms prior to 1981. (Although there is a heated debate about the origin of HIV, professional virologists agree that HIV did not fall from the sky in 1981 and was not created by "communists" working in U.S. government laboratories, as so many rumors have had it.)

Individuals are diagnosed as carriers of HIV if they possess antibodies to the HIV gp120 protein. Persons with such antibodies are HIV-positive. Of all HIV-positive persons, about 1% have AIDS and about 9% have **AIDS-related complex** (ARC). The vast majority (90%) of HIV-positive persons shows no symptoms of AIDS or of ARC, and it is not known what percentage of HIV-positive persons will develop AIDS or ARC. One informed estimate is that 20% to 30% will develop AIDS within five years after infection, but clearly the interval between infection and the development of AIDS varies widely among those with AIDS.

As an immune system is weakened by the loss of T4 lymphocytes and macrophages, there is a corresponding increase of susceptibility to viral, bacterial, and fungal infections, and, in the case of AIDS, to **AIDS dementia complex (ADC),** a gradual deterioration of the central nervous system, and vulnerability to a kind of cancer called **Kaposi's sarcoma (KS),** which spreads without metastasis. Most who die of AIDS die of secondary infections (see Table 2.3), the most common of which is *Pneumocystis carinii* pneumonia. *Candidiasis* (thrush), caused by yeast, is common; *Herpes zoster* (shingles), *Herpes simplex* (ulcers of the mouth, anus, and vagina), and *cytomegalovirus* of the lungs are the three most common viral infections. About 4% of AIDS-related deaths are diagnosed as tuberculosis.

ADC is first recognized by a gradual loss of ability to concentrate, forgetfullness in routine tasks, loss of ability to understand simple directions, onset of apathy, and social withdrawal. By use of a devise that measures the speed of auditory signals (sound) from ear to brain, it appears that a majority of AIDS patients eventually experience a degeneration of brain cells. However,

Table 2.3 People with AIDS usually die of secondary infections. HIV kills T4 lymphocytes, which are key to a functional immune system. A dysfunctional immune system opens the door to secondary fungal, bacterial, and viral infections such as those listed.

Secondary Infection	Site
Pneumocystis carinii pneumonia	lungs
Candidiasis (thrush) — yeast	mouth
Cytomegalovirus (CMV)	lungs
Herpes simplex (HSV) ulcers	mouth, anus, vagina
Herpes zoster (shingles)	mouth, anus
*Kaposi's sarcoma (KS), a rare cancer	

*KS *is not* caused by a distressed immune system but appears early during the course of HIV infection.

it is not always clear whether the degeneration is caused by HIV or by secondary infectious agents. There is evidence that HIV enters certain brain cells (neurons and glial cells that "nourish" neurons) that have the CD4 protein on their outer surfaces. The precise course of ADC is unknown, but it seems certain that it is AIDS related.

The AIDS Epidemic

Since 1981 the numbers of persons with AIDS and the numbers of deaths from AIDS have increased rapidly (the rates of increase have increased; Table 2.4). The increase, in fact, qualifies as an **epidemic.** The study of epidemics is called epidemiology, a most complex field of study. In the case of the AIDS epidemic, for example, blood from blood banks in the African country of Zaire indicates that HIV infections date back to 1959 and that in two communities, two very different trends of infection emerged. One trend showed a slow increase, and the other a rapid increase in the number of people with AIDS. It is speculated that these differences reflect societal differences of sexual behavior. In countries without blood banks, analogous data are nonexistent.

In the United States the disease was first described among homosexual males, then among recipients of blood transfusions (mainly hemophiliac males), among intravenous drug users who share needles, and lastly among children born to women with AIDS. About 120,000 U.S. citizens died of AIDS during the first ten years of its recognition, and about 200,000 had AIDS after the first ten years.

Two sets of facts emerge from the epidemiological studies that have been made. First, HIV infections were on the increase prior to 1981; and second, during the first few years after 1981, little was done by blood banks, hospitals, and the federal government to curb the spread or to investigate the health care technology needed to treat people with AIDS. One of the few rapid responses to the disease was that of a few molecular biologists who quickly, by 1983, isolated HIV. Even the gay communities were slow to adjust their lifestyles as a step toward curbing the spread of the virus, and still little has been done within the intravenous drug-abuse sectors of society, where the incidence of AIDS is rising rapidly at the time of this writing.

Populations at highest risk of HIV infection are those that are most sexually active (multiple partner sex) and that are prone to intravenous drug abuse,

Table 2.4 Facts and figures concerning the AIDS epidemic.

Year	Infections	Cases Reported	Deaths
Dec 1981	50,000	8	—
1982	—	approx 450	—
1983	250,000	approx 1,500	—
1984	—	approx 4,500	—
1985	—	approx 14,000	—
1986	715,000	approx 26,000	—
1987	—	49,743	27,909
1988	—	82,764	47,474
1989	—	115,786	72,880
Apr 1990	1,050,000	158,287	97,792
June 1991	—	182,834	104,687

(1) The numbers of infections represent a rough estimate, a lower estimate than made during the 1980's.

(2) Cases reported to the Centers of Disease Control in Atlanta, GA.

(3) Deaths reported to the Centers of Disease Control, including the 50 states and Puerto Rico, all ages, both sexes, and all modes of infection. The data prior to 1988 were not assembled in this way.

which is to say that young people are at higher risk than old people and that poor people are at higher risk than rich people. Since in the United States African Americans, native Americans, and Hispanics are disproportionately poorer than the ethnic majority, the effects of racism must be factored into the sociological and political aspects of epidemiology, treatment, and cure.

In short, attention must be paid to the fact that two of the high-risk populations, homosexual males (most of whom are ethnic white and middle class) and ethnic minorities (most of whom are young African Americans and Hispanics), have been and are subjected to discrimination and prejudice by affluent members of the ethnic majority. Developing treatments and searching for cures is one issue; dealing with historic prejudices and modern forms of bigotry is another. Rarely is this latter side of the coin included within scientific and political discussions of the AIDS epidemic.

Among the issues of social disease that are being addressed within the political arena (e.g., global warming, pollution, shortage of fresh water, destruction of rain forests, etc.), one issue tends to be ignored, namely relationships among the many diseases. There is a clear conflict of interest within the health professions between individual health and environmental decay, which, in turn, destroys individual health. Relationships among environmental decay, prejudice, bigotry, and compromised health remain ambiguous, in part because health practitioners and researchers tend to focus upon one disease and to use intense focus as an explanation for their reluctance to become involved in discussions of the effects of many diseases upon our global well-being. The good news is that more professionals are beginning to turn their attention toward the larger issues of health.

Summary

A big step toward discovering the chemical composition of genes was the proof that genes are integral to chromosome structure (Perspective 2.1). Between 1930 and 1952 several experiments implicated DNA as being the genetic material. Results from three of these are (1) the wavelengths of UV light absorbed maximally by DNA are also maximally mutagenic; (2) the transforming substance of *Pneumococcus* is DNA; and (3) the "chromosomes" of bacteriophage are DNA. Once the chemical structure of DNA was deduced (by Watson and Crick from chemical and genetic facts and generalizations), it was only a short time before DNA functions were determined. From deductions based on the structure of DNA, the processes of DNA replication and transcription needed only to be verified by experiment; the base-pairing rules and the complementarity of the two strands were powerful insights in the design of these experiments.

An unexpected finding (i.e., not deduced from the structure and function of DNA) was the internal structure of genes (i.e., transcribing segments of DNA are larger than mature, mRNA transcripts). The information carried by exons is recorded within the base sequences of mature mRNA transcripts; the information carried by introns is excised from the pre-mRNA molecules as they are processed to become mature transcripts (Perspective 2.3).

The β gene, its alleles, Hb-A, and Hb-S illustrate the main features of molecular genetics—the chemical composition of genes; mutant forms of genes; the processes of replication and transcription; the process of protein synthesis; the relationships between the primary structures of genes, their transcripts and proteins; and the relationship between mutant proteins, errors of metabolism, and disease. Indeed, the comparison of primary structures of Hb-A and Hb-S contributed greatly to the discovery of the *genetic code*.

The concept of inborn errors of metabolism was refined by the DNA → RNA → protein breakthrough. Hundreds of genetic diseases exhibit the same mutant gene–mutant enzyme relationships that explain sickle cell anemia, relationships that show specific correspondence between base-pair changes in DNA, amino acid changes in enzymes, and altered metabolic pathways that lead to diseases.

The hopes and promises of genetic engineers are premised upon the knowledge of molecular genetics summarized in this chapter. However, even in the absence of gene engineering, a few inborn errors of metabolism can be ameliorated by supplying diseased persons with the missing substance (insulin), by deleting damaging substances from the diet (phenylalanine), or by supplying the missing enzyme (still on the drawing board). In animal models it is possible to engineer the insertion of normal alleles into the chromosomes of tissue and organ cells carrying a mutant allele to correct a diseased condition; however, genes transplanted into somatic cells are not passed on through germ line cells to progeny generations. Trials are now underway to treat cystic fibrosis in this way. Sickle cell anemia is on the drawing board. But the ultimate objective is to transplant normal alleles into early embryos so that they not only will *cure the disease,* but will *be transmitted to future progeny* as well.

During the course of evolution genes are exchanged among species without help from geneticists. Viruses often serve as vectors of species gene exchange. Retroviruses, upon entering host cells, immediately program the formation of DNA proviruses, which in turn become integrated within host chromosomes. HIV is such a virus. Its mode of entry into certain white cells, the integration of its provirus into human chromosomes, its transcription of new HIV genomes, the translation of HIV proteins from these new genomes, the subsequent emergence of new HIV particles, and the lysis of the host cells are discussed in some detail because the ultimate consequence of HIV infection is AIDS, a disease that is spreading in epidemic proportion. Since the first reports of AIDS in the United States emphasized its spread by homosexual males and later by drug abusers, the social policies erected to curb the spread of AIDS were long delayed and ill conceived. However, the scientific studies of HIV and AIDS include the most sophisticated molecular biology known to date — a sharp contrast with the primitive social planning for an epidemic predicted to become the most devastating in all of human history.

Molecular biology often is described as a scientific revolution every bit as spectacular as the revolution in physics that led to the development of nuclear power. While it is as meaningless to argue over which was the bigger revolution as it is to argue who was the better baseball player, Willie Mays or Mickey Mantle, the fact is that *the Atomic Age* changed societal and individual values and mores in significant ways. It also is fact that the new *Age of Reproductive Intervention,* made possible with knowledge of molecular biology, will alter societal values and individual mores to an even greater extent. Once it has been discovered how information is transmitted from a source to a particular destination and once the information code has been discovered, then decoding, recoding, and rerouting the information become matters of (1) sophisticated engineering, and (2) social planning. A question raised by these facts is: Who *ought* to participate in the social planning?

Study Problems

1. What is meant by the phrase "information is stored in DNA?" How is information stored in DNA? Is information stored in both strands of a DNA molecule? Explain.
2. Describe the main features of a Mendelian gene. Defend the thesis that Mendelian genes are composed of DNA.
3. Explain how you would demonstrate the existence of genes. Then define the gene whose existence you revealed. Can you also demonstrate that such a gene encodes the skills needed to understand probability problems?
4. The experiments that provided evidence that genes are made of DNA did not explain what genes are. Explain why this is so.
5. The genetic information carried within your mother's chromosomes was transmitted in two directions; one in the direction of a new generation, that is, to you, and the other within her own cells, tissues, and organs. Explain *how* genetic information is transmitted in these two directions.
6. Show the life cycle of a chromosome; include both semiconservative replication and one mitotic cell division.
7. What is the best explanation for the fact that observed phenotypic differences between SCA and non-SCA children have their origin in one base-pair difference between the β^A and the β^S alleles?
8. Refer to the first 18 base pairs of the β^A allele (Figure 2.22), and show the kinds of mutation necessary to produce the following amino acid sequences of the first six amino acids in the β protein: glu-his-leu-thr-pro-glu; val-his-leu-thr-pro-val; val-his-phe-thr-pro-glu; and, val-his-leu-thr-his-glu.

9. Fill in the open boxes of the following table:

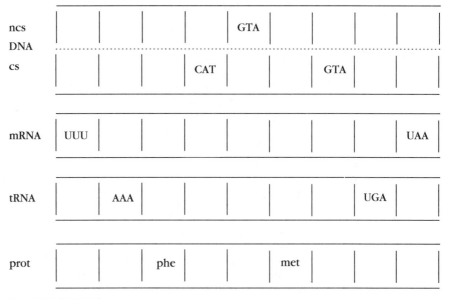

ncs DNA					GTA				
cs				CAT			GTA		

mRNA	UUU								UAA

tRNA		AAA						UGA	

prot			phe			met			

ncs = noncoding strand
cs = coding strand

10. Sketch an instant during the synthesis of a protein: show the mRNA transcript, the required tRNA molecules, and the relevant portion of the growing protein. The first two amino acids are attached to the ribosome; the second two are attached to their respective tRNA molecules, which are attached to the ribosome and to mRNA; and the third two are attached to tRNA molecules that have not yet become attached to mRNA codons. In the order of their appearance onto the ribosome, the amino acids are met, trp, lys, phe, his, and pro. Show the 18 mRNA bases in the correct order, the anticodons of the six tRNA molecules, and the base pair sequence of the segment of the gene that stores this information.

11. With the P gene, the *P* and *p* alleles, the enzyme phenylalanine hydroxylase, and the phenotypes PKU and non-PKU, illustrate the general principle, stated earlier, "that we inherit DNA from our parents and that our phenotypes develop after fertilization." In your illustration be especially careful to differentiate between the roles of DNA and proteins in the two processes of inheritance and development.

12. Refer to Figure 2.35. If the gene that encodes enzyme 4, which catalyzes the transformation of molecule D to molecule E, mutates such that enzyme 4 is inactive, then molecule E, which is essential for life, will not be synthesized. How might the lives of individuals homozygous for this mutant gene be saved? If molecule E is provided and if it prevents premature death, what other alterations in these metabolic pathways might contribute toward ill health?

13. Briefly, what would have to be done to *cure* genetic diseases like PKU or SCA? Would the technology needed to cure SCA be effective in curing a disease like alcoholism or schizophrenia? Explain.

14. Explain why it is easier to control diseases caused by lytic viruses than it is to control diseases caused by latent viruses?
15. Describe how best to protect yourself from becoming infected with HIV.

Suggested Reading

1. Benson, P.F. and A.H. Fenson. 1986. *Genetic Biochemical Disorders.* New York: Oxford University Press. This is a good reference book for a wide variety of human diseases that are correlated with Mendelian and molecular genes. You won't want to read this book from cover to cover, but if you want to investigate a particular disease, this is a good place to begin.

2. Brown, M.H. 1990. *The Search for Eve.* New York: Harper Perennial. A good science-mystery story describing the use of mitochrondrial DNA to discover our biological origins.

3. Cummings, M.R. 1991. *Human Heredity.* 2nd ed. St. Paul, MN: West Publishing Co. This book is described at the end of Chapter 1.

4. *Discover* magazine. Most issues have a short update, called AIDS-Watch, of the progress being made in the struggle to understand AIDS.

5. Freifelder, D. 1985. *Essentials of Molecular Biology.* Boston: Jones and Bartlett Publishers, Inc. This is a small, easy-to-read book that provides an up-to-date account of the material covered in this chapter. While the coverage is more detailed than in this text, it is not difficult to read. This book is recommended to the student who wishes to explore the subject a bit further.

6. Gallo, R.C. 1986. The first human retrovirus. *Sci. Am.* Gallo R.C. 1987. The AIDS virus. *Sci. Am.* These two articles describe human retroviruses in a language that non-specialists can understand.

7. McKusick, V.A. 1988. *Mendelian Inheritance in Man.* 8th ed. Baltimore: The Johns Hopkins University Press. It would be difficult to exaggerate the role that Victor McKusick has played in the study of human genetics. In addition to his extensive research, he has taken the time to catalogue all of the known human genes and to describe the human phenotypes associated with them. His is "The Dictionary" of human genetics.

8. Noonan, D. 1990. Genes of war. *Discover Mag.* October. This is a short, fascinating article about geneticist Mary-Claire King who is tracing the families of children orphaned in Argentina during the 1970s (children of mothers who were slain by the military rulers). Dr. King searches for small segments of specific DNA molecules. Some 50 children have been returned to their biological families. King explains how science and society are but parts of a whole.

9. Stilts, R. 1987. *And The Band Played On: Politics, People, and the AIDS Epidemic.* New York: St. Martin's. Stilts is a journalist, the only journalist in the United States to cover the AIDS epidemic during the 1981 to 1984 period. His book is in journalistic style but with a touch of the mystery book added. It follows the scientists, the physicians, the blood banks, many major politicians, and many of the gay communities. After reading the first page, the book is hard to put down.

10. Werth, B. 1988. The AIDS windfall. *N. Engl. Monthly.* June. This article is about AIDS from the perspective of the community of scientists. Some of the in-fighting is described, especially in relation to the suffering of people with AIDS, their families, and the health care delivery systems.

3

Genes, Sex Chromosomes, and Sexual Dimorphism

You have may have noticed that Chapters 1 and 2 are about sex, and that in fact, genetics is about sex. This is probably not the reason why young biologists choose genetics as a profession, yet once they begin their professional careers they are quick to understand that it is nearly impossible to talk about genetics without talking about sex. Indeed, sex is a serious and perplexing subject of biological research, and many biologists consider sex to be near the top of a long list of unresolved questions of evolutionary biology. "How" and "why," they still ask, did sexual reproduction evolve?

The word sex is an integral part of the vocabularies of nearly everyone over the age of twelve. The word is no stranger to the vernacular, but to say that genetics is about sex is to refer to a very different view of sex than that associated with the popular usage of the word.

One explanation for these differing meanings is that the "liquid properties" of common English have dissolved biological sex into a sea of images, dreams, and sexual fantasies. This is more the stuff of **sexuality,** sexuality being sex behaviors that figuratively or literally lead minds and bodies toward copulation and orgasm. (In other books you will find explanations of the socialization processes through which patterns of sexuality are learned [e.g., Ruth Bleier].)

While biologists disagree about many issues pertaining to biological sex, there is more agreement among them about biological sex than there is about sexuality. In fact, in the strict sense of meaning, human sexuality is as much a psychological as a biological phenomenon, and like some other psychological phenomena, it is learned through socializaton. Sexuality includes personal feelings, emotions toward self and other, imagery, hopes, and fears, all within a social/psychological context.

When we define sex in biological terms, however, we begin to use words unfamiliar to many, words such as meiosis, genetic recombination and sexual dimorphism. For non-biologists the word meiosis denudes the word sex of its excitement, which probably explains why high school biology "is boring," to quote hundreds of first year college students. But the word meiosis rather precisely describes one of the biological meanings of sex.

Meiosis, though, is only one meaning of biological sex. The union of gametes is another, and the mixing of genes from two parents is another. The development of two different sex types **(sexual dimorphism),** often called female and male, is a fourth meaning. No one definition or description of the word sex could possibly satisfy the many scientific investigators who are trying to make sense of the nature and evolution of sexual reproduction, of sexual dimorphism, of the mixing of genes and, in some cases, of the relationships between sexual reproduction and human sexuality.

While some biologists do study the hormones excited by sexual fantasies, the biological meanings of sex have been purged, for the most part, of their capacity to arouse hormonal excitement. However, biologists are not dedicated to purging biology of excitement; but what they have done, wittingly or unwittingly, has been to substitute "intellectual excitement" for "hormonal excitement." While most people may prefer hormonal excitement to intellectual excitement, including, we suppose, biologists, at the same time it would be shortsighted of anyone to knock intellectual excitement without trying it.

Today the biological definitions of sex are very different from the vernacular meanings of sex, yet the latter have always influenced the former, and there was a time when the two usages were indistinguishable. In part this has been the case because until recently almost nothing was known about reproduction. Aristotle left us with the notion that mothers provide the substance of their offspring while fathers provide the forms, and this notion lasted for more than 2,000 years. Yet even in the 18th century biologists argued about the role of eggs and sperm in the process of reproduction. The argument was whether the egg or the sperm carries the "preformed" embryo. Many 18th century biologists believed that Adam and Eve (from the Christian tradition) carried in their sex cells all of the embryos of all of the people who would ever inhabit the earth. In other words, the "ovists" insisted that each of Eve's egg cells carried an embryo, and that the female embryos among these each carried an embryo, and so on to include all who have lived, are living, and who will live. The "spermists" insisted that the "preformed" embryos reside in sperm cells, that sperm cells deposit embryos into the wombs of women, and that the nurturing of the embryo is one of the biological functions of the mother. One conclusion that we can draw from this is that ideas that exist for hundreds and even thousands of years become very resistant to modification. Another is that popular notions about sex are always interacting with and influencing knowledge of biological sex, and vice versa.

In Chapter 2 several tightly bound, cause-effect relationships between the primary structures of DNA, RNA, and proteins were discussed. The high probability of certainty with which scientists regard the physical structures and chemical coding relationships among these primary structures is rare in biology. Usually relationships between biological structures and functions (e.g., relationships between body type and success in farming or between the structure of your chromosomes and the grade you will receive in your next examination) are far more uncertain. When it comes to cause-effect relationships between what we know and how we behave, there is even more uncertainty. Even so, there may exist cause-effect relationships between our knowledge of sex and our behavior. In the September 15, 1990 issue of *Science News,* it was reported in an article entitled, "U.S. populace deemed 'sexually illiterate'," that less than 20% of the nearly 2,000 respondents to an 18-question questionnaire answered 12 or more questions correctly and that only 4, all of them women, answered 16 or more correctly.

The 18 questions were taken from a pool of questions that individuals throughout the country commonly put to the Kinsey Institute at the University of Indiana. The director of the institute, Jane Reinisch, has described the results of the poll as "devastating for the nation's health." Indeed, she describes the nation's political leaders as "dangerously burdened with carnal misinformation." About half of the respondents mistakenly believed that AIDS is acquired by way of anal intercourse even if the participants are not infected with HIV. Another misconception: most women were ignorant of the fact that between one quarter and one third of U.S. males have had sexual experience with another man and therefore of the fact that monogamous wives are correspondingly at risk for becoming infected with HIV. We should consider whether this kind of ignorance about sex is related to facts such as these: a previously uninfected U.S. teenager acquires a sexually transmitted disease every 13 seconds; a new teenage pregnancy occurs every 30 seconds; and babies are being abandoned at the rate of one per minute.

It is not the point of this discussion to suggest that if the general public knew more about meiosis this would solve social problems that resist change because of ignorance. Indeed, there is precious little evidence that social ills can be ameliorated by technical knowledge. But there appears to be a general tendency for those who understand some biology to have a correspondingly greater than average appreciation for individual organisms, species, and the earth's ecosystem. Biologically knowledgeable persons will not all make the same proposal for the improvement of individual, species and ecosystem health, but in general they will tend to think of proposals. However, while technical knowledge is *necessary* to the success of programs for improving individual, species, and ecosystem health, technical knowledge is not *sufficient* in and of itself. In addition to technology, concern and caring are necessary. The key is that it is impossible to be concerned about phenomena that appear not to exist because of ignorance. Therefore, the first step toward achieving awareness and concern is to acquire some technical information. Why has this been so difficult in the case of sex, the most talked about and least understood subject in human history?

Confusion also exists regarding the relationship between the study of genetics and a whole array of cultural beliefs about so-called biological similarities and differences between the sexes (and between so-called races, as will be discussed in Chapter 4). Some popular writers describe genetics as the study of biological differences. This is only partially true. Genetics is the study of biological similarities as well. While biological differences among individuals, groups, and species capture our attention more quickly than similarities within groups and species, our biological similarities are as important to our survival as are our differences. It is similarity that contributes to stability (security) within families, groups, and tribes. It is through similarities that we see ourselves in the grand scheme of biology, our relationships to other species, and our place within the history of life on the planet.

Differences and similarities exist in harmony in the world of biology. Birth and death are as much a part of the harmony as eating and sleeping. The famous French biologist Andre Lwoff said, "Problems do not exist in nature. Nature only knows solutions."

It is within the context of our cultural languages and self images that misconceptions about biological differences arise. Socially indoctrinated attitudes have led to immeasurable human tragedy and suffering, mistreating and denying certain social groups while building privilege into the social fabric for other groups. As the famous geneticist Theodosius Dobzhansky explained in his book *Genetic Diversity and Human Equality,* observations of real biological differences within and between populations of people have been confused, sometimes purposely, with societal inequality. Is it not possible for beauticians and cosmetic surgeons to acquire somewhat equal psychological and material rewards from their work while at the same time being as biologically different as adult females are from adult males? Is the fact that the civil engineer's secretary is of lower social status than the engineer evidence that her genotype is inferior to his? One philosopher, in a discussion of the infamous philosopher Bertrand Russell, asked if Russell was rich because he was smart or smart because he was rich. Just as surely as we live in a real universe of atoms, molecules, and living creatures, we live in a world of ideological warfare much of which appears contradictory to the real world. Some of us may choose to ignore the real world because of these contradictions; others may choose to try to resolve the contradictions; it still is not possible to predict how different individuals will respond to these kinds of contradictions.

Many people, including many scientists, feel that scientists should not discuss or participate in these so-called societal/political matters. At the same time, not all of the people who feel this way ask whether appointed and elected managers of our social systems should discuss or participate in science, that is, to learn as much as they can from science about how to manage social institutions. Then there is the question of whether all of us, beauticians, surgeons, secretaries, scientists, etc., should combine our efforts to integrate all knowledge into a pool of knowledge from which to devise strategies for our ecosystem's survival?

It is difficult to think in global terms, not because human brains are small but because it is not a societal tradition to teach the necessary skills. Many of us have been taught, instead, that it is someone else's responsibility to "mind the store" or that there are no meaningful answers to societal questions, only battles among those who seek power. The relationship between these socially indoctrinated attitudes and genetics is this: we also have been indoctrinated to believe that social roles and divisions of labor are created out of genetic differences among individuals within societies.

In this chapter all of the biological approaches to a definition of sex are mentioned, if only briefly. These definitions may appear more descriptive than definitive, but this reflects both the state of the art and the introductory nature of this book. To get a flavor of the general contradiction before getting into detail, consider a few biological facts (Lewontin, Rose, and Kamin, 1984) and a reaction to these facts by women scholars (Bleier, 1986). On average, slightly more boys than girls are born, and at all ages males have a somewhat greater chance of dying than females do. In Britain and the United States the average male life expectancy is about 73 years, while that of females is about 79. Most elderly people are women; there are more than three women to every man in the 85+ age group, for instance. Males have larger brains, though not when considered in proportion to body weight, and men are physically stronger in terms of performance on the sports field. A high proportion of women are in paid labor outside the home, but the jobs they do tend to be different from those of men. "Men are more likely to be cabinet ministers, business executives, Nobel Prize-winning scientists, doctors or airline pilots. Women are more likely to be secretaries, laboratory technicians, office cleaners, nurses, airline stewardesses, primary school teachers or social workers."

These biological and job related facts have aroused two opposite responses: Women in greater numbers since the 1960's have questioned the social policies that favor men in the workplace, and they have questioned the ideology that has been used to rationalize those social policies. While many male scientists have argued that the social inequalities are justified by the biological differences between the sexes, there are women scholars who disagree: "The 1970's saw a new wave of biological determinism committed to the renaturalization of women; to an insistence that, if not anatomy then evolution, X chromosomes, and hormones were destiny; and to the 'inevitability of patriarchy.' Where the (women's) movement brought women from nature into culture, a host of greater or lesser [biologists] joined eagerly in the effort to return them to whence they had come."

The task of explaining social roles within a society as complex as ours today is not easy. Cause and effect relationships are far from clear. However, it is clear that the members of our society are not equal with respect to opportunity to succeed; and it is clear that biological differences among us fail to explain the differences of opportunity. One question we will address as we go along is whether knowledge of biology is of any help in resolving the contradiction. You can try to answer this question as we proceed.

We begin with a short discussion of sex differences with respect to genetic segregation and inheritance patterns, centering on the fact that most genes are distributed to XX and XY zygotes according to laws of chance and equal numbers of autosomes; this is not the case for genes located on X and Y chromosomes. That females possess 2 X and no Y chromosomes, and males one X and one Y chromosome accounts for sex-specific inheritance patterns of X- and Y-linked genes. This discussion is supplemented by information on the effects of extra sex chromosomes, and of a few mutations, upon the development of sex morphological differences.

Sex Differences With Respect to Sex-Specific Inheritance Patterns

The Discovery of X-Linkage in Drosophila

Mendel introduced the results of his experiments to a small gathering of biologists in 1865, but there is no evidence that anyone in the audience understood what he had done. Between that time and 1900, chromosomes were discovered and described in detail. Of particular significance was the discovery that mitosis and meiosis are different kinds of cell division and that during meiosis chromosomes segregate into gametes, as do alleles of Mendelian genes. Another aspect of this discovery was that chromosomes within genomes are different in size and shape. Within a decade after 1900, genes and chromosomes of many different kinds of plants and animals were being studied; indeed, the small fruit fly, *Drosophila melanogaster,* gained notoriety within biology after 1910, in part because of easy access to its genes, chromosomes, and other sex characteristics.

In 1910 a member of the famous "fly lab" at Columbia University, headed by Thomas Hunt Morgan, discovered a male fly with white eyes—a rare mutant fly. The white-eyed male was mated to his red-eyed sisters (red is the normal eye color) from which an F_1 population emerged; all of the F_1 flies had red eyes. F_1 males were then mated to F_1 females, and among the progeny were 3,470 red-eyed and 782 white-eyed flies. This is a very significant departure from a $3:1$ ratio (from a total of 4,252 flies, an exact $3:1$ ratio is 3,169 red:1,083 white). Even more startling, however, was the fact that all of the white-eyed flies were males! Of the red-eyes flies, 2,459 were females and 1,011 were males (Figure 3.1).

In this experiment, then, there were 1,011 red-eyed and 782 white-eyed males. Why? The red-eyed, F_1 females had two X chromosomes, a W allele on one and a w allele on the other, so why was there not a somewhat equal number of red- and white-eyed males? A test-cross provided a clue: a cross between F_1 (red-eyed) females and white-eyed males gave rise to 129 red-eyed females, 88 white-eyed females, 132 red-eyed males, and 86 white-eyed males (Figure 3.2). From these results it is obvious that there are fewer white-eyed flies of both sexes, a fact interpreted to mean that white-eyed flies are less viable than red-eyed flies. This interpretation turned out to be correct, *but did not explain why, among the F_2 flies, only males have white eyes.*

Both female and male flies have 8 chromosomes in each body cell and 4 in each sex cell. Egg cells are of one type, judged by the sizes and shapes of their chromosomes. Sperm cells are two types: half carry four chromosomes that look exactly like those found in egg cells, and half are missing one of the large chromosomes carrying in its place a smaller chromosome. The large missing chromosome is an X chromosome, and the small replacement is a Y chromosome. Both are called **sex chromosomes** (Figure 3.3).

Females carry three pairs of autosomes (nonsex chromosomes) and one pair of X chromosomes in their diploid cells; males carry three pairs of autosomes, one Y, and one X chromosome in each of their diploid cells. The gene for eye color, W, is located on the X chromosome; therefore females carry two copies of W while males carry one.

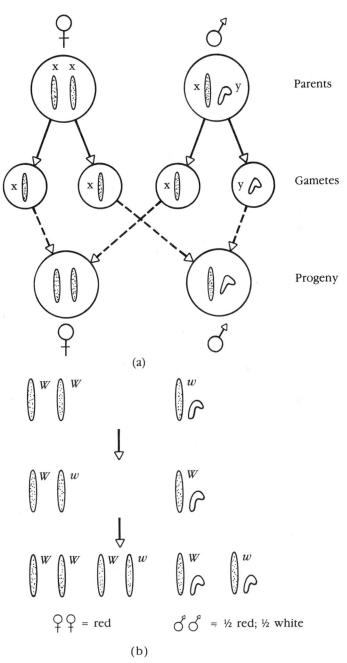

(a)

(b)

Figure 3.1 (a) The inheritance patterns of sex chromosomes. (b) An X chromosome-linked gene.

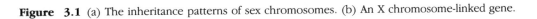

The original white-eyed male fly carried the recessive allele of the W gene (i.e., w). Since males carry only one X chromosome, this male's genotype was X^wY. The original mating, then, was $X^WX^W \times X^wY$. The female progeny of this mating received one X from each parent, and hence were genotype X^WX^w; their phenotype was red-eye. The male progeny received an X chromosome from

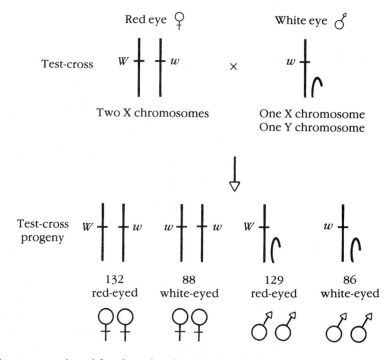

Figure 3.2 A test-cross between a red-eyed female and a white-eyed male, showing reduced viability of the white eye phenotype.

their mothers and a Y chromosome from their fathers, and hence were genotype X^WY and phenotype red-eye.

The brother-sister matings that produced the F_2 generation, $X^WX^w \times X^WY$, gave rise to X^WX^W and X^WX^w daughters, all red-eyed, and X^WY (red eye) and X^wY (white eye) sons, "half" with red eyes and "half" with white eyes, except for the fact that fewer white-eyed flies survived to maturity. The startling observation was that the only white-eyed flies among the F_2 progeny were males and each of these received the w allele carried by their grandfather and their grandfather's daughters (i.e., their mothers; therefore, Grandfather → daughter → grandson). **X chromosome-linked** inheritance patterns are sometimes described as "crisscross" inheritance.

X-Linkage in Humans

The X-Linkage inheritance pattern has been observed for centuries in human families. For example, the Talmud, a book written some 1,500 years ago to summarize earlier oral laws of the Jewish religion, tells of a Rabbi who instructed a woman not to have her son circumcised after three of her sister's sons had bled to death following circumcision. If a woman had a sister who sons had bled to death following circumcision, that woman's sons were exempted from circumcision, but her sons were not exempted if her brother's sons had died of a similar fate. This may be the first recorded description of the inheritance pattern of **hemophilia A,** or bleeders' disease (Figure 3.4).

Today we know of three forms of hemophilia: a rare form initiated by an autosomal recessive gene and two commoner forms associated with two genes on the X chromosome. Of these commoner forms, about 20% of hemophiliacs have Christmas disease, or hemophilia B, and about 80% have hemophilia A.

Hemophilia A is characterized by a deficiency of a protein called **clotting factor VIII,** a protein that is necessary to the blood clotting process (Figure 3.5). In the absence of normal factor VIII, an individual may bleed to death following a minor cut or from internal bleeding. Until it became possible to administer clotting factor VIII to persons with hemophilia A, 75% of hemophiliacs, all males, died before reaching 25 years of age. Today, at an expense of about $12,000 per year, hemophiliacs may enjoy a relatively normal life span. (Recall that hemophiliacs experienced an added risk between 1980 and 1986 because of their dependence upon blood transfusions, and it was during that time that many blood banks refused to examine donated blood for HIV antibodies, antibodies to the human immunodeficiency virus (HIV) believed to cause AIDS. As a result, many hemophiliacs became infected with HIV.)

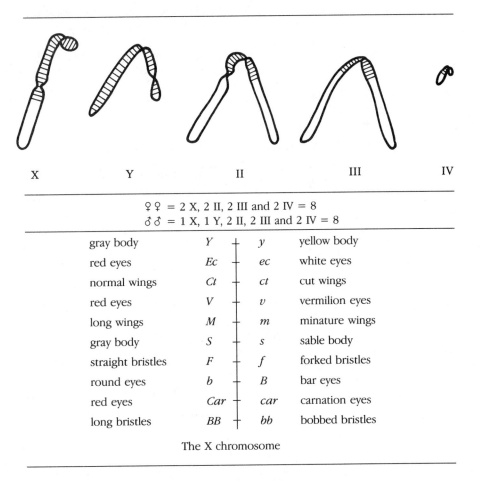

	X	Y	II	III	IV

♀♀ = 2 X, 2 II, 2 III and 2 IV = 8
♂♂ = 1 X, 1 Y, 2 II, 2 III and 2 IV = 8

gray body	Y	y	yellow body
red eyes	Ec	ec	white eyes
normal wings	Ct	ct	cut wings
red eyes	V	v	vermilion eyes
long wings	M	m	minature wings
gray body	S	s	sable body
straight bristles	F	f	forked bristles
round eyes	b	B	bar eyes
red eyes	Car	car	carnation eyes
long bristles	BB	bb	bobbed bristles

The X chromosome

Figure 3.3 The chromosomes of Drosophila melanogaster (top). A few X chromosome-linked genes (bottom).

A rough illustration of the rabbi's observations

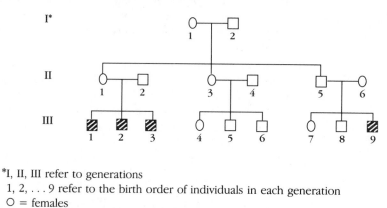

*I, II, III refer to generations
1, 2, . . . 9 refer to the birth order of individuals in each generation
O = females
□ = males

The rabbi's logic:
1. III-1, III-2, and III-3 died because of heredity through II-1.
2. III-5 and III-6 were exempted from circumcision because of common heredity through I-1 to II-1 and II-3.
3. III-9 died after circumcision because of heredity through II-6.
4. I-2, II-2, II-4, and II-5 cannot be carriers; otherwise they would be bleeders.

♀♀ (2 X's)	♂♂ (1X)
HH − normal	H − normal
Hh − normal	
hh − bleeder	h − bleeder

Figure 3.4 A pedigree of hemophilia corresponding to the Talmud, which contains the "first" recorded description of its inheritance pattern.

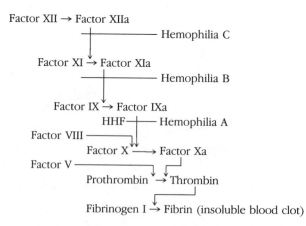

Hemophilia C is an autosomal recessive mutation.
Hemophilia B and hemophilia A are X-linked recessive mutations.

Figure 3.5 The blood-clotting protein *fibrin* is the end product of a cascade of chemical reactions. This series of reactions can be stopped at any of three points by gene mutations that result in hemophilia.

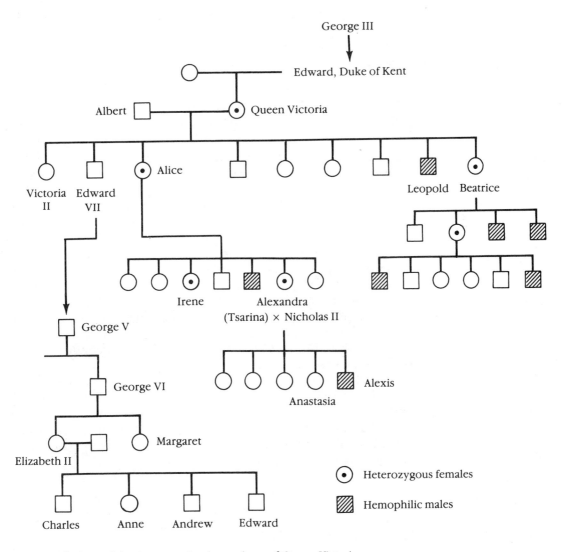

Figure 3.6 Hemophilia A among the descendants of Queen Victoria.

Hemophilia A was made famous because of its presence in some of the descendents of Queen Victoria (1819–1901, Figure 3.6). Apparently one of the Queen's two clotting factor VIII genes mutated (in her germ line cells), and she passed the mutant gene to one son (Leopold, Duke of Albany) and two daughters. Her daughter Alice's daughter, Alexandra, married Tsar Nicholas II of Russia and gave birth to Alexis, who had hemophilia. When he was 14, Alexis, his mother, and the Tsar were killed during the early stages of the communist revolution of 1917.

Among people of Northern European descent, about 1 of every 10,000 male babies has hemophilia A. You can calculate that if the probability of inheriting one recessive allele is 1 in 10,000 the probability of inheriting two is 1 in $(10,000)^2$, which is to say that about 1 female baby in 100 million is expected to have hemophilia A. But for a female to inherit two mutant genes, her father must be hemophiliac, and until recently most male hemophiliacs did not live long enough to become fathers, further reducing the incidence of females with hemophilia A. In the exceedingly rare event of the birth of a hemophiliac female, it is highly unlikely that she would survive her first menstruation.

Red-Green Color-impaired Vision

The genes that encode the proteins that permit us to see red and green colors are X chromosome-linked. Inherited inabilities to see red and green colors are common, and the majority of people with color-impaired vision are males who cannot distinguish red from green, hence the original name *red-green colorblindness.*

People with normal vision "see" three colors, red, green, and blue; these are the three primary colors, as defined by the biology of vision. All others are secondary colors, which are mixtures of primary colors. We see these colors because of **visual pigment proteins** (or cone pigments) located on cone-shaped **photoreceptor cells** in the **retinas** of our eyes. Pigment proteins are called **opsins,** and each opsin absorbs maximally at a characteristic wavelength (see Figure 2.5). The blue sensitive pigment absorbs maximally at 420 nm, the green sensitive pigment at 530 nm, and the red sensitive pigment at 560 nm. There is a fourth pigment, **rhodopsin,** which mediates vision in dim light and absorbs maximally at 495 nm. This is a *rod opsin,* so called because it is found in rod-shaped photoreceptor cells.

The opsins that register green and red are encoded by closely linked genes located at the distal end of the X chromosome (Figure 3.7). A mutation in either gene may result in one or the other mutant opsin. Males carrying one mutant gene will see both red and green as red (**deutan** red-green colorvision impaired), and males carrying the other mutant gene will see both red and green as green (**protan** red-green colorvision impaired).

In the United States among people of Northern European descent, about 8% (1 of 12) of males are red-green colorvision impaired. Of these, 75% are deutan and 25% are protan. From these numbers we calculate that 6% (75% of 8%) of all males are deutan and that 2% (25% of 8%) are protan. Since males carry only one X chromosome and hence one copy of each gene, this means that about 6% of all D genes, which encode green opsin, are in the *d* allelic form, and about 2% of all P genes, which encode red opsin, are in the *p* allelic form (Perspective 3.1).

To illustrate the frequency differences between the sexes, 8% of males (6% deutan plus 2% protan) and only 0.4% of females (0.36% deutan and 0.04% protan) are unable to distinguish between red and green. In fact, these percentages led to the prediction that two genes influenced red-green color vision. Consider the following: if red-green color vision impairment is the consequence of one mutant gene, we would predict that 0.64% of women are homozygous recessive for that mutation (8% × 8% = 0.64%), whereas the observed percentage is only 0.4%, significantly lower. But if two genes act to initiate the phenotype, and if the frequency of *d* is 6%, the percentage of dd females will be 0.36% (6% × 6% = 0.36%); if the frequency of *p* is 2%, the percentage of pp females is expected to be 0.04% (2% × 2% = 0.04%). The addition of 0.36% and 0.04% is 0.4%, the frequency observed.

It is one thing to discover that the genes encoding the red and green pigment proteins are located on the X chromosome and that 20 times as many males as females are red-green colorvision impaired. It is another thing to discover that

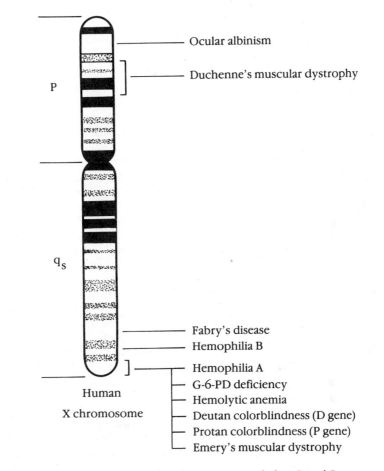

P

q_s

Human
X chromosome

— Ocular albinism

— Duchenne's muscular dystrophy

— Fabry's disease
— Hemophilia B
— Hemophilia A
— G-6-PD deficiency
— Hemolytic anemia
— Deutan colorblindness (D gene)
— Protan colorblindness (P gene)
— Emery's muscular dystrophy

Figure 3.7 The locations of a few genes on the human X chromosome, including D and P.

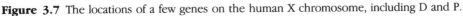

the two genes are nearly identical as measured by base-pair sequences, and that the two genes are very closely linked. In addition, it was discovered that nearly all X chromosomes carry one P gene and, closely linked, from one to five D genes. Next it was found that the blue opsin and rhodopsin genes are similar to the P and D genes, even though they are located on different chromosomes. The conclusion was reached that all of the pigment genes have a common ancestor gene (Figure 3.8). Indeed, this group of genes is a **multigene family,** a family of like genes that arises from an original pigment gene by duplication (unequal crossover) and subsequent mutation. (As discussed in Chapter 4, genetic variation arises from chromosomal and gene mutations.)

Members of this multigene family are carried not only by other primates but also by other vertebrates. Many species of birds see the same primary colors that we see, as well as ultraviolet (UV) light, which no primate sees. Chimpanzees and gorillas, like humans, cannot see UV light, but they do see red, green, and blue, and they may possess more than one D gene. With two or more D genes, one of them may mutate while the other encodes functional pigment proteins. Eventually a mutation may give rise to an opsin that registers a different wavelength of light. (Imagine having improved night vision by being able to register short UV waves or X-rays.) However, neither new

Perspective 3.1

The X chromosomes carried by males now living within any given population represent a "sample" of one third of the X chromosomes within that entire population. This fact is not intuitive for beginners in genetics. But consider the following: males carry one, and females carry two X chromosomes. The X chromosomes carried by males are not "male" X chromosomes, as evidenced by the fact that every male received his X chromosome from his mother and every father transmits his X chromosome to his daughters. If X chromosomes, through the generations, move from fathers to daughters and from mothers to both sons and daughters, it is fairly certain that the X chromosomes currently residing within male cells, including germ-line cells, are an accurate and representative sample of the entire population of X chromosomes.

If this is so, and if X chromosome-linked genes can be "observed directly" simply by observing the phenotypes of males, then the frequencies of the alleles of X-linked genes can be determined simply by counting the numbers of males expressing specific contrasting phenotypes. Consider a somewhat simplified example illustrating this point: the population of the metropolitan twin cities of Minneapolis and St. Paul is about 2.4 million. Round this figure to 2 million, and assume that half the population is female and half male. One million females will carry 2 million X chromosomes, and 1 million males will carry 1 million X chromosomes, for a total of 3 million.

If 6% of the males are deutan colorvision impaired, then 6% of the X chromosomes carried by them will carry the *d* allele. Extrapolated to the entire population of X chromosomes, this means that 6% of the X chromosomes carried by females also carry the *d* allele. *But far fewer females are deutan colorvision impaired.* Indeed, the frequency is 0.36%. Seventeen times as many males as females are deutan colorvision impaired. This is so because *d* alleles are transmitted from one generation to the next according to the rules of chance. If 6% of the X chromosomes carry *d*, and males inherit but one X, then we expect that 6% of males are deutan colorvision impaired.

Now refer back to Perspective 1.3 on page 42. If the frequency of the *d* allele is 0.06, then the frequency of the *D* allele must be 0.94, which is to say that $p = 0.94$, and $q = 0.06$ From this we calculate that within the female population $p^2 = 0.8836$, $2pq = 0.1128$, and $q^2 = 0.0036$, since females carry two X chromosomes. In other words, among the females, slightly more than 88% are genotype DD, slightly more than 11% are Dd, and only 0.36% are dd. In contrast, 94% of the males are genotype D, and 6% are d.

If 6% of the X chromosomes carry the *d* allele, then about 180,000 of the 3 million X chromosomes will carry *d*, while 2,820,000 will carry *D*. Of the 180,000 X chromosomes that carry *d*, only 60,000 of them reside within the male population (6% of 1 million). The remaining 120,000 must reside within the female population (if females carry two thirds of the X chromosomes, they must carry two thirds of the *d* bearing X chromosomes). Therefore females are twice as likely as males to inherit a *d* allele, but only one seventeenth as likely to express the deutan colorvision-impaired phenotype. There are two significant points to be made here. First, the inheritance patterns of X-linked alleles are sex specific; and second, through male phenotypes it is relatively easy to calculate the frequencies of X-linked alleles and the phenotypes initiated by those alleles. (More on this subject in Chapter 4.)

mutations nor the evolutionary trends which they might set in motion can be predicted. Nevertheless, at some time in the future a new mutation may permit us to see another segment of the light spectrum by encoding a new pigment protein.

To view this possibility in hindsight, consider that among modern primates New World monkeys (living in Central and South America) cannot distinguish between red and green colors, while Old World monkeys (living in Africa and Asia) can. These two groups of monkey became separated some 40 to 50 million years ago, after which the gene encoding the red pigment protein duplicated (resulting in more than one copy of the gene on each X chromosome); after that, one of the two copies mutated to encode a pigment protein that absorbs green light. Old World monkeys are ancestors to modern chimps, gorillas, and humans, all three groups of which possess the genes necessary to see red and green.

Another interesting fact about these genes and the way they have been studied by molecular genetics techniques is that the gene encoding rod opsin (rhodopsin) in cows was used to locate the rhodopsin gene in the human genome (See Perspective 3.2). That is, the bovine rhodopsin gene is similar enough, by its base-pair sequence, to the human rhodopsin gene that it can be used to **probe** the human genome by its tendency to form double helices with the human rhodopsin gene. Under slightly different conditions of probing, the bovine rhodopsin gene can be used to locate the blue, red, and green opsin genes in the human genome. In other words, members of this multigene family exist in a wide variety of species, and during the course of evolution the base-pair sequences of these genes have been highly conserved, yet have mutated to produce pigment proteins that have expanded the capacity to see wavelengths of light between about 420 and 560 nm (Figure 3.9).

The G-6-PD Gene

Another gene, linked closely to P and D at one end of the X chromosome, is called the G-6-PD gene. This gene encodes the enzyme *glucose-6-phosphate dehydrogenase,* an enzyme that is found in the membranes of red blood cells

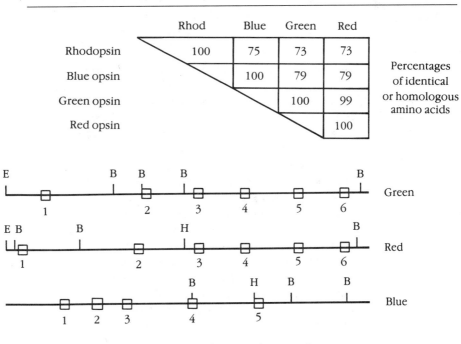

	Rhod	Blue	Green	Red	
Rhodopsin	100	75	73	73	Percentages of identical or homologous amino acids
Blue opsin		100	79	79	
Green opsin			100	99	
Red opsin				100	

The three pigment genes (the genes that encode
the three opsins found on retinal cone cells)

Figure 3.8 A multigene family showing the gross structural similarities among the three pigment genes. The letters above each gene symbolize the sites that are cut by restriction enzymes. The boxes symbolize the coding sequences, or exons.

Perspective 3.2

A probe is a molecule of DNA (a gene) that is synthesized from an RNA template in the presence of the enzyme, reverse transcriptase. After it was discovered that retroviruses carry the gene that encodes reverse transcriptase, it was but a short step to isolate the gene and induce it to make large quantities of the enzyme. The other trick to making a probe is to isolate messenger RNA (mRNA) from cells that are known to synthesize large quantities of a particular protein such as rhodopsin. Rod cells in the retinas of cows' eyes are such cells. Presumably, most of the mRNA isolated from these cells encodes rhodopsin. Like the genomes of retroviruses, this mRNA, in the presence of reverse transcriptase, serves as a template upon which a complementary DNA strand is synthesized. This complementary strand of DNA, in the presence of DNA polymerase, then serves as a template upon which a sense strand of DNA is synthesized. Copies of this double-helical DNA molecule are in fact copies of the cow's rhodopsin gene. Many copies of the gene can be made quickly in the presence of radioactive molecules that subsequently allow detection of its presence among large popula-

tions of nonradioactive DNA fragments. These radioactive rhodopsin genes are called probes.

The reason the cow's rhodopsin gene is called a probe should be clear after discussion of the next step, which is to isolate human DNA, cut it into small fragments with restriction enzymes, and then to spread the fragments out by size and by electric charge on a gel slab within an electric field (Figure 3.10). The DNA fragments are then blotted onto a nylon membrane and exposed to the probes. The probes (i.e., rhodopsin genes of cows) will become attached to segments of human DNA with like (not necessarily identical) sequences, which in turn can be spotted among the thousands of fragments by the presence of radioactivity. It is a simple matter to isolate the cow DNA-human DNA doublets, to separate the human DNA from the cow DNA, and to make many copies of the human DNA. In this way all four of the human pigment genes are isolated, after which their base-pair sequences are determined. Pigment genes of many species have been isolated and compared for their evolutionary relationships to one another, rather like species "fingerprinting."

and that catalyzes the formation of the *high-energy phosphate molecules* needed for many metabolic reactions in cells and tissues (Figure 3.11). Certain mutations of the G-6-PD gene lead to susceptibility to **hemolytic anemia,** a disease characterized by the bursting of red blood cells within the blood vessels. If functional G-6-PD is missing, the membranes of red cells become fragile and burst from the internal pressure of the cells.

Not everyone with mutant G-6-PD experiences hemolytic anemia, but everyone with hemolytic anemia carries a mutant gene and a mutant enzyme; in addition, such persons are red-green colorvision impaired. At first the story seemed to be complicated, but now it is understood to be rather straightforward. To develop hemolytic anemia, a mutant G-6-PD gene is necessary but not sufficient. In addition, individuals carrying the mutant gene must become exposed to raw or partially cooked fava beans (broad beans). Sometime ago hemolytic anemia was called favism, a disease affecting mostly males after eating uncooked fava beans. A chemical found in the beans *induces* the disease, while the mutant gene creates within the red cells the capacity to be induced. The phenotype is as much dependent upon the chemical inducer as upon the mutant gene.

After the relationship between gene and chemical inducer became clear, the mutant gene was used as a marker of susceptibility to uncooked fava beans. Even more to the point, red-green colorblindness could be used as a marker of the mutant G-6-PD gene, a marker that makes it possible to eliminate 92% of males from the risk group. Even though nearly 200 million males carry the mutant G-6-PD gene, 2 billion males can be excluded from the risk group by having normal color vision. In addition to raw fava beans, however, there are other chemical inducers of hemolytic anemia, including antimalarial and sulfa drugs, aspirin, and vitamin K. Therefore, even though the task of identifying the potential for hemolytic anemia is enormous, it is generally agreed that this genetic marker-screening technique has been of immeasurable value to those whose lives have been extended by it.

That more males than females express the phenotypes of recessive, X-linked alleles does not mean that women are carriers of genetic diseases that victimize men. Women homozygous for X-linked, recessive alleles exhibit the same associated phenotypes as men do. The differences between the sexes has to do with the different numbers of X chromosomes carried by

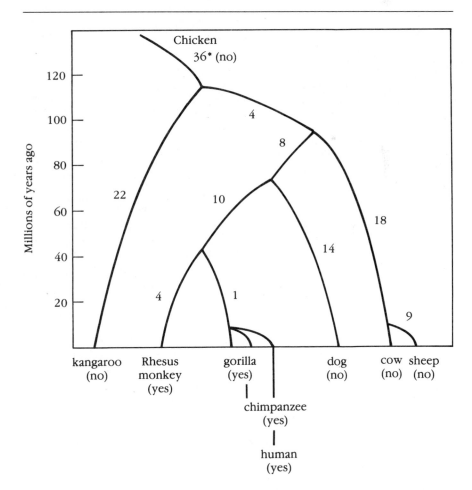

Figure 3.9 Time, amino acid substitutions in α globin, and the evolution of the pigment proteins. The ability to see red and green has appeared during the past 70 million years. Time is measured as a function of amino acid changes in the α protein of hemoglobin from species to species, e.g., the α proteins of gorillas, chimpanzees and humans are identical (no changes in 5 million years) and they differ from Rhesus monkey α protein by 5 amino acids, the number of changes that occur over ~ 40 million years.

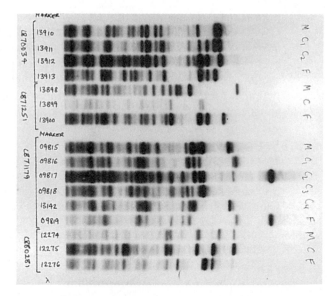

Four sources of DNA yield four-, three-, six-, and three-fragment patterns, each created by a specific restriction endonuclease. This technique creates a DNA fingerprint (no two patterns are identical). The vertical dark bands are DNA fragments.

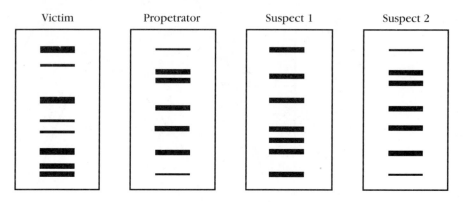

These four samples of DNA came from the blood cells of four different people. The fragment pattern of Suspect 2 and the Perpetrator are identical, which means Suspect 2 did it!

Figure 3.10 Breaking long segments of DNA into short fragments with restriction endonucleases.

each sex and with different frequencies of the phenotypes initiated by X-linked genes.

A few of the more than 200 genes known to be located on the X chromosome are shown in Table 3.1. The frequencies of phenotypic expression of these genes are sex specific; for example, dominant alleles are expressed twice as often within the female population, and recessive alleles are expressed much more often within the male population. However, these statistical differences are not known to be correlated with any so-called qualitative differences between males and females.

Sexual Dimorphism and Sex Chromosomes

While it is interesting that genes located on sex chromosomes give rise to sex-specific inheritance patterns, the information doesn't explain the **development** of sexual dimorphism, that is, the processes which lead some fertilized eggs to develop into females and others into males.

As discussed earlier, biological development includes two major processes: (1) cell division, which results in increased numbers of cells in embryos, fetuses, and children, and in the maintenance of cell number in adults; and (2) cell **differentiation,** which guides the transformaton of dividing cells into the many different kinds of specialized cells found in late embryos, fetuses, children, and adults (e.g., blood, skin, muscle, and hair cells) (Figure 3.12). One of the tenets of faith of developmental biologists is that an increased knowledge of cell division and cell differentiation will lead toward an understanding of how the two sex types, female and male, emerge from basically similar zygotes.

The literal meaning of sexual dimorphism is "two morphologially different sexes." The external and internal genitalia of males and females are different; the

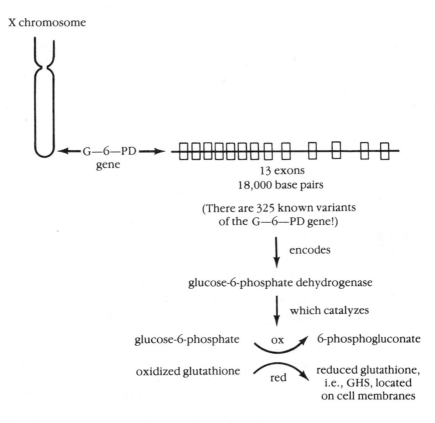

Figure 3.11 The G-6-PD gene and the function of glucose-6-phosphate dehydrogenase, the enzyme encoded by it. Normally the G-6-PD gene produces enough GHS, but rapid oxidation of GHS causes the membranes of red blood cells to rupture, a process exacerbated by (a) mutant forms of G-6-PD and (b) many drugs, including one found in uncooked fava beans.

Table 3.1 A few other genes located on the human X chromosome. Nearly two hundred additional genetic diseases are associated with recessive, X-linked alleles.

Genetic Disease	Symptom
Albrights osteodystrophy	mental retardation
Anophthalmia	absence of eyes
Börjeson's syndrome	mental retardation
Brown tooth syndrome	defective tooth enamel
Cerebellar ataxia	atrophy of cerebellum
Christmas disease	hemophilia B
Combined immunodeficiency	no immune defense
Deutan colorblindness	see red and green as red
Diffuse cerebral sclerosis	hardening of brain
Fabry's disease	renal dysfunction
G-6-PD deficiency	hemolytic anemia
Gout	high uric acid in blood
Hemophilia A	no clotting factor VIII
Hunter's syndrome	growth retardation
Ichthyosis vulgaris	fish scale disease
Lesch-Nyhan syndrome	self-mutilation
Lowe's syndrome	mental retardation
Manic-depressive	mental disorder
Muscular dystrophy	four kinds
Norrie's disease	blindness
Nystagmus	involuntary eye movement
Ocular albinism	no eye pigments

secondary sex characteristics are different; the sex chromosomes are different; and the concentrations of sex hormones are different (Table 3.2, Figure 3.13). It is one thing to describe the morphological differences between the sexes, but quite another to describe how these differences emerge during development.

The Early Events

The starting point of a narrative description of a cycle is arbitrary, in a sense, and we are discussing the human life cycle. Often textbooks begin discussing life cycles by discussing fertilization, the union of eggs and sperm which forms zygotes. But fertilization is only one of the many events that characterize the human life cycle. Where do eggs come from? Where do sperm come from? This discussion begins with these two questions. The terminology used to answer the questions may seem unnecessarily complex, but some of it has been used in the earlier chapters, though in a different context. **Gametogenesis** is the general name given to the processes of gamete formation; **oogenesis** is the name given to the processes of female gamete (egg) formation; and **spermatogenesis** is the name given to the processes of male gamete (sperm) formation (Figure 3.14). And, of course, what sets the processes of gametogenesis apart from other biological processes is **meiosis,** the type of cell division that reduces chromosome numbers from diploid to haploid numbers and mixes ancestral genes into unique combinatorial patterns. These processes are described briefly in Perspective 3.3.

(a) Spermatids differentiate into sperm cells

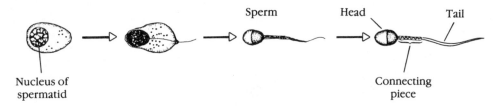

Sperm   Head   Tail

Nucleus of
spermatid

Connecting
piece

(b) Different kinds of cells found in adult bodies

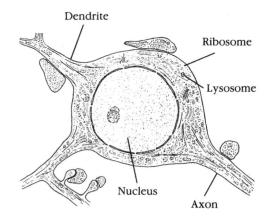

Dendrite

Ribosome

Lysosome

Nucleus

Axon

Neuron

Rod Cell

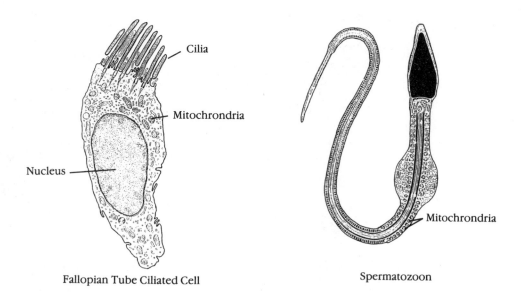

Cilia

Mitochrondria

Nucleus

Fallopian Tube Ciliated Cell

Mitochrondria

Spermatozoon

Figure 3.12 Cells differentiate during development, acquiring new shapes and functions.

(*continued*)

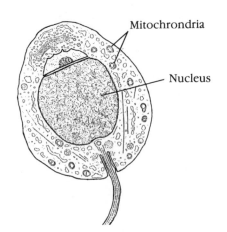

Oocyte

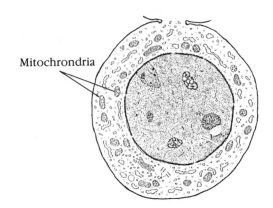

Mitochrondria

Spermatocyte

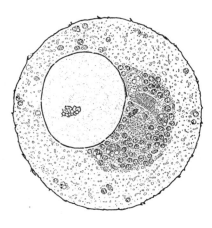

Mitochrondria

Nucleus

Spermatid

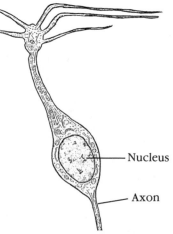

Nucleus

Axon

Olfactory Cell

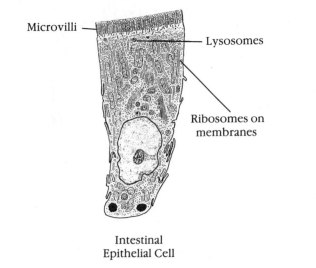

Microvilli

Lysosomes

Ribosomes on
membranes

Intestinal
Epithelial Cell

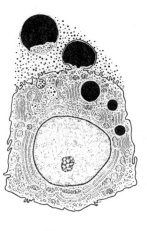

Mammary
Gland Cell

Figure 3.12 (*continued*)

Table 3.2 Sexual dimorphism.

Criteria	♀	♂
Chromosomes	44XX	44XY
Gonads	ovaries	testes
Hormones	estrogens >	< testosterone
Internal genitalia	vagina	prostate
	uterus	seminal vesicles
	oviducts	vas deferens
External genitalia	labia	penis
	vagina	scrotum
Secondary sex characteristics	breasts	voice
	hair patterns	muscle development
	body shape	body shape

Human egg cells are fertilized by one or the other of two kinds of sperm cell. One kind of sperm cell carries 22 autosomes and one X chromosome, and the other carries 22 autosomes and one Y chromosome. Fertilized eggs, then, exhibit sex-chromosome dimorphism (the X and Y chromosomes are morphologically different), which is to say that there are two kinds of human zygote, one with 44 autosomes and 2 X chromosomes and one with 44 autosomes, one X and one Y chromosome. Other than the sex chromosome difference, all zygotes appear to be identical. However, zygotes with two X chromosomes develop into females, and zygotes with one X and one Y chromosome develop into males. (Exceptions to this rule are discussed below.)

At the age of about 30 hours, zygotes multiply to become 2-cell embryos; about four hours later the 2 cells become 4 cells, then 8, 16, 32, 64, and so on. Mitotic cell divisions continue for about two weeks without changing the size of the embryo, which is to say that each new generation of cells is smaller than its parent's generation (Figure 3.15). At the age of 10 to 14 days the embryo (called a **blastocyst** at 5 days) attaches to the upper wall of the uterus where it eventually integrates within the lining of the uterus and continues its development for about nine months. Blastocysts are tiny balls of cells; each blastocyst is the same size as the zygote from which it emerged; all of the cells of a blastocyst look alike, except for tiny "hook cells" which bind to the lining of the uterus. Hook cells apparently are the first to differentiate, at least the first to become morphologically different from the others and to perform a different function.

Cell Differentiation

The puzzling problem of cell differentiation can be stated as follows: *Cell divisions that occur during embryonic development are mitotic. Progeny cells of mitotic cell divisions are genetically identical. How do cells with exactly the same genes and chromosomes differentiate into cells with different morphologies and functions?*

A complete answer to this question cannot be provided here. For one thing, the subject of biological development is not fully understood; in addition, the arguments favoring one view over others are probably more complicated than the actual processes, and certainly too complicated for a book of this type.

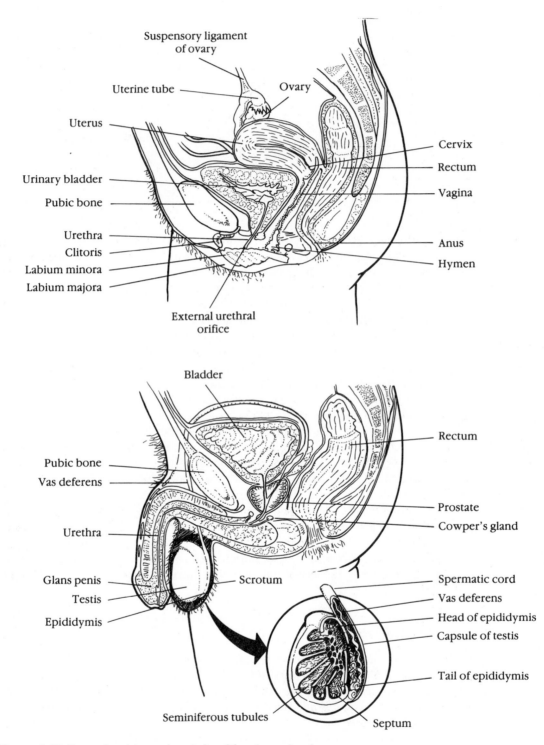

Figure 3.13 External and internal genitalia of females and males.

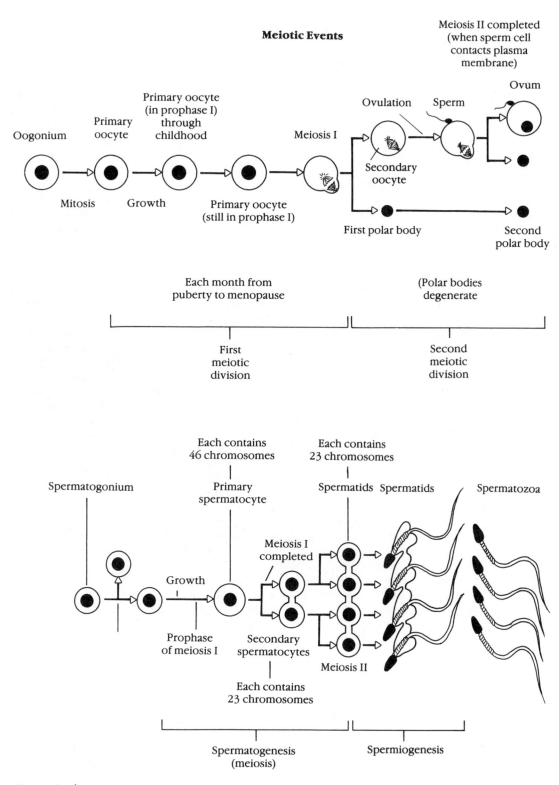

Figure 3.14 Gametogenesis. Oogenesis (top). Spermatogenesis (bottom).

Perspective 3.3

Gametogenesis occurs in gonadal cells, oogenesis in cells of ovaries, and spermatogenesis in cells of testes. The gametes generated within ovaries are egg cells, and the gametes generated within testes are sperm cells. In both cases the processes begin with diploid cells (46 chromosomes in the human species) and end with haploid gametes (23 chromosomes). However, the cellular events of oogenesis and spermatogenesis are markedly different (Figure 3.14).

The cells within ovaries that ultimately give rise to egg cells are called **oogonium** cells. By way of mitotic cell divisions, these cells proliferate, yielding many oogonia within which the first stage of meiosis begins. However, the mitotic divisions that lead to oogonia and the initial meiotic steps that occur within them stop prior to the births of baby girls. Meiosis proceeds to the cell stage called a **primary oocyte**. These early events of oogenesis take place between the seventh and ninth months of fetal development. Baby girls are born with all of the primary oocytes they will ever have, and all of these become suspended at the same stage of the first meiotic division, and all will remain at that stage for many years, until the onset of puberty. The first meiotic division does not resume until an egg cell matures, after the age of sexual maturity.

As an egg cell matures, the primary oocyte completes the first meiotic division, yielding two cells of unequal size. The large cell is called a **secondary oocyte;** it contains most of the cytoplasm of the primary oocyte. The small cell is called a polar body, which may or may not undergo further division. The secondary oocyte will not begin the second stage of meiosis unless the egg is stimulated by a sperm cell; then the secondary oocyte will complete the second meiotic division which results in the emergence of two cells, again of unequal size. The large cell is the egg, the small one another polar body which will be digested without further ado. Since the mature egg cell is not formed until after the arrival of a sperm cell, it becomes fertilized almost immediately, after which the developing zygote delays the meiotic processes of the remaining primary oocytes for 9 to 12 months.

Spermatogenesis, while genetically similar to oogenesis, is a very different cellular process. First, spermatogenesis does not begin until after the onset of puberty, the age of sexual maturity. After it begins, it continues for a lifetime, and each meiotic event yields four functional gametes, not just one.

The inside of a testis is a maze of tubules, called **seminiferous** tubules. The inside "lining" of each tubule is a layer of cells of many types. One type is called a **spermatogonium,** different from the others in that it is capable of undergoing meiosis. Spermatogonia undergo both mitosis, which produces more spermatogonia, and meiosis. A first step of meiosis yields a **primary spermatocyte.** Primary spermatocytes see the completion of the first of the two meiotic divisions, giving rise to two **secondary spermatocytes,** which after the second meiotic division give rise to four **spermatids.**

A process of differentiation called **spermiogenesis** follows, during which spermatids undergo dramatic morphological changes to become mature sperm cells. The entire process of spermatogenesis, from spermatogonium to mature sperm cell, requires about 64 days. However, since the seminiferous tubules of a single testis are lined with millions of spermatogonia, each in a different stage of the 64-day process, hundreds of millions of sperm are produced every few days, a striking contrast to one egg cell every 28 days over a time span of 30 to 35 years.

However, two kinds of conditions seem necessary to growth and differentiation: first, at *specific times* during development and, second, at *specific locations* within developing organisms, chemical signals are released that have the effect of "turning genes off and on." For example, opsin genes are present in blastocyst cells, but they do not initiate the synthesis of pigment proteins in blastocysts. Not until after eyes are fully developed are opsin genes, in photoreceptor cells of the retina, "turned on" to transcribe mRNAs that determine the amino acid sequences of opsins.

In a few organisms a few signal chemicals have been implicated as "messengers" that turn genes on and off, but in terms of development in general we know very little. The position of a cell in an embryo, tissue, organ, or organism has something to do with its morphological development and its function, which means that the morphology and function of some cells are determined in part by the cells surrounding it. What we don't know, either about the signals that regulate gene activity or about the effects of cell position upon cell shapes and functions, is how the individual events are coordinated into processes that occur through time. In the present discussion we describe mainly

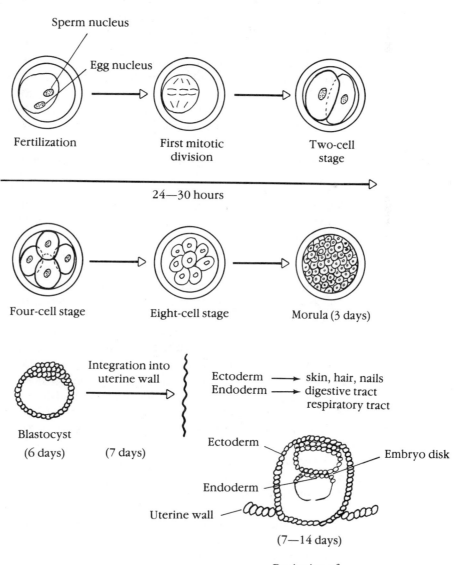

Figure 3.15 Embryogenesis.

the *state of the embryo* at particular moments in time and only rarely *how that state emerged from a prior state.*

By the age of 18 days, embryos have begun to experience rapid and active cell differentiation, but no differences other than chromosomal differences between XX and XY embryos are evident by that time. By age 9 weeks, embryos are less than two inches long and begin to look like miniature mammals, and XX and XY embryos have developed many internal morphological differences.

Sexual Dimorphism

At age 5 weeks, human embryos are about three fourths of an inch long. Small as they are, however, it is possible to identify *primordial internal genitalia.* However, at this age the primordial genitalia of XX and XY embryos are identical (Figure 3.16), and indeed both kinds of embryo are hermaphroditic (the genitalia within both are made up of cells and structures characteristic of both sexes). By the end of the seventh week, however, XX embryos will have lost some of their male characteristics, and the XY embryos will have lost some of their female characteristics. At age 12 weeks (at which time embryos have developed into fetuses about 3 inches in length) miniature, adultlike urogenital structures, femalelike in XX fetuses and malelike in XY fetuses, are present (Figure 3.16). (The embryonic period ends and the fetal period begins around the eighth week after fertilization.)

Following Figure 3.16 closely, notice that a number of things happen during the emergence of sexual dimorphism from hermaphroditism. At 5 weeks the hermaphroditic gonads contain cells of two types, a type found later in primordial ovaries and a type found later in primordial testes. Attached to each hermaphroditic gonad is a **Müllerian duct,** which later will become a fallopian tube in female embryos, and a Wolffian duct, which later will become a **vas deferens** in male embryos. In 5-week embryos it is also possible to identify the cells that later develop into the urinary bladder, uterus, cervix, and clitoris in females, and the scrotum and penis in males.

By week 7 the Wolffian ducts in female embryos and the Müllerian ducts in male embryos have degenerated. In female embryos the Müllerian ducts have developed into primordial fallopian tubes, the hermaphroditic gonads have developed into primordial ovaries, and the urogenital region shows the beginning of uterine and cervical development. In male embryos the Wolffian ducts are on their way to becoming vas deferens, and the hermaphroditic gonads are emerging into testes. Neither embryo, at this age, has developed a urinary bladder.

5-week-old hermaphroditic gonads are very simple structures with just two layers of cells. The inner layer, called the medulla, will degenerate in female embryos as the outer layer, the cortex, develops into an ovary. In male embryos the cortex degenerates as the medulla develops into a testis (Figure 3.17).

What some biologists are seeking to discover is what "triggers" these events to take place. A trigger implies three things: a signal is sent, the signal is received, and a response to the signal is made. When you begin to see the puzzle you may ask, what triggers the trigger? At the moment our knowledge of development is sparse, but this ignorance is eroding fast because development is attracting the attention of many technically skilled biological scientists today.

Since female zygotes carry two X chromosomes and male zygotes carry one X and one Y chromosome, male zygotes possess genetic information not found

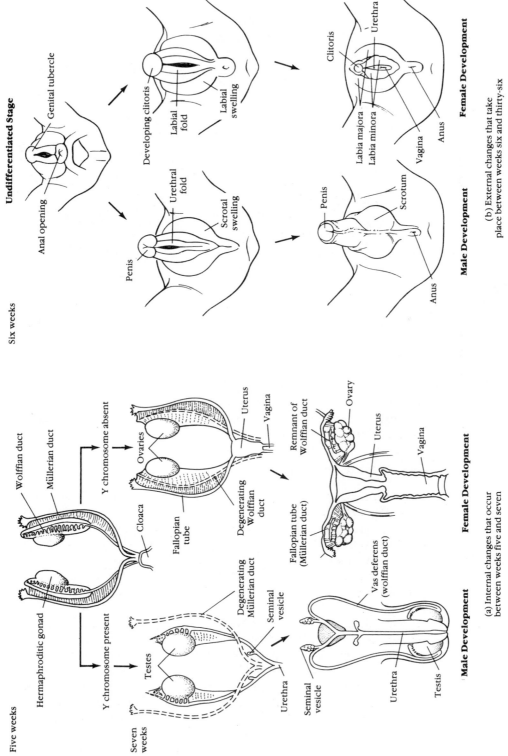

Figure 3.16 Stages of sexual dimorphism.

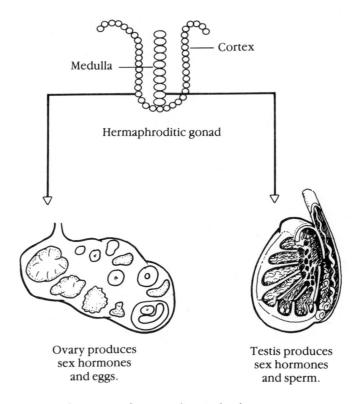

Figure 3.17 The early (five to seven weeks) stages of ovary and testis development.

in female zygotes. We now ask whether this uniquely male chromosome carries genes that encode trigger molecules which in turn participate in the development of sexual dimorphism.

Genes on the Y Chromosome

An article entitled "Sex and the Single Gene" by Jeremy Cherfas, published in *Science,* May 10, 1991, bore a subtitle which read, "British scientists find that all it takes to make a man is a tiny fragment of DNA." *Science* is one of the two most prestigious scientific journals; the other is *Nature,* published in Britain. Even though both journals exert stringent requirements upon scientists whose research is reported, both tend to allow their science reporters to use rather exotic descriptions of scientific research, such as the title, "Sex and the single gene."

The subtitle of the Cherfas paper is even more exotic in its reference to what it takes to make a man. The paper would not have been written or published at that time had it not been for research on changing the sex of embryonic mice, but the subtitle of the article would have us believe that the research was done with humans. The research with mice demonstrated that it is possible to isolate "tiny fragments of DNA" from mice Y chromosomes, to engineer their insertion into the genomes of XX, embryonic mice, and thereby to change the developmental destiny of XX embryos from female to male adults. Female embryos with the appropriate Y chromosome fragment integrated within their X chromosome(s) develop into male mice. The Y fragment carries what is called the Sry gene, which triggers XX hermaphroditic gonads to develop into testes; without Sry, XX hermaphroditic gonads develop into ovaries.

A few years ago it was reported that there is a gene on human Y chromosomes that triggers development toward maleness. This gene, called TDF (for testes determining factor), was alleged to encode a protein that acts as a trigger for the first and most important step of male development, the development of testes (Figure 3.18). The TDF gene was said to be located on the short arm of the Y chromosome and to encode a *testis-determining factor;* this report appeared to be the end of a 30-year hunt for the "switch" that "derails" development toward femaleness and sets it in motion toward maleness. The report was refuted just months after it appeared in a scientific journal.

The search for TDF was a "near miss." The kinds of experiments that were carried out with mice cannot be carried out on human beings; that is, the researchers did not isolate a fragment of the human Y chromosome and engineer its insertion into the X chromosome of XX human embryos. While the technology to do so is in the wings, fortunately it is considered to be unethical to do this kind of human biology. Therefore, the role of the TDF gene during human development was postulated from studies of XX males and XY females, exceptions to the rule that XX embryos develop into females and XY embryos into males. (At that time the Sry gene in mice was called Tdy and was considered to be the mouse counterpart of TDF.)

XX human males appear to be identical to XY males, except that they are sterile. Their appearance as males was believed to result from the attachment of a TDF gene onto one of the X chromosomes. XY females, on the other hand, appeared to carry a Y chromosome that is missing its TDF gene. The fragment of DNA found missing on the Y chromosomes of XY females was isolated from a normal Y chromosome, and this was called the TDF gene. Unlike the work with mice, however, the fragment could not be identified by its activity because it

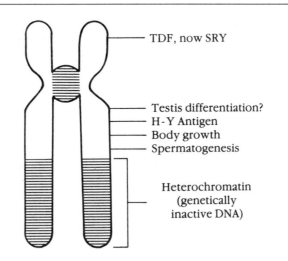

The Y chromosome

Figure 3.18 Genes that initiate development toward maleness. The TDF gene (testes determining factor) was thought to be the gene that "triggers" development toward maleness; today the SRY gene is known to be the "trigger" gene. At least four other genes are important to the development of maleness.

could not be engineered into a human X chromosome. Soon after it was discovered that the TDF fragment is not to be found on the X chromosomes of XX males. The TDF gene hypothesis turned out to be wrong.

The precise identification of the Sry gene in mice led researchers to rename the human counterpart SRY, and it now seems fairly certain that SRY is the Y chromosome fragment found on the X chromosomes of XX males. The new Sry is the old Tdy in mice, and it seems that the old TDF fragment is located near the new Sry gene in human Y chromosomes (Figure 3.18).

All of this work is based on the assumption that both males and females carry the genetic information necessary to transform hermaphroditic gonads into testes, and that males carry a gene that triggers these genes to actively encode the proteins necessary for the transformation. Without the trigger the genes remain dormant and the hermaphroditic gonads become ovaries; with the trigger the genes are activated and the hermaphroditic gonads become testes. SRY is the gene that "pulls the switch," so to speak. (There may be other "switches" within XX embryos, since it is relatively common for XX embryos to develop into adult males.)

There are other genes located on the Y chromosome that encode proteins that are uniquely male but that do not become activated unless the hermaphroditic gonads become testes. One gene initiates the synthesis of the male **H-Y antigen,** an antigen which plays a role in the development toward maleness and which may initiate spermatogenesis, the formation of sperm cells from primary spermatocytes (Figure 3.18 and Perspective 3.3). Shortly after the testes have begun to develop, some of the cells within them begin to synthesize hormones, one of which triggers the disintegration of the Müllerian ducts, and another, **testosterone,** which induces the masculinization of the developing embryo. In the absence of testosterone, the epididymis, vas deferens, and seminal vesicles fail to emerge from the Wolffian ducts; testosterone also induces the development of the prostate gland, the Cowper's glands, and ultimately the penis and scrotum. All of these developmental processes appear to be dependent upon testosterone.

The overall development of sexual dimorphism (Figure 3.19) as summarized above indicates that maleness is induced but that femaleness develops without inducers. One gene on the Y chromosome is said to initiate the cascade of steps that lead to the development of maleness, but so far an analogous trigger has not been posited for female development. Actually it was only in 1959 that the Y chromosome was implicated as a "trigger" of development toward maleness; then nearly 30 years went by before a sighting was reported of a Y chromosome gene (TDF) that triggers these developmental events. These conclusions were reached from inconclusive evidence, so we must not regard the cascade outlined in Figure 3.19 as being the final word. The "TDF fragment" of the Y chromosome was close to the SRY fragment, but in 1991 SRY was shown to be the "triggering device;" who knows how the process will be understood in the year 2000? The history of the studies of sexual dimorphism illustrates one pathway of scientific advancement, that is, by fits and starts, false leads and eurekas, mistakes and breakthroughs, and by continuous repetition of past experiments coupled with innovative new experiments. The result of these processes is never certainty, but rather ever-better approximations of reality.

Researchers in this important area of genetics and developmental biology are continually engaged in two important activities: they are doing more research,

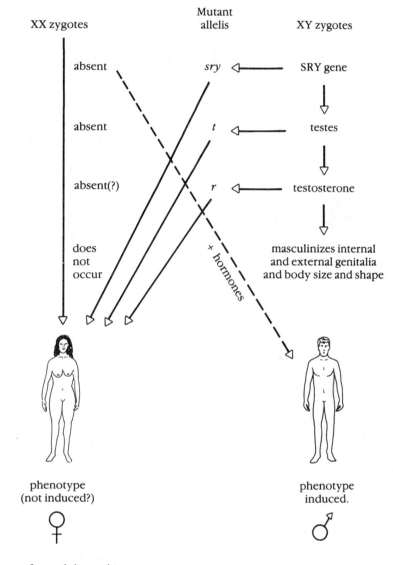

Figure 3.19 Development of sexual dimorphism.

and they are engaging in critical evaluations of the meaning of their results. Now we discuss a few additional complications of sexual dimorphism.

Variations of Female and Male Development

An old person of Troy
Is so prudish and coy
It doesn't know yet
If it's a girl or a boy

Ogden Nash

While a cliché affirms that "variety is the spice of life," history shows that some societies and some individuals disagree. Prejudice against phenotypic differences is common. This is particularly obvious in the case of female and male phenotypes. In most societies women experience more narrowly defined social roles and more limited access to political power than do men, and in most cases the social stratifications are rationalized by stereotyped images of biological and psychological differences between females and males. These stereotypes are flexible, and change sometimes rapidly through time, but the inevitable outcome of stereotyping is the masking of the *real* qualities and characteristics of our species. Stereotyping also accomplishes the social function of rationalizing and reinforcing social classes and social roles (e.g., males versus females, white versus nonwhite, rich versus poor, owners versus workers). Indeed, the real biological potential of individuals within any of the social classes may be obscured by ideas used to rationalize social stratification and maintain social class structures (for example, stereotypes in so-called western societies have it that women are biologically unable to meet the demands of professional football, higher mathematics, political decision making, and so on). The objective of a "good stereotype" is to convince the members of each class that their social status is *natural* (which means, loosely, that DNA programs social status). In some cases stereotyping gets support from *natural* variation (see Perspective 3.4).

Testicular Feminization Syndrome

Adult females who exibit the phenotype called **testicular feminization** (Figure 3.20) are indistinguishable from the XX norm as judged by body size and shape, fat and hair distribution, and breast development. But internally the story is very different. Testicular feminized women possess testes, 44 autosomes, one X and one Y chromosome, and XY levels of testosterone. The syndrome is initiated by the Tfm gene, located on the X chromosome. In the presence of mutant *tfm,* the

---

Perspective 3.4

The words normal and abnormal are used in so many different ways that they have lost much of their meaning. In popular discussions of phenotypes such as female, male, Euro-American, African American, and Asian, the natural variation that exists in nature is belied by the stereotypes used to describe it. In the present discussion, the phrases female norm, male norm, and any other population norm, are **referents** of the variation observed within the targeted populations, and they symbolize something like a collective phenotype of the population. For example, an XX-female norm includes the characteristics of at least 95% of all females with 44 autosomes and 2 X chromosomes, all of whom have a vagina, uterus, two ovaries, two fallopian tubes, and secondary sex characteristics (distribution of body hair and body fat, breast development, and body size). As discussed below, women with 44 autosomes and 3 X chromosomes fit into this "normal" distribution of female phenotypes, except for the fact that they possess an extra X chromosome. But women with 44 autosomes and one X chromosome do not. 44XO women fall outside the 44XX norm in body size and shape, the absence of ovaries, underdeveloped secondary sex characteristics, and no reproductive potential. In other words, a population norm does not mean a normal population, or normal as used in the popular vernacular. Rather, it is both a simple description of population variation and a referent from which individual phenotypes are compared with population phenotypes.

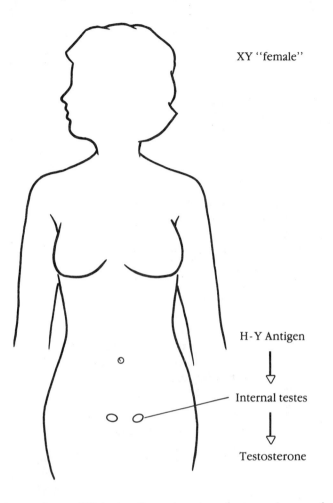

XY "female"

H-Y Antigen

Internal testes

Testosterone

Figure 3.20 Testicular feminization syndrome. X-linked *tmf* mutation gives rise to an absence of receptor sites for testosterone.

SRY gene, the H-Y antigen gene, and the genes necessary for the two testes hormones function normally; the hermaphroditic gonads differentiate into testes, spermatogenesis is initiated, and testosterone is produced in quantities expected for males. However, not all of the expected masculinizing effects of testosterone occur. Testosterone is present, but the receptor cells within certain body tissues do not respond to it; therefore the body develops femalelike external characteristics. *Tfm* alters the hormone receptor sites on cells that in normal XY males recognize testosterone; that is, a signal is sent, but not received.

Testicular feminization is rare, but it illustrates that male genotypes are capable of initiating the development of external female phenotypes. The only genetic difference between the XY male norm and testicular feminized XY females is a gene that prevents cells from responding to testosterone. The biological criteria used to define female and male include external genitalia, internal genitalia, gonads, sex hormones, and sex chromosomes (Table 3.2). Most people are **concordant** with respect to these criteria, that is, most people with a vagina also possess ovaries, more estrogen that testosterone, and two X chromosomes; likewise, most people with a penis also possess testes, more

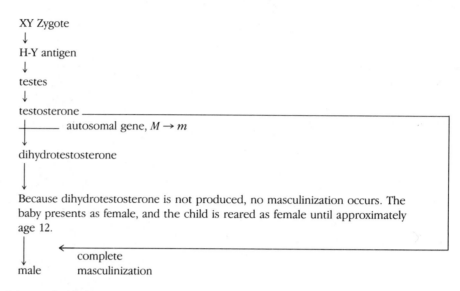

Figure 3.21 Pseudohermaphroditism.

testosterone than estrogen, and an X and a Y chromosome. Testicular feminized females are **discordant** with respect to these criteria; their external genitalia and secondary sex characteristics are female; their internal genitalia, hormones, and sex chromosomes are male.

Pseudohermaphroditism

A related syndrome, called **pseudohermaphroditism,** is relatively common in the village of Salinas in the Dominican Republic, where two dozen pseudohermaphroditic children have been studied (Figure 3.21). Pseudohermaphrodites appear to be females at birth and are raised as females until, at the age of puberty, they develop into normal males. They possess 44 autosomes, one X and one Y chromosome. Actually the prepuberty sex criteria of these individuals are ambiguous. The scrotum appears as labia majora, the vagina is "blind" in that it forms a pouch without a cervix and uterus, and the penis is about the size of a clitoris. At puberty, around the age of 12 years, male musculature develops in a typical male fashion, the voice changes, the "labia majora" develop into a scrotum, testes develop and descend into the scrotum, and the "clitoris" develops into a penis. As far as is known, these individuals become psychologically adjusted to being male, they assume male sex roles in society, and they are fertile.

Pseudohermaphroditism is explained as follows. Before testosterone begins to function as a masculinizing hormone in embryos, it must be transformed, chemically, into dihydrotestosterone. This transformation is catalyzed by an enzyme encoded by an autosomal gene. In persons homozygous recessive at this gene locus, the enzyme is not produced, testosterone is not transformed into dihydrotestosterone, the typical masculinizing processes do not occur, and a female morphology develops. At the age of puberty, testosterone, in its original form, is active as a masculinizing hormone. At the present time it is not known why the cells of pubescent children suddenly become receptive to testosterone or what chemical changes facilitate this development.

It appears that the genotypes of XX and XY embryos possess equal potential to initiate development toward female and male morphology. The SRY gene is sex specific and plays a key role in that it triggers the genotypes of XY embryos to develop testes. A few other genes in mutant form derail development, but clearly the processes leading toward maleness and toward femaleness are similar in kind, but as yet, unknown in detail.

The Female : Male Sex Ratio

Another aspect of sexual dimorphism that is not fully understood is the sex ratio, that is, the ratio of males to females. A diagram of meiosis illustrates that homologous chromosomes pair, then move to opposite poles of the cell such that we would predict that exactly half the sperm cells would carry an X and half a Y chromosome. Without specific information to the contrary, we would further predict that the frequency of the two sperm types would determine the frequency of the two embryo types, XX and XY. These predictions are not borne out by actual ratios of XX to XY zygotes or of XX to XY babies at the time of birth.

There are three developmental stages at which sex ratios are characteristically measured: the primary sex ratio is the ratio of XX to XY zygotes; the secondary sex ratio is the ratio of boys to girls at the time of birth; tertiary sex ratios are those of women to men at various ages of adulthood, most commonly the ages between 20 and 30 years (Table 3.3).

It is estimated that the primary sex ratio is 1.2 : 1, that is, following fertilization there are about 120 XY to every 100 XX zygotes. The secondary sex ratio is near 1.06 : 1, that is, for every 100 baby girls born there are nearly 106 baby boys. The tertiary sex ratio at age 20 to 30 years is close to 1 : 1. Thereafter the ratio changes in favor of women, who in the United States enjoy an average life span nearly eight years longer than that of men. The primary sex ratio greatly favors males, but by the age of 20 years the ratio of males to females is about 1 : 1. Toward the end of the life span, the sex ratio greatly favors females. How is this explained?

The primary sex ratio is poorly understood, and nobody knows for sure what it is. There is no way to examine representative samples of human zygotes. But it is known that the ratio greatly favors XY over XX. It has been suggested, sometimes with tongue in cheek and sometimes seriously, that Y-bearing sperm cells swim faster than X-bearing sperm cells (because of a lighter cargo), and that therefore Y-bearing sperm are more likely to reach egg cells first. Since the only "natural racetrack for sperm" is from the vagina, through the cervix and uterus, into a fallopian tube and to an egg cell, it would be most difficult from our outside observation booths to verify this speculation. Even if it were possible to observe sperm cells with a good microscope as they travel from vagina to egg cell, it would be difficult to distinguish between XX and XY types.

Table 3.3 Primary, secondary, and tertiary sex ratios.

Prediction	Actual		
XX × XY → ½XX : ½XY	Primary	(zygotes)	120XY : 100XX
↓	Secondary	(birth)	106XY : 100XX
½X : ½Y	Tertiary	(age 20)	100XY : 100XX
		(age 65)	83XY : 100XX
		(age 75)	66XY : 100XX

If the estimates of primary sex ratios are correct, then the secondary sex ratios are telling us that a higher percentage of XY than XX types die between fertilization and birth. Within a period of less than nine months, the XY:XX ratio changes from roughly 120:100 to 106:100. The only direct evidence of differential death rates comes from observations of miscarried embryos and fetuses (stillbirths), the majority of which are XY.

Males are lost in excess of females neonatally and through infant death, and the difference continues through adolescence until the age of maximal sexual activity and the age of reproduction, about the age of 20 years, when the sex ratio reaches 1:1. But the ratio doesn't level off at 1:1; males continue to die faster than females. By age 65 only about 16% of females have died, compared to 30% of males.

In one study of 5,000 adults in Alameda County, California, the male:female death ratios varied from nearly 4:1 for homicide to 2:1 for lung cancer, suicide, pulmonary disease, accidents, cirrhosis, and heart disease. Between ages 30 and 50 the higher frequency of cardiovascular disease among men accounts for most of the sex gap. It has been suggested that high and low incidence of cardiovascular disease between ages 30 and 50 is correlated with high and low concentrations of "bad" cholesterol and the actions of male hormones, including testosterone.

Between birth and puberty females and males have the same levels of blood cholesterol (of the two kinds of cholesterol, the high-density lipoproteins, HDLs, are not dangerous, while the low-density lipoproteins, LDLs, are). Between puberty and menopause, however, women have more HDL and less LDL because the female sex hormone estrogen discourages the accumulation of LDLs in the arteries. Indications are that testosterone has the reverse effect, increasing LDL levels and thereby increasing the risk of cardiovascular problems. Following menopause women experience increasingly higher risks of cardiovascular disease, which is correlated with increasingly higher concentrations of LDL. (Postmenopause women are more likely than men in their age group to die following cardiac disease, and they are less likely to be listened to by physicians upon complaining of chest pains. Our understanding of the relationships between cardiac disease and sex ratios is beginning to change rapidly.)

There is one additional point to be mentioned, not because a lot is known about it but because a lot is said about it: males carry one X chromosome while females carry two. A recessive allele on the X chromosome will express itself in male phenotypes more often than in female phenotypes (recall that 20 times as many males as females are genetically unable to distinguish between red and green colors). Since many recessive alleles exert a deleterious effect upon phenotype, more males than females will experience them. One reason for the higher death rate of XY embryos and fetuses may be that many deleterious recessive alleles are expressed during this developmental stage. Again, conclusive evidence is sparse.

The suggestion that it is a "single dose" of X-linked genes that increases the death rates of XY embryos and fetuses must be viewed in light of the discovery made by Mary Lyon that within the cells of XX individuals only one of the two X chromosomes is active (Table 3.4). For about one week after fertilization — to the blastocyst stage — both X chromosomes are active in all cells. Then, until the late blastula stage, the paternal X chromosome becomes inactive while the maternal X remains active within cells of nonembryonic tissue. Within the cells

Table 3.4 Individuals with more than one X chromosome per cell have only one active X chromosome per cell.

Cell nucleus	Number of sex chromosomes per cell	Number of inactive X chromosomes per cell	Morphological sex
XY	1 X 1 Y	0	♂ normal male
X	1 X	0	♀ Turner's syndrome
XX	2 X	1	♀ normal female
XXX	3 X	2	♀ Three-X female
XXY	2 X 1 Y	1	♂ Kleinfelter's syndrome

X→ ● indicates X chromosome inactivation

of the embryonic tissues, from this point on, X inactivation is random, i.e., it is as likely to be one as the other X chromosome that becomes inactive.

One outcome of X chromosome inactivation is that the cells of males and females possess but one active X chromosome. However, this does not explain why more XY than XX embryos and fetuses abort. The differential death rates of XY and XX embryos and fetuses may be due to the fact that the random inactivation of maternal and paternal X chromosomes allows the genes of both chromosomes to become active within most body tissues and organs, that is, each cell will carry only one of each pair of alleles, but each tissue and organ will carry the gene products of both alleles of all X-linked genes.

44X, 44XXX, 44XXY, 44XYY, and Other Chromosomal Variations

During oogenesis and spermatogenesis (Perspective 3.3), homologous chromosomes may fail to segregate, that is, they may fail to *disjoin*. When this happens, both members of a homologous chromosome pair will move to the same meiotic pole and appear together in the same secondary oocyte, or spermatocyte (Figure 3.22). Such meiotic mistakes are called **nondisjunction.**

If an egg cell with two chromosomes of the same kind, for example two of the 21st type, is fertilized by a sperm cell with only one 21st chromosome, the resulting zygote will carry three of the 21st chromosomes and two of the remaining 22 types. The zygote will carry 47 chromosomes, not the usual 46. The name **trisomy** is used to designate this specific chromosomal condition (e.g., the zygote described here is called trisomy 21).

Of the 22 possible autosomal trisomics, only three survive beyond birth, and two of these, trisomy 13 and 18, die shortly after birth. Both trisomy 13 and 18

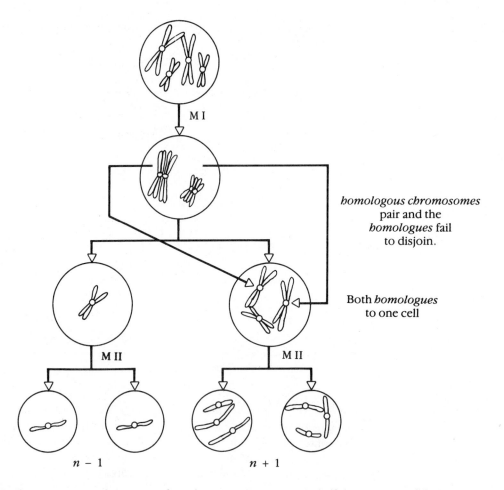

Figure 3.22 Chromosome nondisjunction. If nondisjunction occurs at MI, half the gametes will be $n + 1$, and half will be $n - 1$. If nondisjunction occurs at MII, half the gametes will be normal, n, one will be $n + 1$, and one will be $n - 1$.

lead toward such abnormal development that there is no hope for making them right with medical intervention. The other surviving trisomy, trisomy 21, may live for many years, but not an average life span. Trisomy 21 is called **Down's syndrome** after the physician who first described it in the medical literature (Figure 3.23).

Variations of sex chromosome numbers are more common than variations of autosome numbers. In most cases missing and extra sex chromosomes are less deleterious to health. However, variations in the numbers of sex chromosomes do lead to variations of sex phenotype.

Turner's Syndrome

One sex chromosome variation leads to the development of **Turner's syndrome** (Figure 3.24). Turner's females are 44XO, that is, they have 44 autosomes and only one X chromosome (an O signifies the absence of a sex chromosome). Turner's zygotes are formed by one of two kinds of fertilization event: a 22O egg and a 22X sperm, or a 22X egg and a 22O sperm. (Examine Figure 3.22 again, and notice that a nondisjunction event that results in egg cells without an X chromosome also results in egg cells with two X chromosomes; the analogous event during spermatogenesis results, simultaneously, in 22O sperm cells, and sperm cells with 22 autosomes, one X and one Y chromosome.)

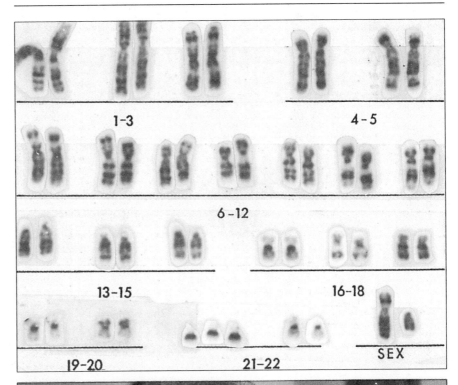

Figure 3.23 Down's syndrome (trisomy 21). Three chromosomes 21 result in a total of 47 chromosomes.

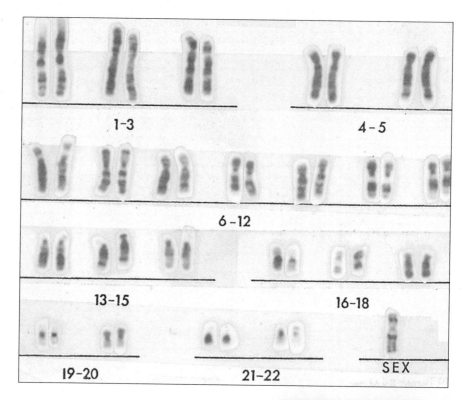

1-3

4-5

6-12

13-15

16-18

19-20

21-22

SEX

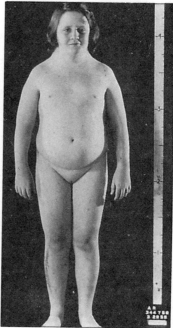

Figure 3.24 Turner's syndrome. The 44X chromosome variation is characterized by short stature, sterility, and underdeveloped secondary sex characteristics.

Turner's syndrome individuals appear to be females as judged by their external genitalia at birth. However, 44XO adults do not have functional ovaries, only streaks of white tissue where ovaries are expected to be. Lacking ovaries, Turner's females do not produce functional sex hormones; therefore they do not experience puberty and the development of secondary sex characteristics following puberty (pubic hair, breast development, subcutaneous fat, body size and shape). Exogenous hormones administered at age 11 to 13 years often enhance the development of secondary sex characteristics of Turner's females, but they have no effect upon ovarian development and sex-hormone synthesis. From these facts it is not unreasonable to conclude that two X chromosomes are necessary for ovarian development, probably to provide a sufficient dose of a trigger substance during the early weeks of embryogenesis. However, the trigger substance has not yet been discovered.

About 1 of every 2,200 female births is a Turner's female, but about 20% of spontaneous abortions of female embryos/fetuses are 44XO. This observation supports the view that embryos/fetuses with only one X chromosome, 44XY and 44XO, abort more frequently than those with two X chromosomes, probably because one X chromosome allows the expression of a few deleterious alleles. It is also true that the frequencies of X-linked phenotypes, such as red-green colorblindness, are the same for 44XO and 44XY individuals.

If a 22XX egg is fertilized by a 22X sperm, the zygote will be 44XXX. There is no syndrome name for this condition, probably because 44XXX females are otherwise indistinguishable from 44XX females. There is some indication that 44XXX females are less fertile than 44XX females, which is consistent with the fact that extra chromosomes sometimes interfere with the syncopation of meiosis.

Klinefelter's Syndrome

If a 22XX egg is fertilized by a 22Y sperm, the zygote will be 44XXY. This condition is diagnosed as **Klinefelter's syndrome** (Figure 3.25). 44XXY individuals are diagnosed as males at birth. Klinefelter's males have longer arms and legs than 44XY males and hence are taller, but they differ from 44XY males in that they develop femalelike secondary sex characteristics. Most 44XXY males are sterile, and many possess juvenile external genitalia. However, most are easily assisted with surgery and exogenous hormones to achieve an external phenotype indistinguishable from the 44XY norm. The frequency of X-linked recessive phenotypes among 44XXY males is the same as it is for 44XX females.

About 1 of every 500 baby boys is 44XXY. Babies with 44XXXY, 44XXXXY, and 44XXYY have been observed, but rarely. The greater the number of extra chromosomes the more the developmental processes deviate from the 44XY norm. However, that such deviations from the norm are viable at all is best explained by X-chromosome inactivation; after about one week of embryonic development, all but one X chromosome in every cell becomes inactive, whether in 44XX females or 44XXY, 44XXXY, or 44XXXXY males.

44XYY (Originally called Jacob's "Criminal" Syndrome)

In 1965 a paper was published under the title "Aggressive behavior, mental subnormality, and the XYY male." The title suggests rather vividly that 44XYY males are more aggressive and less bright than 44XY males. What evidence did the authors use to arrive at that conclusion?

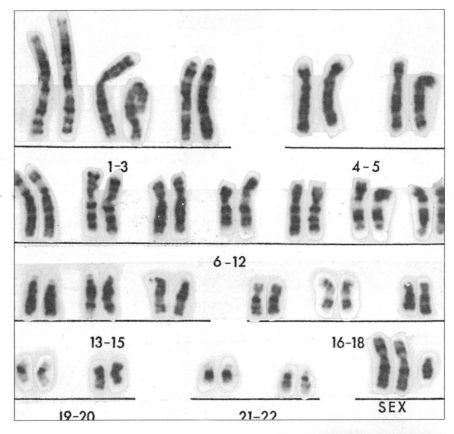

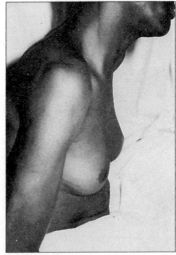

Figure 3.25 Klinefelter's syndrome. The 44XXY chromosome variation is characterized by greater height, sparse beard, female breasts, and frequent sterility.

The paper describes seven inmates found in a maximum security prison in Scotland whose total inmate population was 197. The average height of the seven 44XYY inmates was about 5 cm taller than the average among XY males, their faces were acned, and the acts committed by them for which they were imprisoned appeared to the authors to be more aggressive than was true for their 44XY prison mates. Shortly after the first paper, a second was published based upon observations of nine 44XYY males drawn from a population of 315 inmates of a maximum security hospital. These two papers were hailed internationally for have reported a genetic base from which to understand crime (Figure 3.26).

Within two years following these reports, many prison inmates were examined for an extra Y chromosome, and those with two Y chromosomes were examined for behavior. The psychologists who participated in these studies, believing that males are more aggressive than females (i.e., that the Y chromosome enhances aggressive behavior), were quick to point out that two Y chromosomes must encode even more aggression. Prominent scientists announced that among the many great breakthroughs made recently by geneticists, this discovery of the genetic basis of aggression, criminality, and low intelligence was at the top. (A few geneticists do in fact make important discoveries, but all geneticists are required to justify the public monies awarded to them for research. It is not uncommon to hear some of them "brag" a bit for the purpose of justifying research dollars, but in this case the bragging was short lived).

It turns out that the first reported case of a 44XYY male in the United States was of a "family man" whose outward appearance fit well into the 44XY norm. This 44XYY man displayed no unusual aggressive tendencies. Then it was discovered that about 1 of every 800 to 900 male births is 44XYY, meaning that

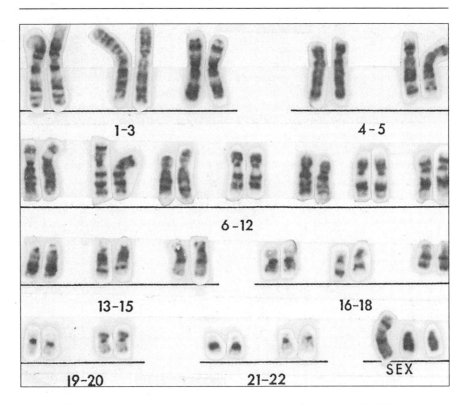

Figure 3.26 XYY syndrome. The XYY syndrome results in acne and greater-than-average height. At one time XYY men were said to be aggressive, antisocial, and unable to fall in love and to establish enduring relationships.

in the greater population there are millions of such males, no doubt most of whom live normal lives. Otherwise jails and prisons would overflow with 44XYY males. Ultimately, of course, someone was bound to ask the question, where do Y-less female prisoners get genetic information that causes them to commit crimes?

Take notice, now, that *the phenotypes under discussion are no longer confined to metabolism, physiology, or morphology.* In the case of XYY males, the claim was made that chromosomal differences lead inevitably to behavioral (aggression) and moral (criminal) differences. This claim was welcomed by those who had always believed that genes encode good and bad behavior and high and low intelligence, just as they encode blue and brown eye color. Indeed, the claim that an extra Y chromosome enhances aggression and diminishes intelligence was used as further evidence that sterilization programs are needed to keep the incidence of crimes in society at relatively low levels (Shockley, 1972).

An Aside

As we shift the emphasis from physical to behavioral phenotypes, we come face to face with the most hotly debated issue of sexual "dimorphism," only the word dimorphism hardly applies. We need a new word that implies two kinds of behavior, female and male behavior. Since females and males are dimorphic (this includes metabolic and physiological as well as morphological differences), some scientists have concluded that female and male norms of behavior are different and that the difference is biological. From this conclusion it is but a short jump to the next conclusion, namely, that biologically-based sex behavior differences serve as the infrastructure of social sex-role differences. The logic is clean and simple—*sex-specific physical differences cause sex-specific behavioral differences, and these, in turn, cause sex-specific social rules.* The "missing link" in this clear and simple scheme is a database—i.e., experimental evidence of a causal link between biology and behavior and between behavior and social sex roles.

Biology, psychology, and sociology are disciplines of study representing three very different levels of structural and functional complexity. Within each level there are clear limits to our understanding. The limits to understanding are even more severe between levels, so much so as to make it impossible to bridge levels using current knowledge found within levels. Even so, some scientists are convinced that there are causal links between levels, and, of course, other scientists are equally convinced that there is no evidence for such links (Figure 3.27). Yet, it is clear that there is more to behavior and societal organization than genetic information. The argument is not, "either genes or experiences," but rather that of primary and secondary inputs. This issue will not be resolved in our lifetime, as is further discussed below. (For an interesting history showing that scientists often change their hypotheses, even in the absence of changes in the database, see the short book by Farley.)

Do Sex-Specific Physical Differences Cause Sex-Specific Behavioral Differences?

Transsexuality

There exists a form of self-image called **gender identity.** Gender identity refers to sex identification, that is, to the sex type that each individual "feels" that she or he is. Many biologists lump the concept of gender with that of biological sex, and often psychologists lump facts about biological sex into their descriptions of gender. It is common even today to hear the phrases sex roles and gender roles used interchangeably. Some scientists are trying to clarify the meaning of gender identity, one motivation being a discordance between biology and gender, a condition known as **transsexuality.**

Transsexuals possess the biological features of one sex and the gender of the opposite sex. Biological males may "feel" as if they are female, and biological females as if they are male. There have been attempts by psychologists and psychiatrists to change the genders of transsexuals to match their biological sex, and there have been attempts by surgeons to change biological sex to match gender. These latter attempts have been widely publicized ever since the "creation" of Christine Jorgensen in the 1950s.

Attempts to change biological sex to match gender are based on the assumption that the developmental processes that give rise to biological sex differences in the first place can be reversed, at least partially. Biological change has been the preferred approach to concordance, since many psychologists have maintained that gender identity is set in place by the age of 3 years, and that once set it cannot be changed. Both assumptions — that biological sex is reversible and

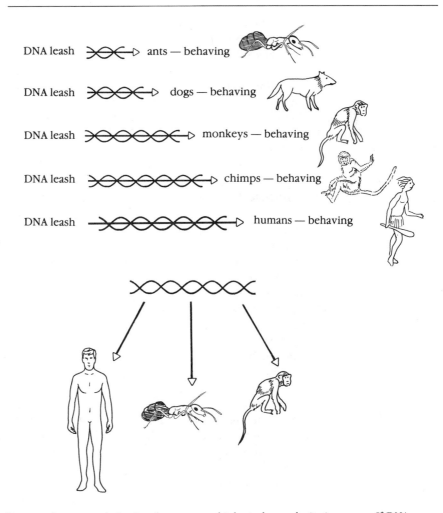

Figure 3.27 The influence of genes on behavior decreases as biological complexity increases. If DNA dictates behavior, and if DNA is a leash, then leash length is species specific.

that gender identity is not reversible—illustrate two hypotheses that lack a database. However, our knowledge of biological sex is based on facts, but our ideas of gender were fairly well set in place before science got into the act (part of the sex mystique). To add insult to injury, the subject about which the least is known (gender identity) is the subject about which the greatest certainty is felt.

It is impossible to reverse the biological sex of human beings. Men made over into "women" are not women; they are infertile, lack ovaries, fallopian tubes, uterus, and an internal source of sex hormones. Women made over into "men" are even less well reconstructed; in some cases there isn't even an attempt to construct a penis. Whether anyone's self-image has been helped by transsexual surgery is a matter that has not been corroborated by science. However, a matter that has been corroborated by science is that transsexual surgery is a one-way action. There is no turning back. After surgery one must try desperately to behave as a member of the opposite biological sex. In real life there is a vastness between these twains that has not been and cannot be touched by surgeons.

Whether gender identity is reversible is still to be determined. But if transsexuals in fact display discordance between gender and biological sex, it is clear that sex-specific biological features, including chromosomes, hormones, and secondary sex characteristics, are not the primary determinants of gender identity and hence of individual sex behavior. But this is not to deny that biology exerts an important influence upon behavior. For the most part, the development of sexual dimorphism can be explained in a formal biological language, but to extend this explanation to gender differences and to sex-specific behaviors that in fact determine social sex roles tortures the kind of scientific logic ordinarily reserved for uniting facts into generalizations.

Are Biological and Social Sex Roles Concordant?

Biological sex roles are easy to recognize and to describe. There are two: *reproduction* and *genetic recombination*. However, neither sexual reproduction nor genetic recombination is a necessary feature of continuity between generations, or of increasing the genetic variation within populations. Asexual reproduction (Figure 3.28) and gene and chromosome mutations (Table 3.5) achieve similar ends but at different tempos, and with less "fanfare."

If all life on the planet earth, from the submicroscopic to the largest whales and redwood trees, is looked upon as one gigantic ecosystem, it would be foolish to argue whether genetic recombination or mutation is the best, or whether sexual or asexual reproduction is the most efficient process for explaining the origin or the future of the ecosystem. Mitosis and meiosis are different kinds of cell division, but both, independently and in concert, achieve generation continuity; recombination is meaningless in the absence of the variation provided by mutations. What biologists want is the most inclusive and general explanation within which all of these phenomena interact and contribute to the *evolution of biological variation* (see Chapter 4).

Within the context of development, whether of humans or peas, both asexual and sexual reproduction are necessary. All cell divisions that occur outside the germ-line cells are asexual, and all cell divisions that result in the formation of gametes are sexual. While asexual reproduction is necessary to the development of individuals, sexual reproduction is necessary to the increase in numbers of individuals. Humans, hamsters, and hyacinths are absolutely dependent upon

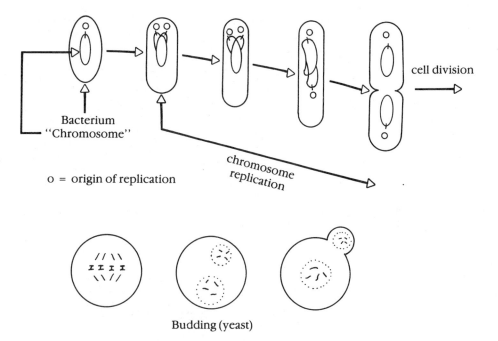

Figure 3.28 Asexual reproduction. In mitotic cell division, the progeny cells are genetically identical to the parent cells.

Table 3.5 Chromosome and gene mutations.

	Chromosome										*Gene*	
Normal	A B C D E F G H I J											A C G A T C T G C T A G
Deletion	A B D E F G H I J									Deletion	A C A T C T G T A G
Duplication	A B C C D E F G H I J J									Duplication	A C G G A T C T G C C T A G
Inversion	A B C F E D G H I J											
Translocation	A B C D E F G H I J X Y											

sexual reproduction for generation continuity. This is not so with viruses, simple algae, bacteria, many plants and some animals. Bacteria increase their numbers via asexual reproduction, as do human embryonic cells; however, bacterial cells are individuals, whereas human embryonic cells are integral to an individual. Bacteria engage in sexual reproduction, and their genomes recombine (Figure 3.29) as do even some viral genomes, but recombination doesn't

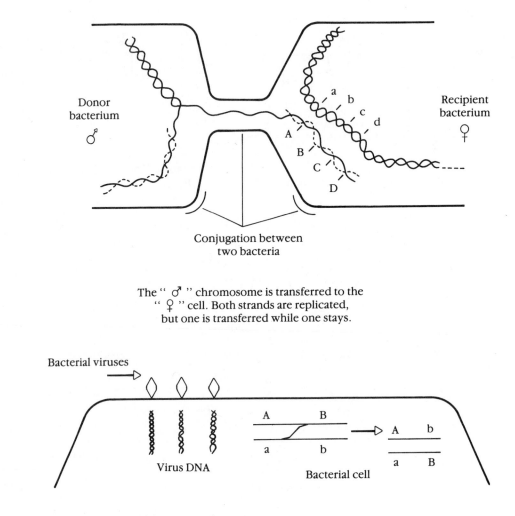

Figure 3.29 Genetic recombination in bacteria and viruses.

appear to be a necessary feature of the process by which they increase their numbers.

Virus, algal, and bacterial sexual reproduction is simple compared to human, hamster, and hyacinth sexual reproduction, and among sexually reproducing plants and animals there are variations intermediate in complexity between these. In the case of humans, however, there is an ensemble of nonbiological sex roles in addition to the biological sex roles. There is no doubt that biological and social sex roles interact and coinfluence one another, but since biological sex roles are essentially the same in all plant and animal species, and since social sex roles are not, it is not self-evident that the latter are encoded by the former.

To illustrate, scientific knowledge gained about sexual reproduction in hamsters explains a great deal about sexual reproduction in humans. But studies of hamster sexuality and hamster social sex roles tells us little if anything about human sexuality and human societal sex roles. Human behavior and human societies are in a class by themselves; other species that reproduce as humans do, do so without judges, priests, psychiatrists, guilt, or ceremony. Indeed, in many modern human societies, the many images of sex derived from sexuality have all but blotted out knowledge of the role of sex in reproduction.

Biological sex roles are sex specific. Males produce sperm cells and females produce egg cells. Egg cells are fertilized within female bodies, and for about nine months the development of new individuals takes place within female bodies. Women give birth to babies, and if a baby is suckled, it is suckled by a woman. For a few days and weeks after birth, relationships between baby and mother are biologically different from those between baby and father.

It is general practice in most modern societies to put diapers on babies, to bathe and clothe them, to feed them, and to attend to them when they are ill. Traditionally women have performed these tasks. But are women genetically programmed to perform these tasks and men not? Furthermore, are women genetically programmed to work at home and men to work away from home, women to socialize and educate children and men to earn a wage? While some professionals will answer these questions in the affirmative and others in the negative, it should be obvious that we do not have enough evidence (scientific data) to decide; there is no scientific proof that social sex roles are genetically determined.

Professionals who feel that social sex roles, like biological sex roles, are programmed by genes, use as (partial) evidence the sex behaviors of other animal species; for example, female birds sit on their eggs while males forage for food; baby gorillas stay with their mothers, sometimes for years, whereas their fathers are out foraging for other females; in some species of fish a male will lead a group of several females, but if the male dies or wanders off, the dominant female in the group will undergo an automatic sex change (become a male) and subsequently lead the group. (Is the take-home lesson that only males can lead?). But not all bird species behave in the same way; in some species males keep the eggs warm. Not all primates behave like gorillas, and not all female fish experience a sex change in anticipation of leading a group of females. In addition, gorillas and fish are not human beings. Finally, not all human societies practice the same societal sex roles.

Professionals who feel that social sex roles are learned rather than genetically programmed use as evidence the fact that children with similar biological histories can be socialized to exhibit very different kinds of behavior. They point to the fact that legal systems, social mores, and religious dogmas dictate and reinforce certain types of behavior and discourage other types. Also, adopted children tend to behave as instructed by their foster parents, not by their genes.

Between these rather opposite views there are agnostics who call for more evidence before closing the question; they understand that both arguments rest on weak foundations of facts. The agnostics agree, however, that societal sex roles *have not* been shown to be instructed by Mendelian and/or molecular genes.

Cultural history, like biological development, cannot be reversed. Our traditions are in place, and they won't easily be changed. For decades and centuries, women's bosses have tended to be men, whether the woman works at home or away from home. Rarely do men have women bosses. Women, then, are more apt to be censured by men than men by women. Since part of a woman's societal contract includes being attractive to men, and since part of a man's societal contract includes being financially successful, censure is likely to be internalized in a sex-specific way. In the world of work and wages, approaches to resolving conflicts are very different for female-male than for male-male

Table 3.6 A brief history of the societal responses to suffrage.

Events in the suffrage and equal rights movements (Thesis: privileges granted by the State are natural, inherent rights)

1848	First women's suffrage convention
1866	First petition to Congress
1869	New England Women's Suffrage Association founded
1916	Both Presidential candidates favor suffrage
	Wilson (D) — ballot by state action
	Huges (R) — Constitutional amendment
1919	Congress sends states an equal suffrage amendment which is ratified by 22 states, including Minnesota.
1920	Equal suffrage amendment ratified by 15 additional states (Alabama, Delaware, Georgia, Louisiana, Mississippi, and South Carolina voted not to ratify, and 5 states were undecided.)
	August 26. Suffrage amendment proclaimed part of Constitution
1963	Equal Pay Act
1964	Title VII of Civil Rights Act (Legal right to equal treatment in work)
1989	Women earn 59–63% of men's wages

relationships. Shall we summarize with "cultural tradition versus DNA coding," or "cultural tradition and DNA coding?"

The social dominance of men has included the denial of human and civil rights to women. Women in the United States could not vote until after 1920, were not hired as truck drivers, police officers, or airline pilots until after 1970 (Table 3.6), and are still excluded by the National Football League, the National Basketball Association, the National Hockey League, major league baseball, and many social organizations. More important to most women than being excluded from boxing and hockey, however, is the fact that the Equal Rights Amendment still has not been ratified in 17 of the 50 states. How shall we force an argument that genes encode the need and the lack of need for guaranteed equal rights under the law? Would it not be as easy to construct an argument that males enjoy their social dominance and will go to some lengths to keep it?

Many men, even those who are not socially dominant, consider sexual privilege to be among their most cherished privileges. Fathers sexually abuse their daughters far more often than mothers sexually abuse their sons; men often rape their wives and lovers, fondle their female coworkers, rape strangers, and sometimes batter and kill women who resist them. It seems as if sexuality can be convoluted in such as way as to justify a dominance relationship, even violence.

Sex objectification is a multibillion dollar industry. Indeed, the objectification of female bodies may explain why so few males understand females as persons, as human beings. The pornography industry (including advertising) explicitly portrays female body parts that both appeal to and help shape male images of what it is to be female. The "porn" industries, in fact, amplify these male images for the whole of society, the result being that some women equate those images with reality. Putting aside for a moment the question of cause and effect, sexuality is a subject about which it is impossible to remain innocent and nearly impossible to become informed.

The list of social sex role differences is long. Some scientists claim that males are "naturally" better at mathematics than women, that mathematics and related skills are facilitated by one half of the brain and that verbal and motor skills and nurturing tendencies are facilitated by the other half. Biological explanations of these social phenomena usually ignore the fact that women have been excluded from mathematics and other primarily male societal activities (e.g., science and sports). It turns out that the societal activities from which women have been excluded are dominant and interesting, whereas the societal activities in which women do excel aren't as interesting, as well funded, or as highly rewarded. But notice that as women enter into the interesting and financially rewarding societal activities, they do well. Florence Joyner Griffith ran 100 meters faster in 1988 than Jesse Owens did in 1936.

As will be discussed in Chapter 4, it is very difficult to "sight genes" within human societies through observations of societal phenotypes, especially within societies that stratify classes and social roles upon economic imperatives, thereby dislodging such phenotypes from their genetic base, if indeed they have one.

Summary

As obvious as the morphological, physiological, and metabolic differences are between males and females, our understanding of how these differences develop after an egg is fertilized is indeed sparse. A few hormonally–induced secondary sex differences are understood, but these have little or no bearing upon the development of sexual dimorphism.

Sex chromosomes are distributed unequally between female and male progeny. Females inherit two X chromosomes, while males inherit one X and one Y chromosome. Genes located on the Y chromosome are expressed only in males. Recessive alleles on the X chromosome are expressed in both sexes, but always in males and rarely (the allele must be homozygous) in females.

Some of the sex changes that take place within early embryos can be described at the cellular level, but little is known about the processes of their development. Most XX embryos develop into females, and most XY embryos develop into males. Four genes on the Y chromosome appear to trigger metabolic events that are necessary to the development of maleness, but no such genes have been found necessary to the development of femaleness. However, variations of male and female development accompany variations of sex chromosome number. The Y chromosome seems necessary for maleness, however variant, and in the absence of a Y chromosome, femaleness, however variant, develops. For example, 44XO leads to variant femaleness (Turner's syndrome), while 44XXY leads to variant maleness (Klinefelter's syndrome).

The X-linked *tmf* allele encodes defective testosterone receptor sites, which in turn prevent testosterone from masculinizing XY individuals. The syndrome, testicular feminization, is recognized by discordance between external body type (female) and internal genitalia (male). A gene that fails to encode the enzyme that transforms testosterone into dihydrotestosterone delays masculinization until the onset of puberty, that is, XY children appear to be females until puberty, at which time testosterone masculinizes, and they become males. This is called sequential pseudohermaphroditism.

Evidence abounds that genes influence the development of morphological, physiological, and metabolic differences among individuals and between the two sexes. But in addition to these differences there are differences in individual and societal behavior, and in particular, in social sex roles. It is contentious whether sex behavioral and sex social role differences are genetically or societally determined. While evidence has been amassed to support both views, a scientific verdict cannot be given. However, social mores and legalistic rules are built on the premise that many forms of behavior, individual and societal, are outcomes of our biological natures, and that they are therefore inevitable.

In short, biological sex roles — reproduction and genetic recombination — are easy to determine and to substantiate. Behavioral and societal sex roles are problematic in that so much modern social practice has been set in motion by historical precedence, kept in motion by political and economic motives, and defended as personal privilege and/or property as to make it virtually impossible to sort out all of the cause-effect events and processes.

Study Problems

1. Explain in what ways the advice of the ancient Rabbi was correct when he exempted a child from circumcision if the sons of his mother's sister bled to death following circumcision; and when he would not exempt a child if the sons of his mother's brother bled to death following circumcision. Describe both circumstances in terms of the inheritance patterns of hemophilia A.

2. Gene D is located on the X chromosome. DD and Dd females and D- males have normal vision, but dd females and d- males are deutan, red-green colorvision impaired. In Metropolitan St. Paul and Minneapolis there are more than 2 million people (round the number to 2,000,000). Close to 50,000 males in this population are deutan, red-green colorvision impaired.
   (a) About how many different kinds of X chromosome are to be found in the Twin Cities?
   (b) About how many *d* alleles are expected within this population of X chromosomes?
   (c) Predict the number of women that are deutan, red-green colorvision impaired?
   (d) What is the probability of a dd × d- mating? a Dd × d- mating? a Dd × D- mating?

3. Explain why red-green colorvision impaired boys never inherit the causative gene from their fathers.

4. If you had a daughter who had been reared as a female and who seemed satisfied being a female, and if you discovered when she was 15 years old that she was genetically a male who had been feminized by an insensitivity to the action of testosterone, would you wish to have her continue to act out female societal and sex roles, or would you wish to change her external sex characteristics to match her chromosomal and gonadal sex type? Explain.

5. A four-day-old embryo is composed of cells of one type, as far as we know. A four-month-old fetus is composed of cells of about 100 cell types, including blood cells, muscle cells, skin cells, and so on. Is it any more surprising to witness the differentiation of a stem cell into a neuron or into a muscle cell than it is to see one embryo develop into a female and another into a male after having been triggered by one or a very few proteins?

6. Identify two known genes, either of which in mutant form explains how an XY zygote might develop into an adult female (female as judged by body form). Explain how each mutant gene alters the course of development.

7. Explain why it is predicted that half of all sperm cells will carry an X chromosome and half a Y chromosome. Compare your explanation with a test-cross: in what ways are they similar?

8. The first 16 44XYY individuals reported to be hyperaggressive and to have subnormal intelligence were discovered in a maximum security prison and a maximum security hospital. It was later said of these men, by psychologists who specialize in gender identification, that 44XYY males are unable to "fall in love," and to "establish enduring relationships." Does this seem to be an objective observation of the 44XYY condition, given the circumstances of their living arrangements at the time their behavior was described? Outline an experiment to determine whether or not there are behavioral differences between 44XY and 44XYY males.

9. Argue for and against the proposition that societal sex roles are determined primarily by genes.

10. Cite a few examples of phenotypes that are known to be influenced by genes but that can be changed by nongenetic means.

11. What is the best explanation for the fact that the majority of elected politicians, doctors, lawyers, and scientists are males?

Suggested Reading

1. Bleier, Ruth, Ed., 1986. *Feminist Approaches to Science*. Oxford: Athena, Pergamon. In this volume, Hilary Rose has an article entitled, "Beyond Masculinist Realities."

2. Cummings, M.R. 1991. *Human Heredity: Principles and Issues*. 2nd ed. St. Paul, MN: West Publishing.

3. Farley J. 1979. *The Spontaneous Generation Controversy From Descartes to Oparin*. Baltimore: Johns Hopkins University Press. This book illustrates the difference between what actually happens during the evolution of science and what straight-line histories of science tell us. It also illustrates how scientific opinion switches back and forth between unchanging databases and changing popular opinions.

4. Haraway D. 1990. *Primate Visions: Gender, Race and Nature in the World of Modern Science*. New York: Routledge.

5. Hubbard, R. and Lowe, M. eds. 1979. *Genes and Gender II*. New York: Gordian Press. This is one in a series of books about aspects of biological and cultural sex, sexuality, and sex behavior written by biologists, psychologists, sociologists, anthropologists, and feminists.

6. Keller, E.F. 1985. *Reflections on Gender and Science*. New Haven: Yale University Press. An exceptionally well written history of social sex roles.

7. Lewontin, R.C., Rose, S.P., and Kamin, L.S. 1984. *Not in Our Genes*. New York: Pantheon Books (in paperback by Penguin). These authors—a geneticist, a neurobiologist, and a psychologist—examine the fundamental flaws of biological determinism as an explanation of human behavior.

8. Nelson, H., and Jurmain, R. 1988. *Introduction to Physical Anthropology*. St. Paul, MN: West Publishing. The chapter on Primate Behavior is relevant to the discussion of sex behavior. However, the entire book is about primate behavior and the evolution of that behavior.

9. Shockley W. 1972. "Dysgenics, Geneticity and Raceology." *Phi Delta Kappan* 53 (5):305.

4

EVOLUTION, POPULATIONS, AND SPECIES

Meet Mr. Charles Darwin

If we visited a popular shopping mall on a Saturday afternoon in any city in North America, polled the shoppers about the differences between biological sex and sexuality, and tried to summarize the wide variety of answers we would get, it might lead us to conclude that scientific knowledge seldom interferes with popular opinion. If we added to our list a few questions about the differences between the biological and the "Christian" views of evolution, we would find additional support for our conclusion. The subject of evolution, like that of sex, is an area in which individual realities, distilled from social, cultural, and religious realities, are seldom disturbed or contested by scientific facts.

This is not a criticism of the shoppers. As individuals they have very little to do with the public confusion about sex and evolution. The confusion is already basic to the society within which each was born. The confusion is not personal but the result of a clash between two powerful social forces or world views that have been at one another since human history was first recorded. One force strives to establish and control a stable and unchanging world view; the second strives for an understanding of how all things in the universe did, in fact, come to be as they are, which assumes that the universe has changed and is changing. Advocates of a traditional and unchanging universe have railed against ideas that suggest tradition is wrong, and those who suggest that tradition is wrong rail against orthodoxy and traditional ideas. Socrates, Galileo, Copernicus, Darwin, and Einstein all got themselves into trouble with the "guardians of orthodoxy" by suggesting that the state of the universe is very different from that being espoused by those guardians.

Darwin was keenly aware of the clash between these two world views. In his autobiography he wrote:

> Whilst on board the Beagle I was quite orthodox, and I remember being heartily laughed at by several of the officers for quoting the Bible as an unanswerable authority on some point of morality. . . . But I [gradually came] to see that the Old Testament was no more to be trusted than the sacred books of the Hindoos. . . . I gradually came to disbelieve in Christianity as a divine revelation. . . . I found it more and more difficult, with free scope given to my imagination, to invent evidence which would suffice to convince me. Thus disbelief crept over me at a very slow rate, but was at last complete. The rate was so slow that I felt no distress.

Darwin enrolled at Cambridge to study divinity and to prepare for the ministry. His family and his childhood were orthodox, and for the most part the whole of England was orthodox. Almost all historians agree that Queen Victoria, who reigned as Darwin penned *The Origin of Species*, exerted a powerful orthodox influence upon England and its colonies. In addition to being orthodox, Darwin was socially timid and very much concerned about not hurting the feelings of others. He married into a conservative and wealthy family, the Wedgwoods (makers of the famous pottery) and he strived hard not to offend that family. After his nearly five year-voyage around the world on the H. M. S. Beagle, he lived a secluded life. He left arguing about the origin of species to others, namely to Thomas Henry Huxley.

Darwin wrote *On the Origin of Species by Means of Natural Selection, or the Preservation of Favoured Races in the Struggle for Life*, which was published in 1859. Within days of its publication his book became a topic of debate and of ideological warfare, and was a major stimulus for biologists to reexamine some of biology's own orthodoxy. One hundred and thirty years later the debates go on, the ideological warfare remains hot, and the theory continues to stimulate biological thought and research. Indeed, the theory of evolution is the most powerful explanatory theory (i.e., it is supported by the widest number and variety of facts) in biology. As will be explained later, the word "race," as Darwin used it in his book, did not mean then what it means in the vernacular today. His usage of the word was based on a mistaken biological notion of within–species populations, and in this century the word race evolved into an equally mistaken but popular description of differences among groups of human beings.

For some 30 years before writing *The Origin of Species*, Darwin had accumulated evidence in support of his theory. He didn't rush to conclusions; in fact, he had difficulty convincing himself that the theory was supported by the evidence. This is understandable. There were, at the time, very few views of human origins other than those provided by the church. But as more facts created an ever stronger case for evolution theory and against "special creation," Darwin eventually gained the confidence to suggest that species were not created as such but change over time, and that present-day species evolved from preexisting ones. Darwin formulated a thesis that expanded these generalizations, and he searched for and presented evidence in support of his thesis.

Darwin's thesis was surprisingly simple, as complex ideas often appear in hindsight. It consisted of three principles, which, summarized, are: **variation**, **heredity**, and **selection**. From his own observations and from the writings of Thomas Malthus, Darwin became convinced that within all species more offspring are "born" than will live to reproduce. Mathematically this means that the numbers of individuals within species will increase faster than their needs (e.g., food, shelter, and space). This suggests that there is competition for survival both within, and between species. In addition there are diseases, predator-prey relationships, accidents, starvation, and other forces in nature that tend to reduce the numbers of offspring that might otherwise reach adulthood. Among the offspring that do reach adulthood, not all will mate and not all will be fertile. In other words, not all zygotes are equally likely to reach adulthood and to leave offspring. Darwin called those that do live to reproduce **fit**, and those that don't, **unfit**, which is to say that zygotes exhibit *differential reproductive success*, or differential **fitness**. He explained differential fitness by way of **natural selection**.

Variation, called phenotypic differences in earlier chapters, had been observed not only by Darwin but probably by everyone who has ever lived, certainly all biologists. Everyone has observed that no two individuals are identical. In the language of biology it is said, simply, that phenotypic variation exists between individuals and within populations. Darwin didn't know about genes, or that genes contribute to a population's phenotypic variation, but he did postulate that for species to change, some phenotypic variation must be hereditary. So without knowing about genes, inheritance patterns, or Mendel, Darwin incorporated the second principle, heredity, into his general explanation of the origin of species. It has been on this principle that geneticists have made so very many contributions to the theory of evolution.

The third principle, selection, seems self-evident: if more progeny are produced than will survive to adulthood and if there are heritable phenotypic differences among the progeny, then some types must be more likely to survive than others. Said the other way, some types will be less likely to survive than others. If there is competition for survival (e.g., for food, space, and escape from predators), each new generation of offspring will come to look more like the successful types within the parent generations. Over long periods of time, gradual slight changes within populations will accumulate, eventually to become large differences between populations. The sum of the forces that give rise to differential survival among phenotypic variants Darwin called natural selection.

Biologists since Darwin have debated the relative importance of natural selection as a motor force of evolution, but most are in agreement that in the main Darwin had identified the three basic principles of evolution. However, these three principles do not tell us all we must know to understand evolution, only how small changes accumulate within species. For example, the three principles do not explain how new species arise. Obviously a single species may continue to accumulate small changes over the years in response to the "pressures" of natural selection, but that species will remain a single species, albeit an ever changing one. To illustrate how Darwin explained speciation, it is necessary to introduce the concept of **geographical isolation** (Figure 4.1).

Imagine dividing a population of inbred mice (the individuals are very nearly genetically identical) into two populations using a random selection method. One population of mice is placed on one island and the other on another island that differs slightly from the first in climate, food supply, and elevation, and is far enough away from the first to ensure that matings between members of the two populations never occur. These conditions allow the processes of variation, heredity, and natural selection to proceed independently in both populations until enough small changes will have accumulated within each population to

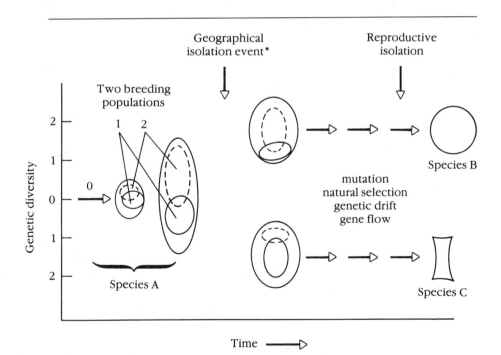

Figure 4.1 Geographical isolation, a necessary condition for speciation. A geographical isolation event is any natural event (earthquake, continental drifting, glaciation, etc.) that separates a species into two or more groups such that between-group matings are impossible.

have produced recognizable differences between them. Eventually the two populations will become so different that a biological "reproductive barrier" will have been established between them, at which time the females of one population will neither mate with nor reproduce by the males of the other population. At that time the two populations will have become two **species**. Darwin suggested that geographical isolation is necessary to **speciation** (Figure 4.1).

Clearly, new species have not been formed by biology teachers who ask their students to perform mind experiments like placing mice on different islands; however, there is reason to suppose that nature has done something similar many times during the course of the earth's history. Islands come and go; continents break up to form sub-continents that drift apart; lakes form and dry up; crashes between continental plates lift sea bottoms to mountain tops, and so on. There is no reason to reject the concept of geographic isolation on the grounds that the earth is calm, quiet, and unchanging. But the fact that the earth has changed and is changing is not proof that geographic isolation plays a role in speciation. Darwin predicted that it did and does, and even though other forces have been found to influence speciation, there is still no evidence to refute Darwin's prediction. What surprises some contemporary biologists when they think back on Darwin's life and work is that he was the first biologist to grasp the idea that *speciation cannot occur if matings, no matter how few, occur between the prespecies populations.*

Toward the end of his life, Darwin entertained the possibility that his theory was an all encompassing explanation of the origin of species, that it might explain more biological facts than any theory theretofore developed, and that he might even possess a streak of genius. But he considered himself rather an ordinary man in matters of the exact sciences (mathematics), politics, and economics. Genius aside, he realized that he had developed an important generalization that would influence the history of biology. Yet it was the custom then for people with outstanding accomplishments to appear modest, even common, so he often retreated from high praise and overstatement of his genius.

Some historians of science have tried to learn enough about the women and men who have left indelible marks on science to reveal their personalities, their private thoughts, and even what it was that gave them their spark of genius. In the long run, however, their personalities were understood and appreciated only by those who knew them personally. What is important to those of us who did not and cannot know Darwin is the stimulus he gave to science through his remarkable summary of the evidence that species come and go and a beginning explanation of how it all happens. John Maynard Smith opens his highly praised book, *The Theory of Evolution*, with the sentence, "The main unifying idea in biology is Darwin's theory of evolution through natural selection." If there is a more unquestioning testimony of Darwin's genius than that, it has been kept secret. The correctness of Darwin's theory has been amplified during the past 130 years by research results in many fields of science, including anthropology, biochemistry, botany, cell biology, chemistry, geology, genetics, physics and zoology.

Like many biologists since his time, Darwin was impressed by the fact that domestic species of plants and animals — wheat, rice, corn, cows, pigs, and cats — can be changed radically by *artificial selection* programs (Figure 4.2). Today, with help from genetics that Darwin didn't have, the success of artificial

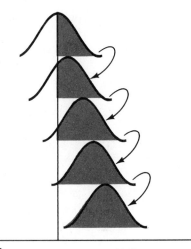

Bulldog	Doberman pinscher
height = 14″	height = 23-27″
weight = 46 lbs	weight = 65-75 lbs

discard   save

Artificial selection has been applied to both animal (e.g., cows, horses, pigs, cats, dogs, and sheep) and plant (e.g., wheat, corn, oats, vegetables, fruits, and ornamentals) species.

The frequency of the desired phenotype can be increased by selecting then breeding individuals of that type, generation after generation.

Figure 4.2 Phenotypic differences achieved by artificial selection.

selection programs can be explained at the levels of genes, enzymes, physiology, and anatomy. It turns out that our current explanations of the effects of artificial selection are much the same as our explanations of the effects of natural selection. In both cases, radical changes of type are possible because most populations possess a great deal of **genetic variability** within their gene pools. What is not known is the actual role of natural selection during the course of time during which modern species acquired their present biological characteristics. On this issue, evolutionists argue among themselves.

The Logic of Natural Selection

Logic is neither evidence, nor data, nor fact. One form of logic, inference, is simply the expression of apparent relationships between things, between events, or between things and events. If one perceives a logical relationship between a child and her pet dog or a logical explanation for why children, in general, benefit from having pets, the logic may take the form of explaining the psychological needs of children that are satisfied by pets or the apparent relationship between satisfying those psychological needs and minimizing juvenile crime. Logic does not provide the evidence, but it often helps to make better sense of evidence.

In Darwin's thesis there is a logic between what he called the forces of selection, whether natural or artificial, and biological change through time. Consider the following: most hairy mammals are good swimmers, and a few, like beavers, swim very well under water. But no mammal, including whales and dolphins, can breath under water. They simply are not able to extract oxygen from water, as fish do. In other words, mammals are **adapted** to extracting oxygen from the air and fish are adapted to extracting oxygen from water. Most birds are adapted to flying, angleworms to living underground, some fungi to living in boiling hot water, and some plants to growing at elevations above 20,000 feet. An adaptation is a special kind of equilibrium between an organism and its environment. Yet the simple existence of an adaptation, like fish in water, is not an explanation of how the adaptation arose in the first place.

What does it mean for an organism to become adapted to a particular environment? Why does it die if it is placed into a radically different environment? In your mind's eye, compare and contrast a duck with a blue whale. The duck is covered with feathers, the whale with very thick skin; the duck has hollow bones, the whale solid bones; the air displaced by a duck in flight is very light, a few fractions of a gram, but blue whales will displace several tons of water as they swim. What's more, the duck will die quickly in the whale's environment, and the duck's environment is so strange to a whale that we can't even image it—a whale sitting on a nest, and flying several thousand miles every year?

The simple observation of a duck living in a certain way does not tell us how the duck came to live in that way. Certainly feathers and hollow bones would seem to aid flying, but bats fly very well, and they do not possess feathers or hollow bones. In fact, bats are mammals and are no more closely related to ducks than are field mice. Any argument we might make, then, to explain how ducks became able to fly (e.g., because they have feathers and hollow bones) will have to include why ostriches do not fly and why bats and butterflies do. *The history of adaptations cannot be understood by way of apparently logical connections between the structures, functions, and environments of organisms.*

What we can do that Darwin could not is to compare ducks and whales in a more detailed way than by watching them fly and swim, or by killing them and dissecting them, or by comparing their physiologies (e.g., both are warm blooded, both possess lungs, hearts, oxygen-carrying proteins in their blood). Darwin was not prepared, either technologically or conceptually, to study *the genetic bases of adaptation.* Today (recall Chapter 2) it is possible to make comparative studies of genes that encode similar proteins and of proteins that perform similar functions, within an almost limitless range of species—studies of the blood proteins of all mammals, for example, or of the metabolic enzymes that convert amino acids into carbon dioxide, water and energy in all species. Every species has evolved a slightly different way of coming into equilibrium with its environment, and now it is possible to understand aspects of that equilibrium through studies of macromolecules and their functions. Closely related species, such as mallard ducks and Canadian geese, have evolved similar equilibria. Whales and geese possess very different, but not absolutely different, equilibria (the genomes and proteins of whales and geese are much more alike than one might suppose from looking at whole animals). It is now possible to compare the genomes of any two species for similarities and differences. (One geneticist "testified" that he heard Darwin chuckle in his grave over the news of comparative biochemistry, but it is rumored that this was the same geneticist who, the week before, reported sighting Elvis.)

The three principles of Darwin's theory—variation, heredity, and selection — imply that some of the phenotypic variation within species is due to genetic variation, and that upon exposure to the vicissitudes of nature *some of the genetic variation will be weeded out and some will survive.* The overall result is that differential reproductive success among the genetic variants *will change the genetic composition of populations through time.*

However, before getting to genetic analyses of adaptation, it must be acknowledged that the theory of evolution has not been easy for all Christian churches to accept. Even to this day evolution is denounced by a few minor denominations, and some individuals within most Christian sects denounce evolution theory.

A Comment About the Controversy Between Evolution and Religion

One student asked me, "Do I change the facts to fit my beliefs? Or do I change my beliefs to fit the facts?" I responded with an unambiguous NO to the first question, and with a qualified and ambiguous Yes, but. . . to the second question.

In the United States, fundamentalist **creationists** have tried to replace the teaching of evolution in the public schools with the biblical doctrine of creation. Their efforts to get evolution out of the schools—the Scopes Trial in Dayton, Tennessee, in July, 1925, was a main event—have failed at the legal level. Today their strategy is to get creationism into the schools.

Creationists counterpoise creation with evolution, so as to pit "creation science" against evolution science. However, the defense of creation is based on selected passages from the Bible, not on science. The evidence for evolution theory, on the other hand, is not found in scripture, but encompasses the legions of facts provided by the research of biologists, chemists, geologists, and anthropologists. Today the database of evolution is comprehensive. Indeed, it is a fact that the processes of evolution have given rise to all contemporary species, including our own. While the theory of how these processes did and do occur is not fully formulated, the theory gets better as more data are incorporated into it. Without doubt the theory of evolution is the most comprehensive summary of biological change through time (Futuyma, 1983).

Most biology teachers fail to explain the many differences between scientific theories and religious beliefs, in particular how scientific theories are evaluated based on the facts that support them (Figure 4.3). The relationships between facts and theory are more complex than outlined in Figure 4.3, but even so, the sketch shows that relationship. For example, new facts can be used to evaluate theory, and theory can be used to generate new facts. Both are in flux, and both are responsive to new information. A theory may indeed "commit suicide" by generating hypotheses which when tested generate facts that in turn disprove the theory. Even facts appear to change as methods for observing them improve and as the world around us changes through time.

Among other things, scientific activity includes the art of asking questions of nature and of listening to the answers. Science is not in the business of discovering absolute truth. Scientists discover how things change through time. In biology the two kinds of change that have attracted the most attention are those that take place during the course of geologic time, called evolution, and those that take place during generational time, called development. Think about it this way: If we took "snapshots" of present-day species or of a tiny baby over a time span of just a few years, we could record evidence of biological change on film. Some species would have died and new species would have emerged; the tiny baby might have become a small violin player or slam dunker.

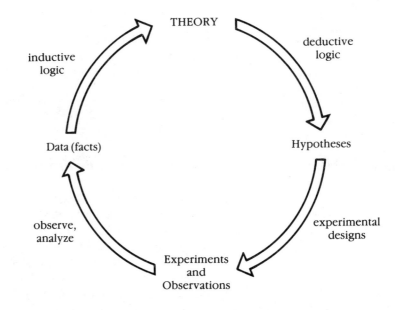

Theory (generalization): a summation of the relevant facts, and the best vantage point
    from which to predict new facts
Hypothesis: a deduction from theory that can be tested by experiment and observation
Fact: a verifiable aspect of the real world

Figure 4.3 Relationships among theory, facts, hypotheses, and experiments.

    The striving (testing hypotheses) to understand change generates knowledge
(facts) that can be used to make better summaries (theories) about the state of
the world now and as it is coming to be. "Snapshots" simply confirm that change
has happened; testing hypotheses provides new information with which to
improve the theory from which new predictions can be made. This ongoing
process of discovery gradually adds strength to our explanations of the world
around us.

    Belief systems (e.g., an organized religion, or social club) and science are
opposites. They have little in common. If one wants to believe a religious
interpretation of a scriptural story of human origins, there is little that facts can
do to dispel the belief. On the other hand, if one wants to get a better
understanding of the changes that have and are taking place in the real
world—whether continental drifting, the ecological effects of acid rain, the
dangers and benefits of genetic engineering, or how acorns become oak trees—
religious beliefs are of little help. Understanding comes only as fast as discoveries
reveal facts and facts are patterned into theories.

    Science is a disconcertingly slow process. But if we wish to ameliorate
diseases, travel in space, discover how zygotes become adults, or preserve the
planet earth for habitation, we have no alternatives other than to proceed at a
pace set by the processes of discovery. This is not to say, "cast off your beliefs,"
not by any stretch of the imagination. It is to say, first, that our acquisition of
knowledge has been and is being delayed by beliefs. The stories of Galileo,
Copurnicus, Darwin and Einstein illustrate this point. Second, beliefs have been
changed radically by scientific knowledge; this can be seen in the history of
discovery and in how discoveries have caused alterations in the dogmas of

Christian religions. Third, today at the dawn of the 21st century, our species has had enough experience with antagonism between science and religion to provide it with a better way of dealing with the antagonism. The first step is to get the issue "on the table." A second step must include checking guns and other Neolithic weapons at the door and encouraging honest dialogue, good will, and the use of evidence.

At this point a beginning attempt will be made to present some of the genetic evidence that supports the theory of evolution as outlined by Darwin.

Breeding Populations and Genetic Variation

What is the origin of phenotypic variation within breeding populations and species? What are the causes of phenotypic variation?

Recall norms of reaction (Figure 1.5). Striking phenotypic variation is possible among genetically identical individuals, which is to say that nongenetic factors contribute to phenotypic variation. One of the tasks of geneticists is to distinguish between the two major kinds of variance, that is, variance caused by genetic differences among individuals within the population and variance caused by nongenetic factors that influence individual development: these are called genetic and nongenetic **components of variance**.

Mendelian genes contribute to phenotypic variation; recall, however, that Mendelian genes can be identified only if they exist in two or more allelic forms, that is, only if the contrasting phenotypes associated with them obey Mendel's rules of inheritance. Mendel did not address the question how genes come to exist in two or more allelic forms in the first place. Earlier, however, we suggested that one of the two clotting factor VIII genes in the germ line cells of Queen Victoria mutated and that *she became heterozygous by mutation,* not by genetic recombination. Today we know that *gene mutations are the major source of genetic variation.* We will now discuss mutant alleles in real human populations.

Lancaster County, Pennsylvania is home to a population of nearly 18,000 Amish people. This population was founded between 1720 and 1770 by a migration of about 200 couples from Western Europe. Primarily because of Amish religious beliefs, this population is reproductively isolated from non-Amish people who live nearby. (Religious isolation is not the same thing as geographic isolation, but it leads to the same biological end point.) Isolating mechanisms allow isolated populations to become genetically different from one another, and from the population from which they arose. However, most within species populations are not totally isolated from one another, but if the frequency of matings within a population is greater than it is between populations, the populations are called **breeding populations**. In other words, breeding populations are identified by mating patterns—frequent mating among individuals within a population and infrequent mating between individuals from different populations. Breeding populations are considered to be the **units of evolution**.

Most people who live in or near Lancaster County are of Western European origin as well, but the Amish differ from the others as measured by the frequencies of alleles of several genes. When homozygous, the recessive allele of one of these genes initiates the phenotype called **Ellis-van Creveld syndrome**, one of several kinds of genetic dwarfism. This particular form of short stature, also called six-fingered dwarfism, is so rare in the general population (only 50 cases in the world had been reported prior to the study of the Amish) that until recently it had been impossible to correlate the phenotype with an inheritance pattern. (Many short people have voiced objection to being called dwarfs, for good reason, but the genetic literature describes all forms of genetic shortness as genetic dwarfism.)

In the small Amish population in Lancaster County, some 50 cases of Ellis-van Creveld syndrome have been found in 26 families (i.e., the same number in a population of 18,000 as had been recognized in a world population of more than 5 billion!). The pedigrees of the 52 parents in these 26 families have been traced back to one couple, a Samuel King and his wife, who emigrated to Pennsylvania in 1744.

This observation, like the one described earlier for the relatively high incidence of Huntington's disease in the villages surrounding Lake Maracaibo, illustrates one aspect of the theory of evolution — a **founder** of a new population may introduce a rare mutant gene into that population, after which the new mutant gene may increase in frequency within that population's gene pool by **chance** and through **inbreeding**.

In Chapters 1, 2, and 3 it was shown how *individuals differ from one another* with respect to heritable phenotypes; in this chapter we see that *populations differ from one another* with respect to frequencies of these heritable phenotypes and of the alleles that initiate their development. Both the Amish and the people living near Lake Maracaibo became measurably different from their parent populations within a time frame of 200 years — less than 10 generations, following the introduction of a mutant gene into those populations by a founder. We may never know whether the founders of these genetic differences inherited the mutant gene from an ancestor or whether the mutation occurred within their own germ cells, as can be said with some certainty about Queen Victoria's mutant clotting factor VIII gene; but in either case mutations do occur, and some of them become established within population gene pools.

Allele frequency changes, within small populations surrounded by rapidly growing large populations, could not happen without an isolation barrier to prevent the "dilution" of the alleles into the greater population. Through matings between members of small and large populations, the population boundaries "dissolve." Here, then, we see two aspects of evolutionary change within breeding populations: after mutations have provided the genetic variation, *mating patterns and chance influence the distribution of that variation.*

Natural selection is discussed below, but consider the fact that the phenotypic variation caused by the Ellis-van Creveld mutation has no potential for survival. Persons with the syndrome do not have children. If homozygous recessive individuals do not have children, then what keeps the allele in the population gene pool? Is the allele kept there by new mutations? Does it stay in the gene pool because the forces of natural selection cannot "see" it when it exists in the heterozygous state?

Forces that Modify Allele Frequencies Within Populations

Darwin's second principle suggests a process of biological inheritance, but he left us without one useful idea of how the process happens. His main point was that for changes to occur within populations through time, some of the variation must be inherited, since only inherited variation can appear in subsequent generations. Darwin was right as far as he went, but he didn't go very far, especially compared with what is known about heredity today.

The phenotypic variation discussed in the earlier chapters is now discussed in a different context, the context of populations. Remember, though, that the *inherited variation that distinguishes one individual from another is the same variation that distinguishes one population from another.* The differences between the Amish population and its neighbor populations, discussed above, included only one allele; in fact, though, many allele differences exist between the Amish and its neighbor populations, as is true for the neighboring Swiss and Italian populations, the Northern European and Chinese populations, and so on. How do population differences evolve? Is evolution continuing this process within contemporary populations, and if so, are these populations becoming different from one another?

First, A Discussion of Genetic Equilibrium

If all the individuals within a population were genetically identical and if mutations did not occur, evolution could not occur. Natural selection could not bring about genetic differences between populations. Things would stay the same. In model populations, things can stay the same even if there are genetic differences among individuals within the population. A model gene pool can be made to "stand still," to remain in equilibrium.

For a gene pool to reach equilibrium — on paper, not in the world — only the forces that bring about change need be eliminated, for example, mutations. A simple illustration of a population in genetic equilibrium is the following: Consider one gene, A, two alleles, A and a, and an allele ratio of 0.5 A and 0.5 a. This means the probability that an egg cell will carry A is 0.5 and the same for a. This also is true for sperm cells; half will carry A, and half will carry a. The symbols used earlier for these two alleles were p and q; in this example, p = 0.5 and q = 0.5. It is possible to begin a hypothetical population in different ways. For example, all of the females can be AA genotypes and the males aa; or all of the members of the population can be Aa; or 25% can be AA, 50% Aa, and 25% aa. In all three cases, p = 0.5 and q = 0.5.

If we opt for the first example — females are AA and males aa — then all of the eggs will be A, all of the sperm will be a, and all of the progeny will be Aa. In this example, the next generation of progeny will have had heterozygous parents, Aa x Aa, so in every case the genotypic frequencies expected among second generation progeny are $p^2 = 0.25 = $ AA, $2pq = 0.5 = $ Aa, and $q^2 = 0.25 = $ aa.

If we opt for the second example, in which all of the individuals are heterozygous (Aa), then the result reached after two generations in the first example is reached after one generation in this example (i.e., 1/4AA, 1/2Aa, and 1/4aa). Indeed, in examples 1 and 2, the starting point of example 3 is reached by the second generation. Therefore, in all three examples we reach a point at which the populations are identical with respect to genotype frequencies — 1/4AA, 1/2Aa, and 1/4aa — and allele frequencies — 1/2A and 1/2a.

If all three examples are followed from this point into the next generation, there are more combinations of matings, and if the matings are random with respect to genotype, the probabilities of each kind of mating will be as shown in parentheses: AA x AA (1/4 x 1/4 = 1/16), AA x Aa (1/4 x 1/2 = 1/8), AA x aa (1/4 x 1/4 = 1/16), Aa x Aa (1/2 x 1/2 = 1/4), Aa x aa (1/2 x 1/4 = 1/8), and aa x aa (1/4 x 1/4 = 1/16). Figure 4.4 shows how it is calculated that the generation to follow this one is composed of AA, Aa, and aa genotypes in a ratio of 1:2:1 and a 1:1 ratio of *A:a* alleles. If you have the patience to follow these genotype and allele ratios through five more generations, you will find that the genotype and allele ratios are in equilibrium and will remain so until time in the universe runs out.

What this means — and this is counterintuitive to many people — is that sexual reproduction and genetic recombination do not disturb allele frequency equilibria; both genotype and allele ratios persist indefinitely. If you draw from your experiences playing cards, you will know that shuffling a deck of cards does not change the frequencies of the suits {i.e., ¼ hearts, ¼ diamonds, ¼ spades, and ¼ clubs} or the frequency of the numbers of the cards within suits {e.g., ¹⁄₁₃ ace, ¹⁄₁₃ king, ¹⁄₁₃ queen}. Neither does meiosis change the kinds or the frequencies of alleles within gene pools.

Sexual reproduction and genetic recombination act to "shuffle alleles" within genomes into different gametic combinations, but they do not increase or decrease the frequencies of alleles within gene pools. As you see in Figure 4.4, the population gene pool remains stable *until something happens to disturb the equilibrium* (This equilibrium is called the **Hardy-Weinberg equilibrium** after the English mathematician and the German physician who independently recognized it. An American, William Castle, and a Soviet, Tschetverikov, also discussed this equilibrium in their research papers).

What forces do in fact change allele frequencies within population gene pools? Suppose *A* mutates to *a* more often than *a* mutates to *A*; the ratio of *A:a* will gradually change toward 0:1. Allele frequency change by way of mutation is an exceedingly slow process. Or suppose that 99% of the individuals of AA and Aa genotypes survive and have children, while only 90% of aa individuals do so. The ratio will change in the opposite direction, toward 1:0, again very slowly. In other words, mutations and differential survival are potentially capable of changing the frequency of alleles within breeding populations.

Mutation

At any given gene locus, mutation is rare. Most gametes carry the alleles inherited by the men and women who produce them, but once in a while, as happened with Queen Victoria and the mutant allele of the clotting factor VIII gene, a gene will mutate in the germ line and be passed to some members of the next generation. On average, between 1 and 10 of every 100,000 gametes will carry a new mutation of a specific gene (Table 4.1). Put another way, after each gene has been replicated 100,000 times, on average one or a few mutations will have occurred.

In large populations the rarity of mutations at any given gene locus will have little influence upon allele frequency; in small populations the same mutation will have a larger effect. For example, one mutation will change allele frequency by only 0.001% in a population of 50,000 people, but in a population of 50 people, one mutation will change the allele frequency by 1%. (1 mutation per 100,000 genes versus 1 mutation per 100 genes, respectively). However, this is not the main issue of why gene mutations are important to evolution. The main

Assume: Population is infinitely large
      No new mutations
      No selection
      No gene flow
Let the frequency of A = p
Let the frequency of a = q $\Big\}$ p + q = 1
Thus $(p + q)^2 = 1$
Then, the frequency of AA = p^2
           of Aa = $2pq$ $\Big\}$ and $p^2 + 2pq + q^2 = 1$
           of aa = q^2

Population 1: p = 0.5 + q = 0.5 = 1

$\female$ gametes

	A = 0.5	a = 0.5	
A 0.5	p^2 = AA = 0.25	pq = Aa = 0.25	}p
a 0.5	pq = Aa = 0.25	q^2 = aa = 0.25	}q
	p	q	

p^2 = AA = 0.25
$2pq$ = Aa = 0.5
q^2 = aa = 0.25

$\male$ gametes

Population 2: p = 0.4 + q = 0.6 = 1

$\female$ gametes

	A = 0.4	a = 0.6	
A 0.4	p^2 = AA = 0.16	pq = Aa = 0.24	}p
a 0.6	pq = Aa = 0.24	q^2 = aa = 0.36	}q
	p	q	

p^2 = AA = 0.16
$2pq$ = Aa = 0.48
q^2 = aa = 0.36

$\male$ gametes

Population 3: 16% of population is phenotype rh$^-$; i.e., rh$^-$rh$^-$ genotype = 0.16
∴ allele rh^- = $\sqrt{0.16}$ = 0.4; thus Rh$^+$ = 1 − 0.4 = 0.6

$\female$ gametes

	Rh^+ = 0.6	rh^- = 0.4	
Rh^+ 0.6	Rh$^+$Rh$^+$ = 0.36 = p^2	Rh$^+$rh$^-$ = 0.24 pq	}p
rh^- 0.4	Rh$^+$rh$^-$ = 0.24 = pq	rh$^-$rh$^-$ = 0.16 q^2	}q
	p	q	

p^2 = Rh$^+$Rh$^+$ = 0.36
$2p^2$ = Rh$^+$rh$^-$ = 0.48
q^2 = rh$^-$rh$^-$ = 0.16

Figure 4.4 Hypothetical populations within which allele frequencies do not change; that is, allele frequencies are not disturbed by sexual reproduction and genetic recombination.

issue is that gene mutations provide genetic variation.

Most new mutations that can be observed by altered phenotype lead to phenotypic variation that decreases reproductive success. To illustrate, hemoglobin proteins have evolved through time to be very efficient in providing oxygen to body cells. The majority of new (mutant) amino acid changes in these proteins will result in a less efficient oxygen-carrying capability, as is true for hemoglobin S (Hb-S), found in SCA individuals. However, as conditions of environment change, an amino acid change may result in a more efficient oxygen-carrying capacity, as it apparently has among populations of native South Americans living at high altitudes in the Andes Mountains. Among these people

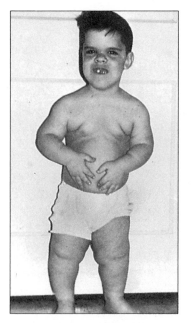

Table 4.1 Mutation Rates. Achondroplasia is a form of short stature due to short leg bones. The arm bones are short as well. Achondroplasia is initiated by a dominant allele. In one study of nearly 250,000 births (to nonachondroplasia parents), 7 children were achondroplasic, i.e., there were

> 7 new mutations per 500,000 gametes, or
> 14 new mutations per 10^6 gametes, or
> 1.4 new mutations per 10^5 gametes, or
> a mutation rate of 1.4×10^{-5}

It is relatively easy to determine mutation rates for dominant alleles, as illustrated by the allele that correlates with achondroplasia. Estimates of mutation rates from recessive to dominant alleles of a few human genes.

Trait	Mutations/10^6 gametes	Mutation rate
Achondroplasia	14	1.4×10^{-5}
Retinoblastoma	6	6×10^{-6}
Osteogenesis imperfecta	11	1.1×10^{-5}
Marfan syndrome	5	5×10^{-6}
Huntington's disease	1	1×10^{-6}

is found a kind of hemoglobin that is a more efficient oxygen carrier than hemoglobin A. Such mutations are rare, but again, over long periods of time they do occur and do accumulate in population gene pools.

Guided by the theory of evolution, we predict that harmful mutations will be "weeded out" of gene pools by natural selection, and that rare, helpful mutations will increase their numbers within gene pools. As we shall see, this interpretation may explain some but not all allele frequency changes that occur within gene pools.

For one thing, deleterious mutations do survive in gene pools. Indeed, human gene pools carry hundreds of deleterious alleles (and probably every human being carries between 4 and 10). Deleterious alleles accumulate in gene pools in part because most new mutations are recessive, and by definition, completely recessive alleles are not deleterious in the heterozygous state (e.g., the p allele of the gene that encodes phenylalanine hydroxylase; the pp genotype leads to PKU, but PP and Pp genotypes lead to healthy phenotypes).

It is possible for frequency equilibria to exist between death rates and mutation rates of alleles (e.g., if the death rate of pp genotypes balances the rate at which mutations convert P to p alleles, then there will always be a few p alleles in the gene pool, a frequency that will be determined by the rate of mutation if all pp individuals die without reproducing). This has not been shown to be absolutely the case with the PKU allele, but the fact that about 1 baby in 12,000 (in Northern European populations) is born with PKU indicates that about 1 of every 110 **P** genes in the gene pool is in the p allelic form. Whether the allele frequency reflects a true equilibrium between mutation rates of P to p and death rates of pp genotypes is not known. The argument about the equilibrium of the two alleles within the Euro-American population, however, does not even touch on the question of why the recessive allele is absent from the gene pools of African-Americans. Allele equilibria can be observed in controlled, nonhuman populations and can be simulated with mathematical models, but in natural populations allele equilibria cannot be determined with precision (Figure 4.5).

Natural Selection

Darwin credited natural selection with being the dominant driving force of evolution. To modernize his language yet remain true to what he said, we say that variation must exist within populations; that some of the variation must be heritable; that differential reproductive success among heritable variants leads to allele frequency changes within breeding populations; and that for speciation to occur, breeding populations must become geographically isolated from one another until such time that the accumulated genetic differences between them will have produced a biological, reproductive barrier between them.

Starting from the beginning, consider sickle cell anemia (SCA), probably the most cited example of differential reproductive success among human genotypes. $\beta^A\beta^A$ and $\beta^A\beta^S$ genotypes are reproductively successful, while nearly all $\beta^S\beta^S$ genotypes die without leaving offspring. (With access to blood transfusions and continuous medical care, sickle cell anemic individuals live longer than they otherwise would, and a few are known to have had children. However, in much of the world in which SCA is important, all SCA persons die without leaving offspring.)

If non-SCA individuals survive and if SCA individuals do not, what keeps the β^S allele in the gene pool? For one thing, mutations from β^A to β^S. However,

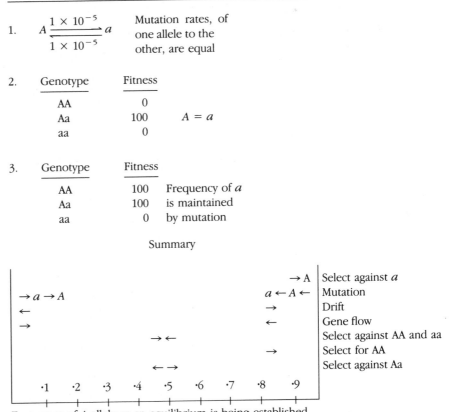

Figure 4.5 Possible ways for equilibria to become established with respect to one gene and two alleles.

mutations are rare (in this case the mutation rate is unknown, but certainly it is no higher than 1 new mutation per 10,000 gametes). In the United States about 1 of every 12 persons of African descent carries one copy of the mutant allele, a much higher frequency than can be accounted for by the mutation rate.

The best explanation so far is that the mutant allele survives in gene pools of populations that live in malarial environments because heterozygous persons are more resistant to falciparum malaria than are $\beta^A\beta^A$ persons. In all other populations of the world, the frequency of β^S in gene pools is probably more a function of an equilibrium between the mutation rate and deaths of SCA children (Table 4.2).

Within genetically controlled populations it is relatively easy to demonstrate the effects of various intensities of selection upon gene pools (Figure 4.6). If a population of bacteria (e.g., *Escherichia coli*, the major species of bacteria found in the human gut) is exposed to an antibiotic (e.g., streptomycin) nearly all of the original bacteria will die, but not quite all. A few cells will live, and will do so because they are resistant to streptomycin. In the continued presence of the

Table 4.2 Relationships between fitness and genotype frequencies and allele frequencies.

Genotype	Fitness	
1. $\beta^A\beta^A$	100	The frequency of the β^S
$\beta^A\beta^S$	100	allele in the gene pool will
$\beta^S\beta^S$	0	be a function of the mutation rate from $\beta^A \rightarrow \beta^S$.
2. $\beta^A\beta^A$	70	The frequency of β^S will
$\beta^A\beta^S$	100	be higher than in population 1.
$\beta^S\beta^S$	0	

For example: Among African-Americans one of every 576 babies is $\beta^S\beta^S$ (SCA); therefore one of every 24 β genes is β^S ($\sqrt{1/576} = 1/24$); therefore one of every 12 persons is $\beta^A\beta^S$. This is a much higher frequency of β^S than can be expected from mutation rates.

Among certain populations in Nigeria, one of every 64 babies is $\beta^S\beta^S$; therefore one of every 8 β genes is β^S, and one of every 4 persons in $\beta^A\beta^S$, an even higher frequency of the β^S allele.

The best explanation of the high frequency of β^S is that $\beta^A\beta^S$ individuals are more fit than $\beta^A\beta^A$ individuals in areas of the world where fulciparum malaria is indigenous, that is, they are more resistant to malaria.

If the frequency of aa is	the frequency of Aa will be:
1/100	1/5.5
1/500	1/12
1/1000	1/16
1/10,000	1/50
1/100,000	1/158
1/1,000,000	1/500

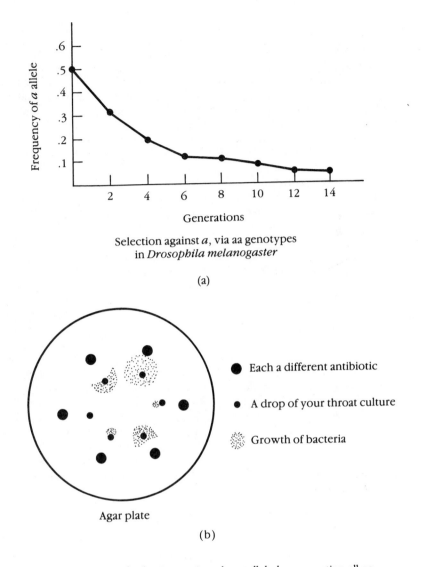

(a)

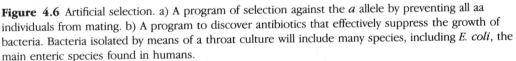

Agar plate

(b)

Figure 4.6 Artificial selection. a) A program of selection against the *a* allele by preventing all aa individuals from mating. b) A program to discover antibiotics that effectively suppress the growth of bacteria. Bacteria isolated by means of a throat culture will include many species, including *E. coli*, the main enteric species found in humans.

antibiotic, a new strain of resistant bacteria will replace the old strain of susceptible bacteria. Within such artificial settings, selection pressures can be manipulated to act fast and effectively. More slowly, but again artificially, plant and animal breeders change allele frequencies within populations of cereal crops and farm animals simply by weeding out undesired types and encouraging the desired types to reproduce.

That there appear to be similarities between artificial and natural selection does not mean that artificial selection is the best model for natural selection or, for that matter, for the sum of the processes that contribute to gene pool changes through time. More and more biologists are beginning to question the heavy role given to natural selection in the past.

Perspective 4.1

A contentious issue among evolutionists centers upon the question, "What does natural selection "see" as it acts to bring about differential reproductive success among genetic variants?" Asked another way, "What are the **units of selection?**" In the case of sickle cell anemia the unit of selection appears to be the individual, not the β^S allele. It appears that the β^S allele cannot be "seen" by natural selection in non-malarial environments if it resides within the genomes of heterozygous individuals. In the case of SCA individuals, the majority are so ill that they do not live long enough to reproduce, but does natural selection "see" the genotypes or does it "see" the biological weakness of these individuals? In either event, when an SCA person does die before reproducing, all of that person's genes—good, bad, and indifferent—die at the same time, except for those genes that are passed into the next gene pool by that person's siblings, cousins and other close relatives.

The issue of units of selection is important. In recent years the idea that natural selection "sees" and acts directly on genes has been pushed hard by several prominent biologists. Such an idea seems silly to many traditional Darwinists. The idea, first proposed in 1966 and first defended with passion in 1976 by Richard Dawkins, assumes that genes are selfish, that alleles compete with one another for survival,

and that by competing, alleles make bodies that are more and more competent, as judged by the survival of the successful alleles. The alleles that beat out their competitors survive. The others fall by the wayside. The selfish gene idea says, or at least Dawkins says it does, that our bodies are but the gene's way of propagating themselves. Dawkins even referred to our bodies as "lumbering robots" carrying "swarms of genes" waiting and plotting to get into the next gene pool. But the gene defined as a unit of selection is not the same gene that geneticists have called a unit of inheritance or a unit of transcription. The selfish gene argument is based on a redefinition of the gene and a rather significant change of the neo-Darwinist theory of evolution. We return to this argument toward the end of Chapter 5. (Dawkins' argument can be found in his book *The Selfish Gene*).

In this discussion it is assumed that the major units of selection are individuals, not genes, and that the units of evolution are breeding populations. Allele frequency changes occur within breeding populations, not within genomes. Genomes may carry mutant alleles, but mutation rates are measured within gene pools. Individuals differ with respect to reproductive success, but it is the sum of the success stories that determines the composition of gene pools, and gene pools are a function of populations, not of individuals.

An example discussed by Maewan Ho is the honeycomb. She brings attention to Darwin's admiration of the honeycomb as an example of a structure so "beautifully adapted to its ends," the outcome of the hive-making instincts of bees so perfected by natural selection. But in 1917, the famous biologist D'Arcy Thompson showed that "the hexagonal cross-section of the cells, as well as their trihedral pyrimidal ends, are both the result of compression due to close packing. In other words, the impressive symmetry of the honeycomb arises from the automatic play of physical forces." Ho cites many examples of phenotypes that arise not by instructions from genes, but out of the physical forces that are integral to their function. In other words, form may not be fashioned as much by natural selection as by function. This is a complex concept made more so by the many examples presented as evidence which demand a knowledge of biology that this book has not provided.

In a book entitled *The Wisdom of the Genes*, Christopher Wills makes a very convincing case for the fact that the old idea that natural selection acts on random mutations is giving way to the idea that evolution is a cumulative process, in that a prior event increases the likelihood of the occurrence of a second, companion event. It is not as if genes are smart and plan their own destinies. Rather, during the course of time, as chromosomes became more complex, and as more genes evolved important contributions to the survival of species, the stage became increasingly better set, so to speak, for species to take advantage of new changes that could occur within their many genomes. (Wills' book is not written as a textbook; it is very easy to read, very informative, and a lot of fun.)

Other Forces that Alter Allele Frequencies

Migration

In addition to mutations and differential survival, population gene pools are changed by migrations of groups of people from one population to another. Armies of one nation often invade other nations, or groups of people from one section of the country may move to another section, as people from the Midwest moved to California during The Great Depression or as African Americans moved from the South to the North early in this century. Small groups of people may leave parent groups to establish new populations. Migrations change the composition of gene pools by **gene flow**. Migrations not only redistribute alleles, they also break down isolation barriers between breeding populations.

The invasion of the Americas by Europeans is an example. Five hundred years ago three ships carrying Portuguese sailors landed on the island of Hispaniola (Dominican Republic and Haiti today) or on a nearby island. Within the next few years Portuguese and Spanish sailors landed on many islands in the Caribbean Sea and along the coasts of South, Central, and North America. Within five years of the first landing, people of African descent were shipped West to farm cane sugar, eventually to become slaves. After the word of easy riches got back to other countries in Europe, migrations became the rule, not the exception. Many isolated populations in Europe contributed genes to new gene pools in the Americas. Later Chinese and Japanese laborers were imported to the Americas to farm and to build railroads. Population gene pools became mixed, and the reproductive barriers between previously isolated populations were brought down.

This mixing of gene pools did not occur in the mode of the fictitious "melting pot." For example, the mixing of European, African, and native American gene pools occurred primarily by white males inseminating African and native American females. In South America the mixing of these three large gene pools occurred in a slightly more democratic manner. In North America the native American gene pool was nearly destroyed (by 1930 the native American population in the United States had been reduced to fewer than one million).

While the human species became divided into many isolated breeding populations prior to the advent of agriculture (10,000 years ago), the advent of written language (5,000 years ago) and commerce (4,000 years ago) paved the way for integrating those gene pools. While the original gene pools have not been thoroughly mixed, few if any of them remain isolated in their original form.

Genetic Drift

Critics of the theory of evolution often assert that grasses, geese, and gorillas are highly nonrandom forms of nature, and that such highly fine-tuned creatures cannot have had their origins in the haphazard random events said to propel evolution. It is true that the next mutation to occur cannot be predicted and therefore will appear to be a random event; fertilization events are random, and climatic changes, if not random, are unpredictable. But the critics ignore interrelationships among phenotypes and their environments, within which evolutionary and developmental processes are not random. Once a new mutation has occurred, it may persist in its mutant form for thousands of generations. Once an equilibrium has been established between a species and its environment, the events that maintain the equilibrium are not random, as were the events that contributed to its origin. The critics of evolution rarely mention the stabilizing forces of evolution.

However, genetic drift is another chance phenomenon in the same mode as fertilization and recombination events. Recall that of the many kinds of egg produced by each of our mothers and the many kinds of sperm cell produced by each of our fathers, each of us inherited genes from but one egg and one sperm. That is, you represent the union of a very small sample of the gametes that might have formed zygotes, all of which would have been very different from the zygote that in fact did give rise to you.

Analogous to the formation of zygotes from small samples of gametes, progeny gene pools are samples of genes carried within parent gene pools. Genetic drift is the name given to the changes in allele frequencies brought about by the sampling process, a process which consists of both the random events that accompany the formation of gametes from genomes, and the chance unions of gametes to form zygotes. Then, of course, there are the chances that some zygotes won't live to reproduce, and so on. Genetic drift is just a fancy name for chance, and it's by chance that a progeny gene pool is somewhat different from its parent gene pool.

A small population that becomes separated from its parent population takes with it a sample of genes from the parent population's gene pool. By chance, the allele frequencies in the sample may differ from those in the parent gene pool.

The rules of chance explain why large populations are affected differently by chance than are small populations. Consider one gene, **A**, and its two alleles, A and a, within a population size of 24—about the size of some hoards of great apes, and possibly the size of early human, mobile families. There are 48 **A** genes in a population of 24 individuals; if half of them are A, we will predict that in equilibrium 6 individuals will be of AA, 12 of Aa, and 6 of aa genotypes. If an Aa individual in the parent group fails to mate for some reason, the ratio of A:a will not be disturbed, since one of each allele will suffer the same fate. But if an AA or an aa individual fails to mate, the ratio of A:a will be disturbed. (The loss of one AA individual represents an 8% reduction of A alleles.) Even if all the adults mate and have children, Aa × Aa matings can, by chance, result in only AA offspring. Chance deviations from the original allele frequencies, from one generation to the next, are said to arise from genetic drift.

In large populations, chance deviations from parent to progeny gene pools won't have much of an effect upon the overall allele frequencies. In a population of 10,000 individuals, with 20,000 **A** genes and an allele frequency of $1A : 1a$, a loss of 40 AA individuals from the matings that will give rise to a new gene pool represents a decrease of less than 1% of the A alleles (i.e., 80 of 10,000 A alleles). For this reason it is speculated that genetic drift plays a minor role in the evolution of large populations, but that it played a big role in the early evolutionary history of our species when population sizes were comparatively small.

Neutral Mutations

The literal interpretation of Darwin's main thesis, that natural selection forces the adaptation of Mendelian genes, may be wrong. It may not be that the fate of every allele is decided by the forces of natural selection. In 1968 Motoo Kimura suggested that new mutations may be **neutral** with respect to natural selection.

Refer back to Figure 2.31, the genetic dictionary. Notice that the genetic code is redundant, that is, some amino acids are called to the sites of protein synthesis by more than one codon. In fact, there are three times as many codons as amino acids, suggesting that two thirds of the changes made in DNA could have no effect upon phenotype. For example, GGU, GGC, GGA, and GGG all code for the amino acid glycine. You can see from this that any change in the third base of any of the four codons will have no effect upon the amino acid sequence of the protein encoded by the gene; any of the four codons with any of the four bases in the third position will call for glycine during protein synthesis. Therefore any mutation of the third base of any of the four codons will be neutral with respect to phenotype. In contrast, if either of the first two bases are changed in any of the four codons, a different amino acid will be inserted into the growing protein, and some of those amino acid substitutions may alter the function of the protein and in turn alter phenotype. (If all third-base mutations were neutral and all first- and second-base mutations resulted in a phenotypic change, then only one third of all mutations would be neutral. This is not the case, however.)

After it became possible to determine base sequences of DNA coding strands, it was reported that in natural populations there are more mutations of third-codon positions than of first- and second-codon positions. Mutations in the first two positions are selected against, which is to say that the original sequence of the first two bases of most codons are conserved, presumably by natural selection, whereas mutations in the last position may change without phenotypic consequences, because they will be neutral. The frequencies of neutral mutations within gene pools, then, should be functions of mutation rates and genetic drift. (Recent evidence, in contrast, supports the view that neutral mutations do not appear in gene pools more frequently than biased mutations. It will be some time before the issue of neutral mutations is cleared up by experiment.)

There is another aspect of mutation neutrality. Proteins whose amino acid sequences are known have been compared in two different but related dimensions. First, distantly related species such as pigs and humans possess closely related proteins, as judged by function and amino acid sequence (e.g., the hormone insulin). The amino acid sequence of swine and human insulin is identical but for one amino acid. This means that the amino acid sequence of insulin is highly conserved over very long periods of evolutionary time; thus the gene that encodes insulin is highly conserved over long periods of evolutionary time (Figure 4.7). Second, not all proteins or short amino acid sequences within proteins are as highly conserved. For example, within some proteins there are segments of amino acid sequences that are highly conserved and segments that are relatively variable (Table 4.3). The highly conserved regions are the **active sites** of protein molecules (regions of proteins that bind to substrates, to other proteins, or to nucleic acids); the least conserved regions are located at some distance from the active sites. In these distant regions it may matter less which amino acid is found in some of the positions, which is to say that in some codons a mutation of the first, second, or third base may cause an amino acid

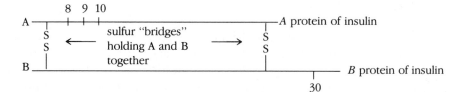

All amino acid positions except A8, A9, A10, and B30 are identical in these organisms:

Organism	A8	A9	A10	B30
Human	thr	ser	Ileu	thr
Swine	thr	ser	Ileu	ala
Rabbit	thr	ser	Ileu	ser
Dog	thr	ser	Ileu	ala
Horse	thr	gly	Ileu	ala
Cattle	ala	ser	val	ala
Sheep	ala	gly	val	ala
sperm whale	thr	ser	Ileu	ala

Figure 4.7 The primary structures of many proteins are conserved over long periods of evolutionary time.

Table 4.3 Hemoglobin beta, delta, and gamma proteins. The active site regions of the hemoglobin proteins are highly conserved among the higher primates (over approximately 50×10^6 years), but other regions of the proteins are somewhat variable.

Source	1	2	3	4	5	6	7	8	9	10	63	64	65	66....	89	90	91	92	125	126	127
human β	V	H	L	T	P	E	E	K	S	A	H	G	K	K....	S	E	L	H	P	V	Q
Chimp β	V	H	L	T	P	E	E	K	S	A	H	G	K	K....	S	E	L	H	P	V	Q
Rhesus β	V	H	L	T	P	E	E	K	N	A	H	G	K	K....	S	E	L	H	Q	V	Q
human δ	V	H	L	T	P	E	E	K	T	A	H	G	K	K....	S	E	L	H	Q	M	Q
chimp δ	V	H	L	T	P	E	E	K	T	A	H	G	K	K....	S	E	L	H	Q	V	Q
Spider mon δ	V	H	L	T	G	E	E	K	S	A	H	G	K	K....	S	E	L	H	Q	V	Q
human γ	G	H	F	T	E	E	D	K	A	T	H	G	K	K....	S	E	L	H	E	V	Q

active
site

Amino acid code
A = ala   H = his   Q = gln   C = cys   K = lys   R = arg   E = glu   L = leu   S = ser   F = phe   M = met   T = thr
G = gly   P = pro   V = val

substitution without altering the protein's function. Such mutations will be neutral, and their frequency within a gene pool may increase in the absence of intervention by unknown factors.

There are other kinds of evidence that genetic variation can accumulate within gene pools without passing the tests alleged to be administered by natural selection. Neutral mutations occur; they may accumulate in gene pools or they may be lost from gene pools by chance. Even more to the point, some mutations may be neutral in one environment but not in another. In the long run the accumulation of neutral mutations will confer versatility to a species as pressures imposed by natural selection change sharply. An example of this in human evolution is difficult to verify, but the following mutation may illustrate a related point.

It has been suggested that people with sickle cell trait are more resistant to falciparum malaria than homozygous non-SCA persons. The nature of this resistance is not fully understood. However, there is another kind of malaria in many parts of Africa called vivax malaria. People who are susceptible to the vivax malarial parasite possess a protein on the membranes of their red blood cells to which the parasite attaches, and that attachment guides the parasite into the red cell. In the red cells it multiplies, and as red cells burst, releasing a volley of young parasites, the victim spikes a fever around 105° F. Vivax malaria is called benign malaria and is not lethal as is falciparum malaria (called malignant malaria), but it is a devastating disease.

People who are resistant to vivax malaria are missing the membrane antigen, called the Duffy antigen. The parasites cannot attach to red cells that are missing the antigen, and therefore cannot enter the cells. The gene that encodes the Duffy antigen is called the Duffy gene. Resistant persons are homozygous for the mutant Duffy allele, and they carry no Duffy antigen. The point is that the function of the Duffy antigen is unknown. Persons lacking it appear to be as healthy as persons with it; the only known difference is that persons with the antigen are susceptible to vivax malaria. The entire gene appears to be neutral; why does it survive if the protein encoded by it is not needed? Very simply, we do not know.

Allele Frequencies in Human Populations

A, B, AB, and O Blood Types

Many people living in North America and Europe will have had their blood typed, that is, drawn and tested for blood proteins. One family of blood proteins is called ABO. This protein family includes two antigens and two natural antibodies. Each individual is one of four possible phenotypes, A, B, AB, and O, and each phenotype is defined by the presence of two, one, or zero blood antigens and zero, one, or two antibodies. Persons with type A antigens also carry antibody B; persons with type B antigens also carry antibody A; persons with neither antigen, type O, carry both antibody A and B; and persons with both antigens carry neither antibody (Table 4.4). One gene, I, encodes the family of antigens, and this gene exists in three allelic forms, I^A, I^B and I^O (Table 4.4). Since each person carries just two alleles of each gene, there are six possible genotypes: $I^A I^A$, $I^A I^O$, $I^A I^B$, $I^B I^B$, $I^B I^O$, and $I^O I^O$.

The frequencies of the three I alleles are known for most human populations (Table 4.5). From the vantage point of population genetics we want to know

Table 4.4 The ABO blood types, the I gene, its three alleles and six genotypes.

Genotypes	Phenotypes	
$I^A I^A$	A	A antigens
$I^A I^O$		anti-B
$I^B I^B$	B	B antigens
$I^B I^O$		anti-A
$I^A I^B$	AB	A and B antigens
$I^O I^O$	O	anti-A and anti-B

Table 4.5 The allele frequencies of I^A, I^B, and I^O in nine populations.

Population	I^A	I^B	I^O
Armenia	36.0	10.4	53.6
Alaska	35.5	4.6	59.9
Belgium	27.0	5.9	67.1
Greece	22.9	8.2	68.9
Russia (Urals)	29.5	19.5	51.0
Russia (Siberia)	13.0	25.1	61.9
Russia (Tadzhikistan)	21.0	37.3	41.7
Nigeria (Ibo)	13.2	9.5	77.3
Nigeria (Yoruba)	13.8	14.6	71.6

whether the allele frequency differences between populations are due to genetic drift, migration, or to natural selection. If natural selection, we will want to know the influence of these blood proteins upon reproductive success within specific environments.

Evidence for gene flow comes, in one case, from the fact that on a line between Central Asia and Western Europe, there is a gradual decrease in the frequency of the I^B allele, from 0.3 in Central Asia, to 0.2 in Russia, to 0.15 in Central Europe, to 0.1 in France and England (Figure 4.8). It is speculated that the frequency of this allele was near zero in Western Europe prior to the Tartar invasions from Central Asia, which ended in 1480. One bit of evidence for this speculation is the fact that the frequency of the I^B allele in populations now living in the Pyrenees mountains is very low (0.05). The ancestors of these people may have escaped the Tartars by fleeing into the mountains. This is not a fact but a prediction based upon facts, including the fact that historians have yet to describe an invasion by an army that was not accompanied by gene flow.

Evidence for differential reproductive success of the three alleles comes out of different kinds of population studies. For example, in Britain, persons of O blood type are 40% more likely to develop duodenal ulcers than are persons of A, B, and AB blood types. Yet the I^O allele is the most common of the three alleles in Britain, and duodenal ulcers usually develop after the age of reproduction—not an "airtight" case that natural selection in Britain works to eliminate I^O alleles from gene pools. Yet even if the numbers of persons with O type blood are greater than the others, selection pressures could be acting to reduce the frequency of the I^O allele today.

Another seeming contradiction: Persons of B and O blood types carry natural antibodies against A antigens, and A antibodies confer some resistance against the smallpox virus. One might conclude from this that smallpox epidemics in the past had some influence upon the frequencies of I^B and I^O alleles. But it turns out that the frequency of the I^B allele is lower than that of the I^A allele in most present-day world populations.

It is an amusing and seductive activity to guess the kinds of environments within which modern alleles became established residents of our ancestor's gene pools. What were the circumstances that favored the survival of those alleles? What role did natural selection play? How many alleles "squeaked" into the gene pool by chance? Were many "useful" alleles lost from the gene pool by chance? Some questions, these among them, may never be answered. It is quite unlikely that we shall ever see an "instant replay" of evolution.

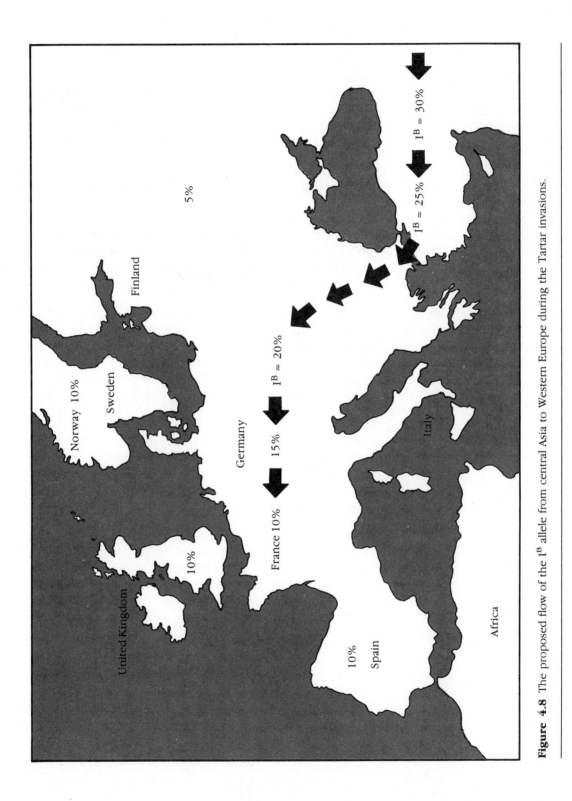

Figure 4.8 The proposed flow of the I^B allele from central Asia to Western Europe during the Tartar invasions.

Rhesus Blood Types

A different family of blood antigens, called Rh because they were observed first in rhesus monkeys, presents an even more puzzling problem of allele frequency changes through time. About 84% of the people of Northern European descent living in the United States are phenotype Rh$^+$, and about 16% are phenotype rh$^-$. The blood of Rh$^+$ persons reacts positively (agglutinates) to Rh antiserum of rhesus monkeys; the blood of rh$^-$ persons reacts negatively (does not agglutinate). (A cautionary note: In most descriptions of gene-phenotype relationships, the language is unambiguous about which is gene and which is phenotype. But in the case of Rh genes and phenotypes, the same names are used for gene and phenotype. The phenotypes are Rh$^+$ and rh$^-$, and the alleles are Rh^+ and rh^-.)

Genes that encode Rh antigens are complicated, so in the interests of pedagogy we discuss only one gene, and we call it the **Rh** gene. Persons phenotypically Rh$^+$ are genotype Rh$^+$Rh$^+$ or Rh$^+$rh$^-$; the rh$^-$ phenotype is always associated with the rh$^-$rh$^-$ genotype. The frequencies of these two alleles are widely different among populations of the world (Table 4.6), and genetic demographers would like to know why.

Rh antigens came to the attention of the medical profession when their relationship to serious complications between mothers and their developing fetuses was discovered (Figure 4.9) If an rh$^-$ woman carries an Rh$^+$ fetus and if the blood of the fetus and the mother exchange across the placenta, Rh$^+$ antigens of the fetus will enter the mother's blood and provoke the formation of Rh$^+$ antibodies. When the antibodies formed in the mother's blood get back into the blood stream of the fetus, the fetal blood cells are destroyed, causing severe anemia. Before the Rh system was understood well, this hemolytic disease of the newborn resulted in death for nearly 1 of every 100 babies in the United States. Today these deaths are prevented by administering anti-Rh to mothers at risk.

Have the deaths of newborns caused by this hemolytic disease changed the frequencies of the Rh^+ and rh^- alleles? It wouldn't seem so because every baby that dies is heterozygous, which is to say that in the case of death, the two alleles depart from the gene pool in pairs. Deaths do not change their relative frequencies. Why then do modern gene pools differ so widely? The rh^- allele is exceedingly rare in populations of native Americans, common in European populations, and more frequent than the Rh^+ allele in the Basque population living in the Pyrenees mountains (Table 4.6). There are no agreed-upon answers

Table 4.6 The frequencies of the Rh^+ and rh^- alleles differ widely among populations of the world.

| | Frequency of | |
Population	Rh^+ allele	rh^- allele
Basques (Spain)	.45	.55 (highest known)
Central Europe	.60	.40
China	.99	< .01 (low)
U.S. Whites	.60	.40
U.S. Blacks	.72	.28
Native Americans	.99	.01 (low)

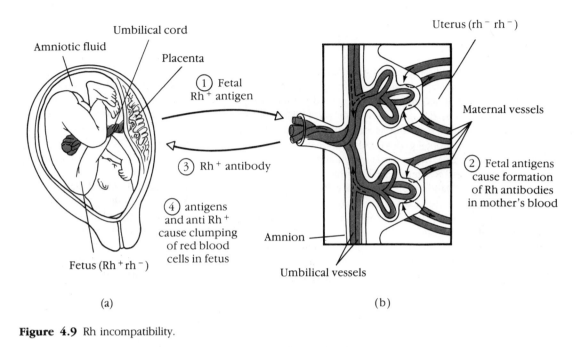

Figure 4.9 Rh incompatibility.

to these questions or explanations of the facts. Genetic drift may have played a role earlier in the history of our species, at a time when human populations were small.

Qualitative versus Quantitative Phenotypic Variation

Mendelian genes are observed within breeding populations just as they are observed within families. One gene and two alleles often are associated with two or three discrete forms of a phenotypic trait, *traits that are **qualitatively** different from one another.* But populations reveal a class of phenotype that cannot be examined and evaluated by studying one or a few individuals. This class of phenotype can be observed and evaluated only by studying populations; in a sense, they are population phenotypes. Population phenotypes that cannot be assigned to individual genes *vary continuously*, not discontinuously, sometimes over rather wide ranges. In the language of genetics, *the variation is **quantitative.*** Since phenotypes that vary continuously cannot be expressed qualitatively, they must be analyzed and described in different ways, that is, quantitatively.

Consider human height (Figure 4.10). Unlike Mendel's observation of long- and short-stemmed pea plants (a qualitative difference), human heights vary continuously (quantitative differences) between short and tall, from one extreme phenotype to the other. While there are human genes that in one or another allelic form lead to short stature, as in the case of the Ellis-van Creveld syndrome, these kinds of genes cannot explain the differences between heights of 5'5" and 5'6", or for that matter, between 5'5" and 6'5".

Said another way, Mendelian genes can be identified only if their associated phenotypes are qualitatively different from one another, and only if the

An Aside

Up to this point the discussions of genetics have been lopsided in one of two ways: The discussions of interesting phenotypes, e.g., the development of zygotes into adults and of sexual dimorphism (Chapter 3) are not accompanied by detailed explanations of the roles of the genes postulated to influence them. On the other hand, detailed discussions of genes, Mendelian (Chapter 1) and molecular (Chapter 2), are not accompanied by detailed explanations of how phenotypes develop. Our current discussion of genes in populations continues the same lopsided view of gene-phenotype relationships because the genes of population genetics are the genes of Mendelian and molecular genetics. The **I** and **Rh** genes are examples.

However, population geneticists ask different questions of these genes than do Mendelian and molecular geneticists, because population geneticists are as concerned about the "genotypes" and "phenotypes" of populations as Mendelian geneticists are about the genotypes and phenotypes of individuals. The assessment of population genotypes is difficult, since it involves allele frequencies, how these frequencies change, and how similar populations evolve into species. But the methods used by population geneticists to study genes add nothing new to our understanding of the characteristics of genes or to our knowledge of how genes influence the development of individuals from zygotes.

These lopsided views provided by genetics technology reflect to some extent the state of the art. Of the estimated 100,000 human genes, fewer than 3,000 have been studied in detail, and these by the methods of Mendelian and molecular genetics. As time goes by, the state of the art improves. More genes will be discovered; more will be learned about how phenotypes emerge during development. But the worry is whether our views of gene-phenotype relationships today will adversely bias our approach toward discovering new genes and studying the processes of development. Recall the old adage, "a little knowledge is a dangerous thing." The danger here lies in the temptation to draw conclusions about all genes and phenotypes from the small number that have been observed. Mendel studied seven genes and concluded from that small sample that genes segregate independently; we now know that this "rule" isn't a rule; it fails to explain inheritance patterns of linked genes.

Mendelian and molecular genes are relatively easy to discover. The geneticist looks for contrasting phenotypic traits, but this methodology is so taken for granted that for some enthusiasts it is easy to conclude that the discovery of contrasting phenotypic traits is evidence for contrasting forms of a gene. This tendency is less dangerous within science than within social settings. Within science the observation of contrasting phenotypic traits can be tested for corresponding contrasting alleles. In society, genes can be assigned to any difference observed from which attitudes, mores, and even legal systems logically follow.

We now consider genetic influences upon phenotypes that cannot be partitioned easily into Mendelian and molecular genes.

inheritance patterns of these qualitatively different phenotypic traits obey the laws of chance. But if phenotypes differ quantitatively (continuously), the relationships between phenotypes and genes are difficult if not impossible to unravel. The branch of genetics concerned with these complicated gene-phenotype relationships is called **quantitative genetics**.

Some of the most interesting phenotypes of plants and animals vary continuously. In humans, skin color, body weight, body shape, metabolic activity, reproductive potential, and behavior vary continuously between outer limits, said to be set by our *species genetic potential*. No adult humans are shorter than 2 feet, and no adult humans are taller than 10 feet; no human has absolutely black skin, and no human, excepting certain albinos, has absolutely colorless skin. What's more, matings between "short" men and "tall" women usually result in children of intermediate height, not tall children who, in turn, if they mated with each other, would produce tall and short children in the ratio of 3:1.

There is a question of whether quantitative genetics provides a weak sighting of different kinds of genes, but most geneticists seem to think not. What quantitative genetics seems to be describing are complex interactions between

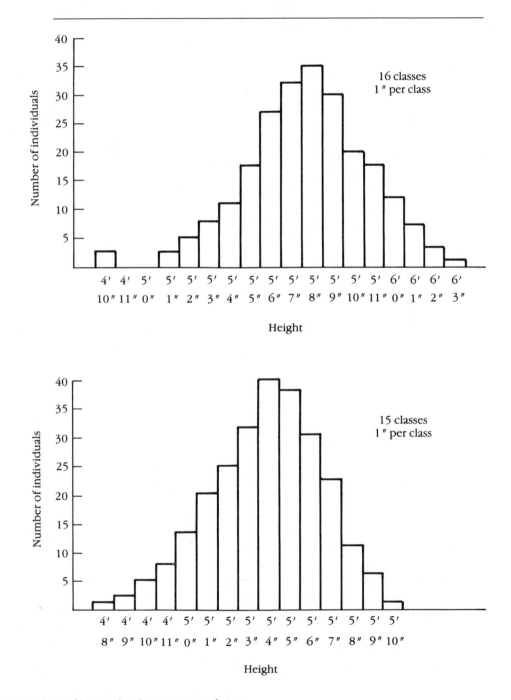

Figure 4.10 Variation of human height in two populations.

genes and their environments. Phenotypes that vary continuously *are influenced by many genes and by their environments.*

The genetic contribution to variation can be illustrated with models showing the influence of one, two, three, four, or five genes upon height (Figure 4.11). As the number of genes increases, the number of phenotypic classes (of height) increases. *As classes of phenotype increase, the differences between classes decrease until, finally, with more than four or five genes, the differences between classes nearly "melt away;" hence, continuous variation.*

The environmental influences are less well known, in part because they are difficult to quantitate and in part because the study of norms of reaction has never been popular. It is difficult indeed to partition phenotypes into genetic and environmental inputs. Given the varied diets and lifestyles (which influence height) that exist within human populations, even within families, the relationships between environmental inputs and phenotypic classes are largely unknown.

Geneticists who study quantitative inheritance often ask the same questions of the genes that influence continuous variation as Mendelian geneticists ask of Mendelian genes, especially within plant and animal improvement programs dedicated to changing phenotypes of economic value. But unlike Mendelian genetics, quantitative genetics requires a far more sophisticated statistical framework within which to plan experiments, gather data, and analyze data. In the present discussion we touch only on a few of the highlights of quantitative genetics. For example, we will ask whether there is any evidence to suggest that continuously varying phenotypes are influenced at all by genetic variation. We also will ask how it is possible to measure the separate contributions of environments and genes.

In Figure 4.11 you see the classes of height and the relative differences between classes, premised on one, two, three, four, and five Mendelian genes postulated—for the purpose of making the point—to influence height. In each of the **histograms** you see a height **distribution pattern**. One height, or height class, is more frequent than the others. The most frequent height class is called the **mode**. On both sides of the mode the numbers of individuals in each height class decrease, and in the models shown, the decrease is symmetrical on both sides of the mode until the least frequent classes are reached. If all of the individual heights are summed and then divided by the number of individuals measured, the **mean** height is determined. This is the arithmetic average height.

One more statistical concept needed for analyses of continuous variation is that of **correlations**. You may have noticed, for example, that tall parents tend to have taller children than short parents. Not all children of tall parents are tall, and not all children of short parents are short, but there seems to be a *correlation between parent height and offspring height.* (Figure 4.12). There are two obvious logical explanations for the observed correlation: tall parents are likely to pass to their children genotypes for tallness, or tall parents are likely to provide environments that lead to tallness. The problem for the geneticist is to discover the processes that lead to height and thereby to distinguish between these two (and other) explanations for the observed phenotypes.

A word about correlations: the fact that correlations between two phenomena exist is not proof of a cause-effect relationship between them. Even if tall parents are genetically different from short parents and even if tall parents have taller children than short parents, the average height differences between their

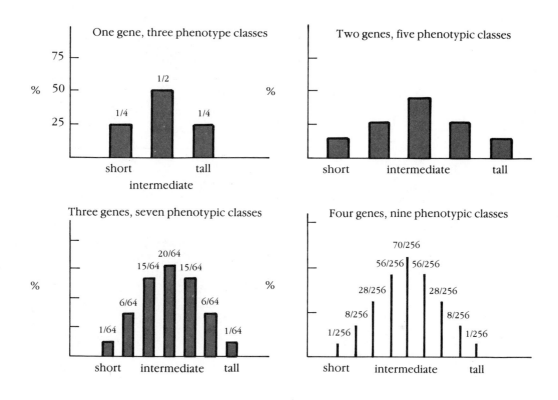

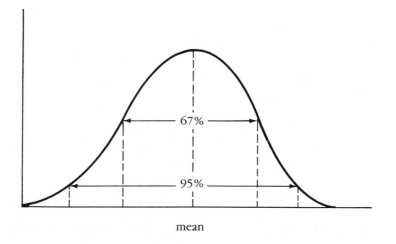

Figure 4.11 Gene number and phenotypic classes. As gene number increases, the number of phenotypic classes increases, and variation becomes more and more a continuum.

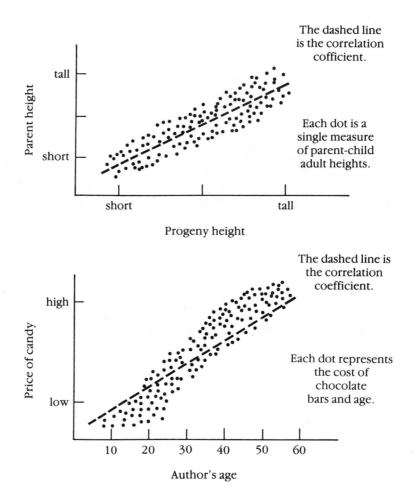

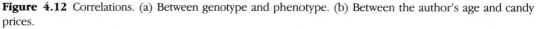

Figure 4.12 Correlations. (a) Between genotype and phenotype. (b) Between the author's age and candy prices.

children may be due to nongenetic factors. In other words, there may be no cause-effect relationship between genes and height even though there is a correlation between them. The price of candy has increased as my age has increased, but my age has had no effect upon candy prices, nor have candy prices had an effect upon my age. So while candy prices and my age are positively correlated, an increase of one is not causally related to an increase of the other (Figure 4.12).

Geneticists have devised several approaches to the problem of separating genetic from nongenetic influences upon continuously varying phenotypes. Only one is discussed here, mainly to illustrate the difficulty of the task and to show that so far it has been difficult if not impossible to separate genetic from nongenetic influences upon continuously varying phenotypes within human populations.

Farm crops and animals are selected mainly for their economic value; pets are usually selected for their esthetic value. An economically important feature of dairy cows is the amount of milk they produce. How are cows influenced to produce more milk? In most cases, they are influenced by various combinations of breeding, selection, and husbandry. Cows that produce lots of milk are used as breeding stock, and bulls born of cows that produce lots of milk are bred to them. (So far the best diet found for increasing milk production is a combination of oats, hay, and a continuous supply of strong ale; but this is not recommended to dairy farmers, for many reasons, one being the cost, another being the difficulty of managing tipsy cows.) From breeding programs, cows that produce the most milk are selected as parents for the next generation and so on (Figure 4.13).

If the progeny of high-producing cows give only an average (of their mother's generation) amount of milk, then the high producers selected (their mothers) as breeding stock must have produced lots of milk because their environments were good, not because of their genotypes. (Their daughters produced less milk.) However, if the progeny cows produce as much milk as their mothers, the

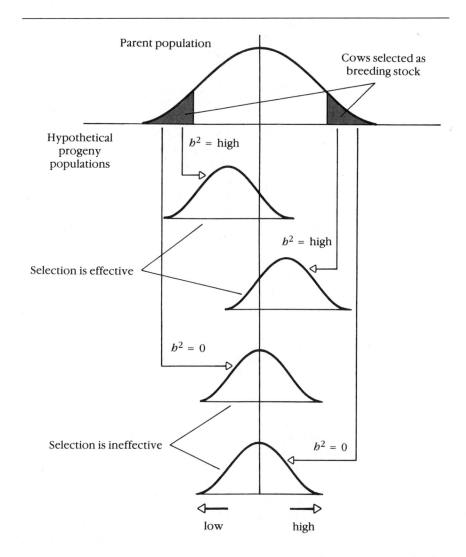

Figure 4.13 Selection for heritable components of phenotypic variation, in this case the quantity of milk. The effectiveness of selection is a measure of the heritability, h^2, of the phenotypic variation.

mothers must have had genotypes that lead to high milk production. On the other hand, if the progeny cows produce intermediate amounts of milk (intermediate between that of their mothers and the average of their mothers' generation), then milk production must be a function of both genotypes and environments.

The amount of milk produced by progeny cows above the average of their mothers' generation is said to reflect the contribution of genotypes to milk production. This genotypic contribution to milk production is called the **heritability** component of milk production variance. In a crude simplification of the problem, the variance of milk production that cannot be ascribed to heritability is said to result from the environmental component of variance, plus an interaction between heritability and environment. Some of the variance is due to genotypic differences among the cows, some is due to environmental differences, and some is due to interactions among genotypes and environments.

To be more specific, if the amount of milk produced by progeny cows is midway between the average (of the parent generation) and the amount produced by their mothers, heritability is estimated to be 50%, or 0.5. This means that half the variation observed in the parental generation is due to genotypic differences among the cows in that generation, and half the variation is due to environmental differences experienced by the cows in the parental generation.

Harking back to norms of reaction, however, it is a simple fact that genotypes do not give rise to identical phenotypes in all environments. For example, cows selected to produce large amounts of milk in Wisconsin do not produce large amounts of milk if they are reared in the dry plains of West Texas; and cows selected to produce large amounts of milk in west Texas do not in Wisconsin. Therefore any estimate of heritability is reliable only within narrow limits of environment. Nevertheless, heritability estimates are useful to breeding programs that are designed to change phenotypes within narrow ranges of environments.

Heritability estimates of continuously varying phenotypes are relatively accurate if the method used for making the estimates includes (1) selection to change the mean from one generation to the next, as outlined briefly above; and (2) comparisons of parent and progeny generations within narrow ranges of environments. *The degree to which a mean can be changed by selection is the degree to which genetic variation influences phenotypic variation.* Compared with Mendelian and molecular genetic methods for sighting genes, heritability estimates are vague. But they continue to be useful in plant and animal improvement programs.

With human populations it is questionable whether meaningful heritability estimates can be made at all, other than by pedigree analyses of the kind that employs simple Mendelian principles. But of quantitative variation within large populations, there is no way to determine whether the heritability estimates that have been made, in fact measure what they are said to measure. For example, it is impossible to use selection methods for estimating the heritability of human phenotypes. It is impossible to produce pure lines of humans, or to design breeding programs to produce tall or short people, or to select from one generation the individuals who will produce the next generation. The study of identical twins allows us to see at least some of the effects of environment upon developing phenotypes.

Twin Studies

About 8 of every 1,000 births are twin births. Some twins are no more genetically alike that siblings born years apart. Such twins result from the fertilization of two eggs, each by a different sperm, and hence are called **dizygotic** (**DZ**) twins. Other pairs of twins are genetically identical. One egg is fertilized, but during early embryonic development (e.g., at the two-cell stage) the genetically identical cells come apart and develop into genetically identical individuals, called **monozygotic** (**MZ**) twins.

If two genetically identical individuals are separated at birth and adopted into very different environments, say, Lagos, Nigeria and Kyoto, Japan, and if the twins never meet during childhood, how different will they become? For sure they will speak different languages and adhere to different social values, allegiances, and mannerisms. They will eat different foods and learn different skills, and it is likely that they will become socialized into different socioeconomic classes. But what about height, weight, intelligence, and aggression? If phenotypic differences between them are observed, then the nongenetic features of upbringing will be credited with causing it.

In the case of height, the two members of the above MZ twin pair are expected to be more alike than pairs of individuals picked at random, one from each city. They may differ slightly in height, but we still expect to find a *high, positive correlation between genetic relatedness and phenotypic similarity*. If this expectation is correct, then genotypes will be a better predictor of height than environmental factors, which is to say that heritability is a more significant component of height variation than are the environmental differences. But the heritability of height variation does not identify the genes that influence height, and such genes might have their effects upon metabolism, physiology, or bone length; correlations between genotypes and height cannot reveal the genes or what the genes encode.

Heritability estimates of height variation cannot predict rather sudden changes in population height, for example, like that observed of U.S. soldiers whose height increased 2 inches between World War I and World War II. The mean height of Japanese people increased nearly 3 inches between 1950 and 1980. These increases of population mean height are probably more a reflection of dietary changes than of new mutations. That is, *even if the heritability of height variation is high within a population, the population's mean height may change radically in response to environmental changes;* this is added evidence that genotypes and phenotypes, while highly correlated within nonchanging environments, are not tightly locked over wide ranges of environments.

In the "real world" it is observed that two populations may be genetically different (as people of European descent are genetically different from people of Japanese descent) and phenotypically similar (as Euro- and Japanese-Americans are in height today), or that two populations may be genetically different and phenotypically different (as were Euro-Americans and Japanese 40 years ago). In other words, high heritability is not a good predictor of mean heights in new environments. In many places in the world mean heights have increased with improved nutrition and changing life-styles. We have no idea how much of the present variation in height observed within the human species would remain if everyone had access to the same diets and standards of living.

And if we could do the proper experiment, as can be done with milk production in cows, we shouldn't be surprised to find that the heritability of height variation is high, even while the mean can be shifted easily by nongenetic influences.

Twin studies do not address this problem. Twins cannot be used to assess the relative influences of genotypes and diets upon height, mainly because members of a twin pair usually are reared in similar environments, either in the same family or in the same social and ethnic class. Only a few cases of twin pairs adopted into strikingly different social and ethnic classes have been reported.

The major controversy stirred by twin studies is not centered upon height variation, but upon variation of intelligence quotient (IQ) and other aspects of behavior. This adds to the confusion, as you will know if you compare the ease and accuracy with which height can be determined with that of intelligence assessment. Intelligence simply cannot be defined or measured neatly, as height can. Indeed, intelligence may be more a social construct than a biological one, but as the news media report, the postulated relationships between genes and social constructs have given birth to one of the most celebrated controversies in science since Darwin. We will return to this subject in Chapter 5.

Human Breeding Populations

A species (e.g., *Homo sapiens*) is a population of organisms that is biologically reproductively isolated from organisms of all other species. Males and females from any existing human population can make babies, but no human male or female can make a baby with a member of any nonhuman population. This assertion is true for chimpanzees, carp, and centipedes. Species are well-defined populations; the biological criterion of sexual reproduction is unambiguous. This is not true for subpopulations of species, especially for subpopulations of the human species. Many names have been given to these subpopulations, and many criteria, some biological and some not, have been employed to describe these subpopulations. Probably the most notorious name given to human subpopulations is **race**.

So-called human races are not well defined; the biological criteria of race are ambiguous and confusing. In fact, species and races are another example of an inverse relationship between the knowledge we have of a subject and the certainty we may feel that we know all about that subject. The word race is used every day in the news media, on the streets, and in schools; the word is used with a confidence that signifies that the users know what it means. But as it turns out, *the popular definition of race has no meaning in biology;* the popular definition, in fact, confuses what we do know about human populations.

The popular meaning of the word race has changed a great deal since it was first used in 1749 to describe the geographical distribution of human populations. Since that time the word race has acquired a pejorative connotation; it often is used to imply "differential worth" of societies of people, ethnic groups, and even religious groups. Newer words such as racism and racist attest to this connotation.

Look back at Figure 4.1, which illustrates the role of geographic isolation in the process of speciation. Following geographic isolation, long periods of time are needed for speciation, enough time for two or more breeding populations to become sufficiently genetically different as to become reproductively isolated, the biological event recognized as the last step in the formation of new species.

The word race implies that at some juncture between the time of geographic isolation and the emergence of genetic reproductive isolation, the two populations will have become sufficiently biologically different as to have become races. How different must that be? At what point between the two isolation events do the two populations become two races? Are the processes of **raciation** simply those that act during the early stages of speciation? Many anthropologists focused their attention upon these questions for many years, but there is an effort now to recover anthropology from that trap (see Nelson and Jurmain, 1988).

Most population geneticists argue that there are no such entities as human races. To argue otherwise requires a definition that specifies the amount of genetic difference that must exist between otherwise similar populations. This has not been done. The word race was coined for a different purpose; its maintenance in the vocabulary serves yet another purpose. As used in the vernacular, the word race is a relic of past ignorance, but it lingers on, probably because of its vulgar political usefulness.

Perspective 4.2

There are many examples of social inequities experienced by different **ethnic** groups— different wages for similar work, different schools, different housing facilities, different access to social and sporting facilities, and so on. An example will show how these kinds of inequities tie in with a belief system that devalues the lives of people who do not belong to the dominant ethnic group.

In the Fall of 1932 in Macon County, Alabama, a study was initiated to follow the course of *untreated* syphilis in African-American males. Through advertising in local churches and drug stores, 600 "volunteers" were picked for the study—400 with syphilis and 200 controls. The men with syphilis were in the latent stage and did not need the kind of immediate treatment administered to men in acute stages of the disease.

At the time of the first Public Health Report about the experiment, there had been four surveys: 1932, 1938, 1948, and 1952. Between surveys the 600 men were seen regularly by a nurse and irregularly by a physician. The most significant aspect of the study was the postmortem examination to determine, if possible, the final result of the infection. Syphilis leads to complications such as heart disease, blindness, insanity, deafness, paralysis, and death.

It turns out that the volunteers did not volunteer for what was to follow but only to submit to a blood test. They were promised burial assistance if they signed a form granting autopsy rights, a promise that looked good to poor people during the worst depression ever in the United States; since the 400 men with the disease had no option but to die of its complications, even after the discovery of penicillin (an antibiotic that effectively stopped syphilis infections), they were in fact tricked into accepting almost any short-term economic concession.

In 1969, nearly 40 years later, a panel was created to investigate the circumstances surrounding "The Study of Untreated Syphilis in the Male Negro" initiated by the United States Public Health Service in 1932. The findings of the panel were chilling. The issue of individual dignity and integrity was pitted against the issue of the freedom of scientific inquiry, and it was the conclusion of many on the panel that "Negroes have less access to dignity than whites," and that science really wasn't the issue in this case, since the methods were so unethical and ineffective that the results had no scientific value.

In 1973 Hubert H. Humphrey made the following statement supporting the establishment of a "National Human Experimentation Standards Board Act": "Mr President, would these physicians care more about the test population if it were composed of U. S. Congressmen? Would they be more caring if the human subject were their next door neighbor? I submit they would."

To obtain a working understanding of the genetic variation within the human species, geneticists have used the concept of breeding population as a way to divide the species into smaller and more easily diagnosed populations. Breeding populations are not clearly defined, because, with a few rare exceptions, the boundary lines between breeding populations become more porous with each succeeding generation, that is, more between-population matings take place each generation. Nevertheless, the frequency of within-population matings determines whether a population is a breeding population or not. Genetic differences between breeding populations can be ascertained; the gene flow between populations can be measured; and more to the point, the future of human breeding populations can be predicted by observing, for example, population mergers and isolation barriers. Biologists who insist that the word race is a useful biological concept usually describe biological differences between races in much the same way that the differences between breeding populations are described. Indeed, the definition of a breeding population is the best biological definition of race, but (1) breeding populations are not defined by use of socially loaded words, only by mating patterns, and (2) the word race contributes nothing to our understanding of human populations. The use of the word race in the vernacular keeps it from acquiring the full meaning of breeding population, and its plethora of nonscientific meanings will not be easy to jettison.

A Short Review of Human Origins

An introduction is needed for this section. While most biology teachers will teach Mendelian and molecular genetics in like ways — because these two approaches to the study of genes are comparatively advanced, and therefore possess a common database and theory — those same teachers may explain human origins in very different ways. The study of human origins by way of the fossil record is not advanced; agreements are few and conjecture is the rule, not the exception. This is said with all due respect for the difficulty of making sense out of a relatively few, well-preserved fossils and in the knowledge that human origins are better understood now than at the time Charles Darwin first took an interest in the subject. The account of human origins outlined briefly below is somewhat of a middle ground and is taken in part from Roger Lewin's *Human Evolution*. No attempt is made to criticize any of the conjectures about human origins or to forge new ground. The purpose here is to place the descriptions of genetic evolution, discussed above, into the context of primate evolution.

Darwin used evidence from the fossil record as support for his theory of evolution. A century and a half ago the evidence was sparse, but nevertheless Darwin made a good case with the evidence available to him. Today the evidence is far less sparse — many of the so-called gaps in the record that existed in Darwin's day have been filled — but still not as abundant as we might desire. A brief outline of the record is shown in Figure 4.14. (As stated above, the details of **primate** evolution are contested at many points, but the question of whether evolution gave rise to the higher primates is not contested. New fossils are being discovered almost daily, and some of them stimulate adjustments of details within the record.)

One current adjustment that is being hotly discussed was provoked not by a new fossil but by analyses of mitochrondrial DNA (mtDNA) taken from many

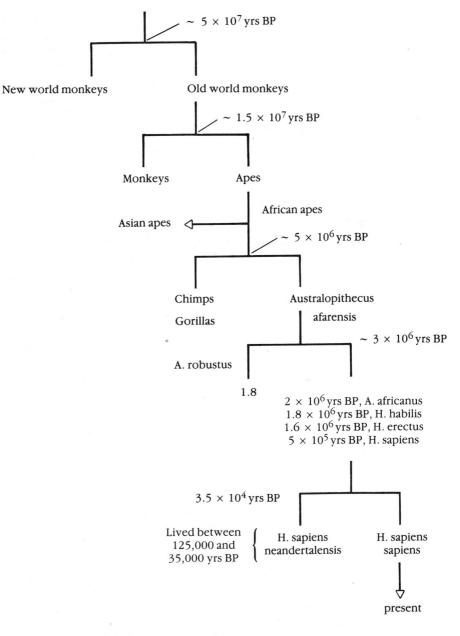

BP = before the present time

Figure 4.14 A very brief outline of primate evolution derived mainly from the fossil record.

types of contemporary humans. Allan Wilson and his colleagues have presented evidence that the history of mtDNA indicates that *H. sapiens* did not interbreed with any of its ancestor species, but can be traced back to one woman who lived some 200,000 years ago in Central Africa. The importance of this claim will appear more relevant after we review the fossil record.

Between 40 and 50 million years ago a species of monkey gave rise to two species. A geographical isolation event may have initiated this split, since one of the new species gave rise to what are called New World Monkeys (in South America today) and the other to the Old World Monkeys (in Africa and Asia today). Some 15 million years ago an Ape species emerged from an Old World monkey species, and at nearly the same time an Asian ape species gave rise to an African ape species. About 5 million years ago the African ape species experienced a split, giving rise to chimpanzees and gorillas (African apes); shortly thereafter the chimp-like species split, initiating the origin of chimpanzees and a species called, *Australopithecus afarensis*. Later (between 3 million and 4 million years ago) *A. afarensis* became two species, *A. robustus* and *A. africanus*. *A. robustus* became extinct about 1.8 million years ago, and from *A. africanus* a new species, *Homo habilis*, appeared about 1.8 million years ago. Following *H. habilis*, *Homo erectus* appeared about 1.6 million years ago, and fossil remains indicate that this species lived until about 0.5 million years ago, the time at which the first fossil remains of *Homo sapiens* have been dated. (For a detailed description of fossil remains, how they are dated, and how confusions are cleared up by the discovery of more fossils, see Nelson and Jurmain.)

During the past 30 years, with the technology of molecular genetics at their disposal, biologists have been able to measure genetic relatedness among modern species of primates by the homology of their DNA base-pair sequences and the amino acid sequences of at least a half dozen proteins (Table 4.7). These studies do not contradict the studies made of fossil remains but rather confirm them with added evidence of evolutionary relatedness. For example, the hemoglobin proteins of chimpanzees, gorillas, and humans have identical amino acid sequences, and no protein of the three species has been found to differ by more than two amino acids. Judged by amino acid sequence homology, African apes and humans are less than one half of 1% different; judged by their antigen proteins they are about 1% different; and judged by their DNA base-pair sequences they are about 1.2% different. The three species are so alike, in fact, that some biologists suggest that they should be classified into the same genus, *Homo*.

More is known of *H. erectus* than is known of the earlier species. The oldest fossils of this species were found in Africa, dating from 1.6 million years ago, but

Table 4.7 Genetic relatedness of humans to other primates as determined by the primary structures of proteins and DNA.

	Amino Acid Sequences	Antigenic differences	DNA sequences
Human—chimp	0.27	1.0	1.8
Human—gorilla	0.65	0.8	2.3
Human—orang	2.78	2.0	4.9
Human—gibbon	2.38	2.6	4.9
Human—macaque	3.89	3.6	—
Human—spider monkey	8.69	7.6	—
Human—tarsier	—*	8.8	—
Human—loris	11.36	11.2	42.0
Human—tree shrew	—	12.6	—
Primates—other placentals	—	12.11-14.91	—
Placentals—marsupials	—	15.83	—

about 1 million years ago members of this species migrated into Asia and into Europe. This is the first species of *Homo* to show geographical dispersion and hence the first species to show both "cultural" and physical divergence. *H. erectus* was the first species of *Homo* to show, over a period of about 1 million years, an increase in cranial size (an estimation of brain size). The increase was from about 600 cc to about 1,100 cc. Modern apes and the fossil remains of earlier apes have a brain size around 400 to 500 cc, but during the 1 million year period of *H. erectus* evolution, brain size doubled. Since 1 million years ago this species inhabited much of Asia, Europe, and Africa. *H. erectus* developed hand tools, established homesites, used fire, and may have been the first *Homo* species to develop communal relationships, evidenced by artifacts found in a few of the home sites. There is no evidence of their having a language or of their attempting to record history.

Breeding populations of *H. erectus* became geographically isolated from one another. From that time allele frequency differences between breeding populations became possible, and the evidence of slight physical differences suggests that this happened. *H. erectus* eventually became extinct; the causes of extinction are unknown, although a very recent discovery (the history of mitochrondrial DNA) has increased the intensity of speculation about the demise of *H. erectus.* This will be discussed further below.

The fossil record is not clear about whether *H. sapiens* arose from *H. erectus*, but many anthropologists claim that the fossil record does in fact prove that lineage. There is a complication, however. One *Homo* species that evolved after the demise of *H. erectus* is *H. sapiens neandertalensis*, a "species" that lived between 125,000 and about 40,000 years ago in Europe. Of course, those who have opposed the entire theory of evolution have had their fun with jokes about the Neandertals. As Nelson and Jurmain have said, few groups of humans have been so brutalized by prejudice and bigotry as have the Neandertals:

> These troublesome hunters are the cave man of cartoonists, walking about with bent knees, dragging a club in one hand and a woman by her hair in the other. They are described as brutish, dwarfish, apelike, and obviously of little intelligence. This image is more than somewhat exaggerated.
>
> While cartoonists' license is not to be denied, the fact remains that Neandertals walked as upright as any of us, and, if they dragged clubs and women, there is not the slightest evidence of it. Nor are they dwarfish or apelike, and, in the light of twentieth-century human behavior, we should be careful of whom we call brutish.

As measured by body size, brain size, and general physical features, Neandertals were not very different from us. In fact, dressed in modern clothes it is unlikely that any of us would recognize a Neandertal woman or man walking through a shopping mall. Their brains, in fact, were slightly larger than ours, and they had a larger amount of lean body mass. There is some evidence that they possessed a language, and quite clearly their tools were more sophisticated than any that preceded them. Fire was used for cooking; they buried their dead with ritual, sometimes with animal artifacts and sometimes with flowers, and they may have had other rituals as well. Their most recent fossil remains date to 35,000 years ago. The cause of their extinction is unknown.

Whether Neandertals are direct descendents of *H. erectus* and/or direct ancestors of *H. sapiens sapiens* remains hotly debated. Either this or they are an offshoot of *H. erectus* with a separate lineage that eventually became extinct, somewhat like that of *A. robustus* from *A. afarensis*. If they are a separate lineage from *H. erectus*, it is possible that *H. sapiens sapiens* also is a separate lineage. Debates abound. In either case, modern humans are best known from about 40,000 years ago, but fossil evidence of their presence in Europe and Africa date to 500,000 years ago. Since 40,000 years ago their dispersal has been worldwide, and a great deal is known of their physical and cultural characteristics since about 30,000 years ago.

The theory of evolution gets supporting evidence from nearly every field of biology. Genetics got into the act early in this century and has been central to the study of evolution ever since, but more recently molecular genetics opened an unexpected door to our past. To get the feel of molecular genetic studies of evolution, make an analogy between comparing skeletal structures of different peoples and, for example, comparing protein structures. The comparison of individual skeletons is one of the many features of comparative anatomy. If chimp and human skeletons are compared (Figure 4.15), you see that chimps have longer arms than humans do and that humans have longer legs than chimps do. Then among humans, some have short fingers, others long fingers, and so on. Molecular genetics can do the same sorts of comparisons, only with proteins and DNA molecules. Indeed, it is through the door of comparative macromolecules - DNA, RNA, and proteins - that evolutionary relationships are being explored.

While the technology of comparative macromolecules appears difficult to the beginner, the steps needed to compare the hemoglobin proteins of any two or more species are routine. With a little practice most people could learn the skills. The bigger point, however, is that comparative studies of protein and DNA primary structures literally magnify, many times over, the resolving power for identifying genetic differences and similarities within and between species.

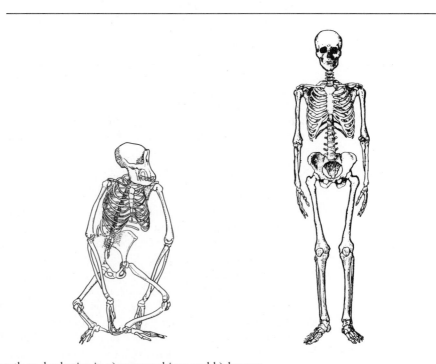

Figure 4.15 Arm length vs. body size in a) pygmy chimp and b) human.

Very few who study evolution question the results of molecular geneticists who study the comparative "anatomy" of proteins and DNA. These studies, for the most part, have fine tuned the work of the anthropologists. However, the claim by Allan Wilson and his colleagues that all present-day humans trace their ancestry to one woman who lived between 150,000 and 250,000 years ago in Central Africa has caused a stir, to say the least. The controversy will not die down soon.

If Wilson is correct, the genes of *H. erectus*, *H. sapiens neandertalensis*, or of any other putative human ancestor do not exist within modern human gene pools. The group to which this first woman belonged may have become isolated from or defeated the other forms of *Homo* without exchanging genes with them. Furthermore, this group must have populated the entire planet during the past 200,000 years without exchanging genes with the extant Homo types in Asia, Europe, and Africa. Even more surprising, if Wilson is correct, is that the variation within our species today must have evolved during the past 200,000 years, beginning with a very small population. None of these consequences of Wilson's discovery is impossible, within the time constraints suggested; the surprise was that these consequences are quite different from those derived from the anthropological evidence, from which we should expect a genetic connection among modern *H. sapiens*, so-called archaic *H. sapiens*, and *H. erectus*.

(Richard Leakey discovered two skulls in Ethiopia, one of which is identical to modern *H. sapiens* and one of which is in a class of fossils called archaic *H. sapiens*. The two skulls came from the same location and are exactly the same age, more than 100,000 and less than 130,000 years old, and they have the same cranial capacity, slightly more than 1,400 cc. The "modern skull" is essentially identical to skulls found in Europe that date between 25,000 and 35,000 years ago, called Cro-Magnon. Leakey's discovery can be interpreted to mean that modern and less modern peoples lived at the same time and in the same location.)

Mitochrondria are inherited through egg cells, and all of us inherited our mitochrondrial DNA from our mothers. During the meiotic events that lead to the formation of egg cells, mitochrondrial "chromosomes" do not exchange segments by crossing over. Therefore mtDNA is passed from generation to generation intact. The only changes that occur in mtDNA through time are mutations - a change in a base pair, or the loss of a base pair. Usually more serious mutations do not survive, although a few have, one of which is found in orangutan, a small southeastern Asian and a small African population. Mutations are rare events, and they do not change mtDNA very much or very fast. It is estimated that between 1% and 2% of the base pairs change during a time span of 1 million years.

If you imagine mutations ticking away like a clock, but much slower, it is possible to imagine telling time by counting the numbers of mutations that have occurred between two time periods. A mother and her children will possess identical mtDNA molecules; a mother and her grandmother will possess identical mtDNA molecules. Indeed, the 18,000 Amish living today in Lancaster, Pennsylvania, probably possess identical mtDNA molecules. But the people living in Tibet may possess mtDNA molecules that are different, by a few mutations, from the mtDNA of the Bantu people from Southern Africa.

It turns out that there is less than one half of 1% difference among mtDNA molecules within the entire human population. According to the mutation clock, this means that the world population of *H. sapiens* is very young. If 1% to 2%

of the base pairs change every 1 million years and if human populations show between 0.18% and 0.4% difference, than all modern humans arose from a population that is less than 400,000 and more than 100,000 years old. The speed of the clock, 1% to 2% change per 1 million years, is calculated from an estimated starting time (e.g., the 5 million years at which chimpanzees and the australopithecines split into different evolutionary trajectories) and proceeds toward the present (e.g., the mtDNA differences found between modern chimps and modern humans, close to 7% difference.)

The mtDNA mutation clock ticks relatively fast, compared with mutation rates in nuclear DNA. But what is difficult to visualize in the mind's eye is the meaning of 200,000 years, or of 1 million years. If we put the history of the earth into a 24-hour time frame, each hour in the time frame represents roughly 200 million years; each minute represents 3.3 million years, and each second, more than 55,000 years. This is to say that the woman who appears to be our "mitochrondrial mother" lived less than 4 seconds ago; the mutation clock has been ticking for 4 seconds, but since it started ticking, two ice ages have come and gone, neandertals have come and gone, and the human population has doubled about 22 times, the last two times since I was born, slightly more than one thousandth of a second ago (Figure 4.16). The final verdict on the mtDNA method of searching our past is not in, but if you can enjoy the show without fretting over the outcome, the next few years of debate about Eve will get a "two thumbs up" rating, possibly in both drama and comedy.

There are two major factors that contribute to heated controversy within the scientific community: (1) as the popularity of a field of study increases, so does the competition among scientists to come up with a final or at least an unusual answer; and (2) as the competition increases, so does the fierceness of fighting among the proponents of different answers. Add to this the self-evident truth that the less known about a subject the more room there is for conjecture, be assured that we have a ready recipe for confusion within the broad and general field of study known as human origins.

The study of human origins is immensely popular, and of the many different approaches to the study of human origins, fossil evidence is by far the most popular. These days the fossil evidence grows rapidly. The numbers of human and prehuman fossils are large today. But still, compared to the sheer bulk and relative ease of obtaining molecular genetic evidence, there is a paucity of fossil evidence. Every new fossil generates conjecture, and every new conjecture generates controversy among those whose previous conjectures are threatened by the new ones. The beat goes on. Within the fracas, however, there is general agreement. The earliest primate fossils date back about 65 million years, and since then speciation among the ancestors of the earliest primates has given rise to modern primates. The details of the events are being refined, but not always within an atmosphere of refined behavior.

Between 30,000 and 2,000 years ago the breeding populations of modern humans were probably more isolated from one another than are contemporary breeding populations. Between-continent populations were more isolated from one another than were within-continent populations, and as a result, differences between gene pools emerged. The subcontinent which today includes China, southeast Asia, and eastern parts of the U.S.S.R. is mostly populated by people who have black straight hair, double-fold eye lids, and skin color ranging between white and tan. Much of Africa is populated by people with black, tightly curled hair, tan to dark brown skin, and a wide range of body types and heights.

One hour = 200 million years
One minute = 3.3 million years
One second = 55 thousand years

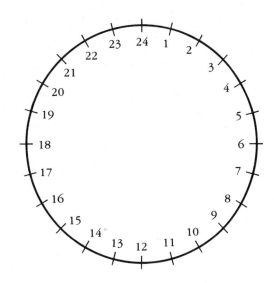

"Time" elapsed since...

WW II	0.0005 sec
WW I	0.001 sec
Civil War	0.002 sec
Magna Carta	0.015 sec
Socrates	0.05 sec
Agriculture	0.2 sec
Mitochrondrial Eve	4 sec
Dinosaur extinction	20 min

Figure 4.16 If the age of the earth is framed in a 24-hour clock, the dinosaurs became extinct about 20 minutes ago.

Europe is populated by people with straight or wavy, light colored hair, a lighter skin color, and blue eyes. The frequencies of many alleles have diverged, sometimes markedly. But, the bottom line is that these major populations carry the same alleles; only the frequencies of the alleles differ.

Studies of human populations during the time since Darwin have provided us with a stockpile of information that exceeds his many fold. But with his information he generated a theory of the origin of species that in turn generated the information gained since, the bulk of which now exceeds his, possibly by orders of magnitude. What causes many biologists, including myself, to stand in awe of Darwin is the fact that his theory has remained pretty much intact during the whole time, that is, it has passed the tests administered to it by all of the information gained since he proposed the theory, far more information than went into making it. If there is a basic, devastating flaw in Darwin's theory, why hasn't someone discovered it? This is not to say that Darwin's theory hasn't been fine tuned by the information and ideas generated since he proposed it; it has, but the general framework remains the same.

Another interesting aspect of the evolution drama involves the word *race*. Whether dressed in its vulgar white hoods and burning crosses, or in its most sophisticated evening attire, the word race has not added so much as a jot or a tittle to our understanding of our species, of past or extant human populations, of family differences, or of differences within families. The question, "Did populations of *H. sapiens* become sufficiently genetically divergent, at any time during the history of the species, as to have developed into different races?" is unanswerable. Would that we could drop race from our vocabulary.

Some of the impetus to classify people into different races comes from a desire to explain cultural, ethnic, and behavioral differences among groups of people. But this desire often is at odds with reality. In the United States, ethnic differences are maintained even after the biological differences said to have given rise to them have dissolved. For example, some 30% of the people who are classified as belonging to the "African race" carry more genes from the European gene pool than from the African gene pool. In the United States, persons with any of the physical features of African ancestry are classified as belonging to the "African race," even if 90% of their genes are from a European gene pool.

Furthermore, people classified as African American are segregated from the other "races" by social policies, not by allele frequency differences — the same policies that have been used to keep members of the "African race" out of the mainstream of "white" societies for 500 years. It is segregation, not genes, that dictates the assignments of people to ethnic groups. But once the ethnic groups are formed and begin to function as social entities, the prosegregationists will begin to advertise that the ethnic groups differ from one another genetically, not socially. The prosegregationist argument falls on its face, however, if it is contested with facts: The skin color of ethnic African Americans varies between black and white. African Americans possess genes from every continent and not because they went out looking for them; the genes came to them in very brutal ways.

The biological criteria used to define races are qualitatively different from the socio–cultural criteria used to create and define ethnicity. Since ethnicity is real and race fictional, the only social function that can be served by "race" classification is that of perpetuating segregation policies, possibly the strongest foe of democracy ever devised within societies that espouse democracy.

Summary

Charles Darwin wrote a book about the origin of species that was published in 1859. The book was widely read during two decades following its publication, but eventually the numbers of its readers dwindled. The book is still widely discussed, but only rarely read. Darwin's book was not written with the intent of insulting Christianity any more than Galileo's descriptions of the solar system were undertaken to insult the Pope. Rather Darwin examined a large number of biological facts and from these facts generated a theory that best explained them. His theory came to be called the theory of evolution.

The theory of evolution is but one of a number of scientific advances that led to splits between science and religion. Sixteenth century astronomers were excommunicated from the Church for suggesting that the earth spins around the sun. Einstein's theory of relativity was a jolt to orthodoxy until the emergence of the atomic bomb. Today creationists are working overtime to denounce evolution and to remove its teaching from the public schools.

At the same time, the theory of evolution explains a greater number of biological facts than does any other generalization in biology. Darwin's main thesis came out of the observations that more progeny are produced than will reach adulthood, that there are heritable differences among the progeny, that some of the progeny are more fit (reproductively successful) than others, and that, thereby, populations change through time by survival of the fittest. Natural selection was said to be the motor force of change.

Darwin did not understand heredity. It remained for 20th century geneticists to discover how Mendel's observations fitted into the grand theory of evolution. Geneticists proposed that breeding populations are the major units of evolution. Through mutation, genetic variation arises, and through differential fitness the genetic composition of breeding populations changes through time. Examples of gene pool differences between human, breeding populations are known, but only rarely is it possible to provide meaningful explanations of the prior histories of these populations that account for the observed gene pool differences. However, natural selection does act to eliminate some alleles from gene pools and to preserve others; migrations (gene flow) break down isolation barriers between breeding populations, and genetic drift explains some allele frequency differences among small populations. The degree to which neutral mutations contribute to genetic variation within gene pools is still being debated.

Quantitative genetics arose in response to continuous variation within populations; it is not helpful to the study of individual genomes. While phenotypes of populations are influenced by many contrasting forms of phenotype attributable to contrasting forms of Mendelian genes, the more interesting population phenotypes are those that vary continuously, for example, height, weight, reproductive potential, and scores on tests of performance, physical and mental. An objective of quantitative geneticists is to partition the population variance into its genetic, environmental, and other components. So far this has not been accomplished within human populations.

The word race is an erroneous concept that has guided the division of the human species into subpopulations for more than 200 years. The concept had been developed and set prior to the discoveries of modern population genetics and evolution; the concept cannot be fitted into this modern knowledge; it has no value as a biological concept. From what is known from the fossil record of human evolution, there is strong evidence for the evolution of breeding populations but only weak evidence for the evolution of Homo-like species. (It is difficult to define species without knowledge of reproduction barriers.) There also is evidence that the isolation barriers of the early breeding populations have been broken and that most modern gene pools are mixtures of the earlier gene pools.

However, the recent approach to human origins by way of mtDNA differences among modern human populations indicates that modern humans have no genetic connection to any of the Homonid types that lived prior to 200,000 years ago. The mtDNA of contemporary human groups is surprisingly similar, and at the estimated rate of mtDNA change via mutation, none of the mitochondrial chromosomes examined could have been more than 400,000 years old, and some of them may have been as young as 100,000 years.

Study Problems

1. Imagine a debate between a creationist and an evolutionary biologist, and then write a brief summary of the main thesis of each.

2. If the *A* and *a* alleles of the A gene are in a Hardy-Weinberg equilibrium and if their frequencies in the population are 0.9 and 0.1, respectively, what percentage of the population members is expected to be heterozygous? Explain.

3. If a large population is isolated from the forces of natural selection, mutation, migration, and drift, to what extent, if any, will genetic recombination change the frequencies of the alleles within its gene pool? Explain.

4. Explain how new mutations may enter and become established in a gene pool in the absence of the "forces" of natural selection. Do your explanations rule out natural selection as a "driving force" of evolution? Explain.

5. Do mutations play an important role during the course of evolution in bringing about genetic differences between breeding populations within the same species? Explain. What is the major role played by mutations in evolution?

6. If a new mutation is absolutely recessive to its common allele (as measured by the fact that AA and Aa genotypes are phenotypically identical and reproductively successful) but in the homozygous recessive state is absolutely lethal (no aa genotypes live to reproduce), how does natural selection act to eliminate the mutant allele from the gene pool? Can natural selection determine that the mutant allele is potentially harmful while it is residing in a heterozygous individual? Explain.

7. Staying with the last question, if one of every 10,000 babies born is aa, what is the predicted frequency of the two alleles in the population? What fraction of the population is predicted to be heterozygous? If one of every 100 babies born is aa, what is the predicted frequency of the two alleles? and of heterozygous individuals? If in both populations all aa individuals die before reaching the age of reproduction, explain how the two populations might have evolved such strikingly different frequencies of these two alleles.

8. Explain why it is relatively easy to define species and difficult if not impossible, to define human races. If human races do exist and if we were asked to provide the very best biological definition of them, what would be the main points of that definition?

9. Explain the main differences between breeding populations and ethnic groups in the United States.

10. Explain why it is difficult if not impossible to make reliable estimates of the heritability of continuous phenotypic variation in human populations.

11. Can norms of reaction be determined by studying phenotypic differences between pairs of MZ twins? Explain.

Suggested Reading

1. Banfield, E. C. 1970. *The Unheavenly City*. Boston: Little Brown & Co. A culture-determinist explanation of inner-city poverty and crime.

2. Boch, K. 1980. *Human Nature and History: A Response to Sociobiology*. New York: Columbia University Press. An easy-to-read expression of the importance of human history to the study of human origins.

3. Cummings, M. R. 1990. *Human Heredity: Principles and Issues*. 2nd ed. St. Paul, MN: West Publishing Company.

4. Darwin, F. 1892–1958. *The Autobiography of Charles Darwin and Selected Letters*. Mineola, NY: Dover Publications.

5. Dawkins, R. 1990. *The Selfish Gene*. 2nd ed. New York: Oxford University Press. This book argues that the gene is the unit of selection, that genes struggle for survival, and that phenotypes are programmed by them to achieve survival success. This concept is controversial among evolution biologists.

6. Futuyma, D. J. 1983. *Science on Trial*. New York: Pantheon Books. This book explains the ideological warfare that has persisted since Darwin between evolution as science and creationism. This book was written for the nonprofessional.

7. Haraway, D. 1990. *Primate Visions: Gender, Race and Nature in the World of Modern Science*. New York: Routledge. This book discusses more than the history of primatology and more than the history of primates. It also discusses the evolution of ways we think about ourselves. One of the most engaging books describing how and why we explain ourselves the way we do.

8. Hirsch, J., McGuire T. R., Vetta, A. 1980. "Concepts of behavior genetics and misapplications to humans." In: Lockard, J.S., ed. *The Evolution of Human Social Behavior* New York: Elsevier. This book is a bit more technical than the text, but its explanations of the misapplications of genetics to human behavior are very readable.

9. Leakey, R. E., and Lewin, R. 1977. *Origins*. New York: E. P. Dutton. This is a fascinating explanation of human origins. The book is more than a history; it examines what it means to be human, and how this meaning might have coevolved with human biology.

10. Lewin, R. 1984. Human Evolution: An Illustrated Introduction. Oxford, England: Blackwell Scientific Publications.

11. Lewontin, R. C. 1982. *Human Diversity*. New York: Scientific American Library. This is a readable account of the diversity of human types in which the significance of between-population and within-population diversity is explained in an easy-to-read format.

12. Lewontin, R. C., Rose, S.P., and Kamin, L. J. 1984. *Not in Our Genes*. New York: Pantheon Books. This is the best critique of "biological determinism as ideology" yet to be published. Informative, provocative and easy to read. (Available as a paperback from Pelican Books.)

13. Nelson, H., and Jurmain, R. 1988. *Introduction to Physical Anthropology*. St. Paul, MN: West Publishing Co. See Chapter 4, pp 96-109; and Chapters 5, 7, 9, 12, 16, and 17.

14. Smith, J.M. 1975. *The Theory of Evolution*. 3rd ed. Middlesex, UK: Penguin Books. This book was written for nonspecialists as well as for specialists. It goes beyond the genetic aspects of evolution to provide nature stories at their best as evidence for the many aspects of evolution that not included in this text.

15. Wills, C. 1989. *The Wisdom of the Genes: New Pathways in Evolution*. New York: Basic Books. This book provides nonspecialists with an interesting perspective of a new way of looking at the genetic aspects of evolution. Very well written, fun to read, informative.

5

Evolution, Individual and Social Behavior

Heredity and Society

This book is entitled *Human Heredity and Society,* but society is not discussed in detail in comparison to the discussions of heredity. We have not addressed the question "What is society?" in the way we have addressed the question "What is a gene?". Indeed, the discussions of heredity and society have been one-sided in favor of heredity.

The history of science teaching explains why it is one-sided in favor of science and opposed to including discussions of society. Many scientists view science as a search for objectivity that proceeds best when society does not intervene; many assume that the properties and functions of atoms and genes are independent of the opinions and politics that propel societal activities. The view that societies and science dance to different beats dictates that science must be taught as science, not as a societal activity.

While many geneticists feel this way about science and society, it is becoming increasingly difficult for them to resist teaching genetics outside the context of society, given the fact that genes are being called upon to explain so many of the characteristics of human societies. But now that genes are being invoked as causal determinants of the characteristics of societies and **cultures**, many geneticists who wish to teach genetics as a societal activity are beginning to realize that their knowledge of human societies is weak compared with their knowledge of genetics.

For example, if we could see into human embryos to determine the sex chromosome composition of each, we could predict rather accurately which ones will develop into females and which into males. And even without the precision provided by knowledge of genomes, we can draw rather accurate outlines of how embryos develop into adults and how different kinds of mutation lead to different kinds of developmental mistakes.

We possess no comparable knowledge of human societies. We have not witnessed the developmental history of any human society. Phrases like "societal illness" and "societal mutations" are simply metaphors. Yet even in our ignorance of society, we can agree that legal systems and religious practices play important roles as influences upon interactions among individuals within societies. At the same time, societal behavior cannot be reduced to a genotype or to any other unit of genetic information; the information that influences societal behavior is qualitatively different from that which influences eye color, hair texture, and the amino acid sequences of proteins. The theory of the gene is not a reliable base from which to formulate scientific models of human society. (Recall from Chapters 1, 2, and 3 that it is relatively easy to demonstrate the phenotypic outcome of a single gene but that it is as yet impossible to demonstrate, at a comparable level of precision, the phenotypic outcomes of concerts of genes [i.e., whole genomes].)

For these and other reasons it is important to learn how to evaluate the alleged evidence that genes inform the structures of societies as they do the structures of cells. Many biologists have tried to make analogies between biological and societal evolution, but before we pass final judgment on these analogies, we should know at least as much about societies as we do about biology. This chapter addresses two kinds of attempts to expand the explanatory power of genes: the first is the abstraction called intelligence and the second is the structure and evolution of human societies and cultures.

A Tradition of Biological Determinism

The view and the social practices born out of the view that individual and societal behavior are direct outcomes of genetic programs are called **biological determinism**. There has never been a time since Darwin in the history of the United States and Western Europe when biological determinism has not been a major, if not dominant, ideological force, that is, a force driven to rationalize certain social practices, school and university curricula, and the economic or political success of one "race" over others. The cliché is "biology is destiny."

Darwin focused his first book about evolution on the origin of species. This book was about the history of biology and his new theory of how natural selection explains biological change through time. Later he wrote *The Descent of Man*, a book about the origin of the human species in which he attempted to describe the human species as accurately as he had described other animal species in his first book. To do so it became necessary for him to include human behavior, clearly one of the most fascinating attributes of the primate *Homo sapiens*.

Before psychology became a formal, intellectual discipline, human behavior was often used as an indicator of a person's worth, race, social standing, and even potential for success within social structures dependent upon divisions of labor. Mean-spirited behavior bespoke a mean-spirited mind; industrious behavior foretold success. Each category of behavior reflected a category of societal performance, occupation, or talent, and each was viewed as the outcome of a particular determinant, whether the grace of God or biology.

Since these descriptions of behavior often did not distinguish between individual and group behavior (i.e., national, ethnic, racial, or species behavior), the descriptions were difficult to fit into the theory of evolution. (Recall that breeding populations are the units of evolution and that the theory of evolution is better at explaining the genetic composition of populations than the genetic composition of individuals.)

Anyone who has carefully watched monkey, chimpanzee, or gorilla behavior has been rewarded with enjoyment, and possibly with the disturbing realization that the similarities between "them" and "us" are too close for comfort, except that their manners are less inhibited than ours. After careful observation we may begin to question why primates behave as they do, and we may notice that the different primate species can be recognized by specific forms of behavior. After a while we may begin to recognize relationships between species-specific behavior and species-specific morphology. For example, monkeys chase one another through the trees at break-neck speed, and they are built to do so. Gorillas, on the other hand, stay on the ground and do a lot of sitting. We may conclude, even without help from experiments that *the biology of higher primates is a determinant of their behavior.*

After at least one course in psychology, we will have have been primed to notice that monkeys and apes appear to have different states of consciousness, somewhat as humans do. One moment they are asleep, the next moment one is chasing another, and a moment later the chaser is grooming the chased. Their reaction to pain "looks" like ours feels. Higher primates learn quickly, remember well, and modify their behavior with learned information. If there are differences among primate species, the differences may be quantitative, not qualitative, with the possible exception of language. Emotions motivate our behavior, and to the extent that emotions motivate other primate behavior, we will conclude that other primates are intelligent. But if we attempt to scientize our feelings about intelligence we discover that intelligence is as difficult to affirm in chimps as it is in humans (Figure 5.1).

After reading Mendel and Darwin, our tendency will be to ascribe the phenotypic differences to allele differences and the similarities to a common origin. Then, once the subject of origins grips our attention, we will want to know whether Darwin's biological theory of evolution explains the history of primate

Figure 5-1 How is intellegence defined, measured, and affirmed?

behavior as it does the history of primate physiology, metabolism, and morphology.

However, upon taking a first course in sociology we will find striking differences between other primate **social behavior** and human **societal behavior**. Parent chimps may teach baby chimps how to retrieve termites from a rotting log with a wet straw, but it is far from clear how chimp "social" behavior is related, if at all, to universities, libraries, computers, and other complex means for transferring information from one generation to the next within human societies. The question of whether societal behavior is uniquely human is discussed in the last half of this chapter.

Human individual behavior may or may not turn out to be related to genes in the way that metabolism, physiology, and morphology are. We simply do not know yet. Behavior cannot be inferred from fossil remains or from the artifactual remains of early populations that failed to leave written records, as it can be, for example, from the writings of the early Greek civilizations. It is possible to infer a great deal about the biology of ancient peoples from their fossil remains, but not about their behavior except in a very general way from artifacts and tools; the problem is that we will never know how close our inferences match reality.

While there is general agreement, then, about Darwin's theory of evolution (including the additions to it that have been made since 1859), there is little agreement about the evolution of human behavior. A major point of controversy centers on the question, *How much of human behavior is determined or influenced by genotypes?* Biological determinists answer this question differently than do historical and societal determinists.

If it were easy to demonstrate cause-effect relationships between genes and behavior, analogous to relationships between genes and the steps in a metabolic pathway, the task would have been completed long ago, since it is clear that more people are curious about behavior than about metabolism, and just as clear that there were more prominent 19th century psychologists than 19th century geneticists.

There are many explanations for the popularity of the first psychologists and for the obscurity of Mendel, both during their lifetimes and during the first half of this century. In their beginnings, psychology and genetics were very different subjects employing very different kinds of scientific methods of study. Mendel's approach to the study of genes was to create pedigrees by engineering specific matings, and to let the results speak for themselves. Mendel's scientific approach to the study of genes forbade him from altering the reproductive processes of peas; the only way he could have gotten false results would have been for him to deliberately alter the numbers.

Sigmund Freud, the primary architect of **psychoanalytical theory**, and John B. Watson, the founder of **behaviorism**, had the difficult task of generalizing from particulars, which consisted mainly of their own observations of human beings. It was impossible for them to design experiments using human beings as subjects that would allow a physical process to take its course without interference from the investigator, as Mendel was able to do with the reproductive processes of peas. The conclusions reached by these psychologists were as much the outcome of their intuitive powers as of the behavior of the people observed by them. Freud's intuition, for example, was rather counter-intuitive for most of his contemporaries and even for many of us today. For example, he claimed that humans do not possess as much conscious control over their behavior as they may imagine; rather, a biologically influenced unconscious mind lies behind most behaviors. What he tried to do was to analyze the motives and drives of the unconscious, and this came to be known as psychoanalysis. His explanations were spellbinding, and they have fascinated generations of psychologists and non psychologists. Many people have read Freud; only a few have read Mendel. But Mendel got it right, and it is being debated to this day whether Freud was even in the right ball park.

Not only did Freud and Mendel explore vastly different research subjects, they employed very different methods of investigation. As one would predict, differences as great as these generated different opinions about the validity of their conclusion and about the application of their conclusions to human biology and psychology. From these opinions it is relatively easy to sketch a "field of battle," and the battle plans of those who concluded that behavior is biologically determined, partly biologically and partly experientially determined, or wholly experientially and historically determined. It turns out that none of these conclusions is overly endowed with "big guns" (facts); thus many of the battles are fought with hot air balloons. Very simply, there are few if any agreed-upon explanations of human behavior.

For that matter, there is disagreement about the taxonomy of behavior, i.e., about the different kinds of behavior and the definitions of these behaviors. What is **aggressive behavior**? What is **shyness**? It turns out that behavior is very complex, almost beyond imagination. The history of behavioral explanations is understandably filled with theories that were too simplistic for the task and that suffered from a lack of unambiguous data. It is not that psychologists are less bright than geneticists and physicists; it is that the subject they are curious to know more about is devilishly more complicated than genes and electromagnetic forces.

Genetics is catching up with psychology in the public opinion polls, however, partly because genes are now being asked to explain behavior. During the latter half of the 20th century genes have emerged from public obscurity to headline news. After DNA came the "universal language of biology," then came a technology that permits the isolation of genes and the transfer of genes from one species to another, the synthesis of genes in the laboratory, and now a "chain-reaction" method for making millions of copies of DNA fragments "during coffee break." We've moved from blue jeans to designer genes during the same period of time that we moved from motor-propelled airplanes to space travel. Today within every modern field of biology there are biologists trained to do genetics research in pursuit of that field's mainstream questions.

Of probably greater social importance, the explanatory power of genes has invaded fields of study that a few years ago seemed "light years" away from mainstream genetics — neurobiology, psychology, sociology, anthropology, and even economics. Today intellectuals in many disciplines are manufacturing genetic explanations for the kinds of "phenotypes" they observe in their corner of the universe — shyness, intelligence quotient (IQ), creativity, aggression, sexual abuse, war, territoriality, altruistic behavior, and criminal behavior. The logic of biological determinism tells us that if humans are capable of doing something, they have their genes to thank or to blame for it.

As the popularity of the gene grows, so do demands on it for more and better performances. As with other celebrities, the gene has been endowed by its fans with talents it does not in fact possess. It is sometimes surprising, sometimes embarrassing, but almost always flattering to geneticists to learn from others — whether theater critics, sports analysts, or "movers and shakers" — that genes possess an impressive repertoire of determinist powers. (For example, Donald Trump, with characteristic modesty, credits his genes for his economic success, just as a century ago John D. Rockefeller credited evolution for his financial empire.)

In these situations it is difficult to distinguish between dangerous misuses of the explanatory power of genes and what instead is simply harmless speculation about what role genes play in forging health, behavior, and culture. The point is that from the geneticist's perspective there are but two experimental approaches to proving gene-phenotype relationships — Mendelian and molecular approaches. One cannot replace these approaches with intuition (gut reactions) without breaking the rules of science. Intuition is useful in designing hypotheses, but experiments and careful observations are necessary for assessing the validity of hypotheses.

Genes and Individual Behavior

Two of the major strategies for sighting genes that influence human behavior arise from (1) the hope of finding a Mendelian gene that causes a specific form of normal behavior such as aggressive or **altruistic behavior**, and (2) the fact that most forms of human behavior exhibit quantitative variation within populations.

If Mendelian-molecular genes do in fact determine normal aspects of behavior, it should be easy to locate, isolate, amplify, and transplant them. If the genes that influence behavior are quantitatively distributed within populations, and if each gene has only a miniscule impact upon phenotype, the only way to demonstrate the fact will be to estimate *the genetic component of the population variance.* This is difficult if not impossible to do with humans. As is discussed below; however, attempts have been and are being made to isolate the genetic component of IQ score variation.

So far, the mutant alleles of Mendelian genes that have been observed to alter behavior exert their effects by altering metabolism, physiology, or morphology. Most of these kinds of studies are done with small animals whose behaviors are relatively simple and whose chromosome maps are well known (Figure 5.2). There are analogous mutations within human gene pools as well, mutations that alter behavior by altering metabolism, physiology, or morphology (e.g., Huntington's disease, phenylketonuria, and sickle cell anemia). Mendelian genes that alter behavior in this way tell us nothing about relationships between genotypes and normal behavior other than the obvious, namely that normal behavior and normal metabolism, physiology, and morphology are positively correlated.

Mendelian genes have not been discovered to encode normal behavior, nor has the behavior variance within human populations been partitioned into its genetic and nongenetic components. A third problem arises from the fact that behavior is very difficult to define unambiguously throughout wide ranges of its variation. Even textbook behaviors such as aggression, altruism, and shyness are difficult to define, especially as they are expressed differently in different human cultures and at different stages of development in the same individual. From the perspective of the geneticist, success in sighting genes is absolutely dependent upon success in sighting (and defining) phenotypes.

What is Behavior?

Physicists will refer to the behavior of atoms, chemists to the behavior of molecules, and geneticists to the behavior of genes and chromosomes. The meaning of behavior, used in these ways, is "what things do." Atoms, molecules, and genes do things; so do organisms, including ourselves. But the explanations of atom behavior are qualitatively different from explanations of organism behavior. Thus, while physicists and geneticists share the common desire to know what atoms and genes do and to discover the causes of atom and gene

Drosophila melanogaster

The dark and light areas symbolize a mosaic condition; the dark areas represent female cells, and the light areas represent male cells. The female cells carry two X chromosomes and the male cells carry one (XO flies are male). Mosaic flies arise from a strategy of including one normal X chromosome and one X shaped in the form of a ring, called XX-R, in a fertilized egg, and allowing the egg to develop into an adult. Because the X-R chromosome does not behave normally during cell division, it is often lost, leaving a progeny cell with only one X. If the normal X carries a recessive mutation that leads to abnormal behavior ($_bX$) all of the male cells will then be genotype b, and the female cells will be genotype Bb.

The mosaic flies reveal the tissue within which the *b* allele must be expressed for a fly to exhibit the mutant behavior. For example, a mutation that leads to premature death in adult flies must be expressed in brain cells. The mutant in any other body tissue or organ has no effect upon age of death. Another mutation causing flies to hold their wings high over their heads leads to the expression of the mutant gene only if the gene resides in the thoracic muscle cells.

So far, all so-called behavioral mutations lead to alterations of metabolism, physiology, or morphology and then to behavioral changes.

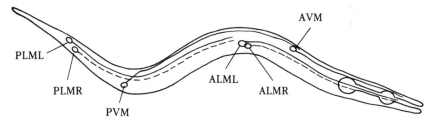

Caenorhabditis elegans

A worm (nematode) about 1 mm long. The worm is composed of 959 somatic cells, 302 of which are neurons. The worm is transparent, permitting the observation of every cell by use of a light microscope. The number and position of every nerve cell is the same in every normal nematode; therefore every mutation that alters number or position (or morphology) of nerve cells can be identified by phenotype. In addition, some of the physical changes to neurons result in behavioral changes.

For example, the six cells identified in the diagram are touch receptors. All six have an easily recognized morphology. Mutations at 12 gene loci are known to alter the worm's response to touch and to alter some morphological feature of these six touch receptor cells and/or the processes (neurons) attached to them.

Many mutations are called *uncoordinated* (*unc* for short); all are either missing muscle cells or neurons or possess morphologically different muscle cells or neurons.

Figure 5-2 a) Mosaic flies are engineered to show the tissue within which a mutant allele must reside in order to program altered behavior. b) In nematodes, mutations have been observed to alter behavior by first altering metabolism, physiology, and nerve and muscle cells.

behavior, they will use very different methods of investigation and they will discover very different causal relationships. Chapter 2 was mainly about what genes do. This chapter touches on the subject of what people do.

Psychology is the study of what people do. The task of the psychologist is not easy, and as a result many models of behavior underlie the research methods that have been employed for studying human behavior. Central to one model (after John B. Watson) is the view that individual human behavior is a window through which the brain can be observed to function—loving behavior is evidence of a loving mind, and criminal behavior, a criminal mind.

Another model assumes that by studying animal behavior and from this learning about animal brains, we can thereby learn about the human brain (Boch, 1980, p 7). One criticism of this approach is that human behavior is more complicated than animal behavior, which is to say that knowledge of other animal behavior is of little value to the understanding of human behavior. A further complication stems from the difficulty of separating so-called **innate behavior** from **learned behavior**, since it is a fact that more of human than of other animal behavior is learned. Counter criticisms assert that it is a false issue to separate innate from learned behavior. For example, one such model assumes "that the development of behavior, from birth, is a continuing interaction between, and integration of, organisms and their environments." Since behavior emerges out of the same developmental processes that give rise to adult bodies, "behavior can no more be partitioned into innate and learned fragments than can our muscles and bones. . .the ways in which genes and environments interact are far too complicated for that" (Slater, in Halliday and Slater, p 82).

These arguments illustrate the complexity of the problem. For example, the claim that human behavior cannot be understood through comparisons with animal behavior is interpreted to mean that humans are not animals and that humans do not "dance to the beat" of DNA. Many people do in fact feel that human behavior is "other-worldly" and not integral to our being animals. But this is not the feeling of many who study human history and who see history as having an influence upon what people do and think. As Boch says, "To whatever extent other animals might have histories, it is obviously much more the case with humans. This is the feature of human existence that is the focus of humanistic inquiry. Whatever light biology might come to shed on humanity, it would be tragic if the historical dimension of human life were left unstudied" (ibid, p 2).

If behavior develops as bodies develop, "as a continuing interaction between, and integration of, organisms and their environments" and is made even more complex by history, then developmental behavior is far more complex than some simplistic models assume, such as those that observe criminal behavior, deduce a criminal mind, and attribute both to criminal genes.

How do we identify, define, and study behavior that is influenced simultaneously by human history, by individual experiences during development, and by our impressive repertoire of biological capabilities? Is it meaningful or misleading to approach the study of behavior believing that each specific form of behavior is the end product of a cascade of cause-effect events triggered by genes?

Let us try to address these questions to a specific behavior, that called gratification delay. It has been alleged that inner-city poor people are poor because they are unable to delay gratification (Banfield, 1970). In other words, their tendency is to spend their money as fast as they get it. Saving is foreign to them, it is alleged, and since this behavioral pattern is uniform among poor people, it must have a biological base. The contrasting "phenotypes," then, are posited as the inability and the ability to delay gratification. As cautioned many times in earlier chapters, contrasting phenotypes are insufficient evidence for contrasting alleles.

The evidence used to support the genetic model of delayed gratification behavior was obtained from comparisons of inner-city poor children with middle-class suburban children based on their responses to the following proposition. Each child was offered the choice whether to get one candy bar that day or two candy bars the next week. The majority of the inner-city children choose to take one candy bar at the time it was offered, and the majority of suburban children made the decision to wait to get two candy bars the next week. These results were believed by some to settle the matter. The reason poor people are poor, they argued, is because they are unable to delay gratification, that is to plan for the future or to understand that "a penny saved is a penny earned."

However, this "experiment" was later repeated exactly but with one added feature. The suburban children, as before, opted to receive two candy bars the following week. When the time came for them to collect, however, the experimenters met with them and said that they had been unwise to accept such an offer without a certified guarantee. They explained that the inner-city children were smart to take their candy when it was offered. Follow-up experiments on the same suburban children and on other children from the same neighborhoods showed that many of the children modified their behavior to match this particular experience. The propensity to delay gratification was changed by a single experience and without help from genetic mutations. (A more meaningful approach to the observed differences between inner-city and middle-class children as measured by gratification delay might have been to study the contrasting social experiences of the children and to discover which of these taught them to delay or not to delay gratification.)

Indeed, the observations that were made may tell us more about the behavioral scientists who decided that gratification delay is a specific aspect of behavior than it does about the behavior of the children who became their subjects. Scientists who approach their experiments with a clear view of the answers they want, e.g., biological proofs of what constitutes normal behavior, can design experiments "that work." This approach, wittingly or unwittingly, does little more than to prop up the view that historical, religious, and moral philosophies, originally designed to both define and prescribe normal behavior, are biological imperatives (laws). It is not uncommon for parents to believe that their children *ought* to save their money and take care of their toys and clothes, and as judged by societal rules and mores, it is not wrong for them to so believe. What is wrong is for professionals to transform moral *oughts* (to save) into biological "realities" (that saving versus not saving is a one-gene, two-allele phenotype).

Human behavior is not simple. Indeed, it is a far more complex aggregate phenomenon than molecular genetics. If individual behavior is an aggregate of prescribed, learned, and biologically based behavior, it will be more difficult to unmix the whole of it into discrete components of behavior than it is to unmix a cup of coffee and retrieve the cream, sugar, coffee, and water in their original forms. This complexity partially explains why there is so much confusion at the interface between psychology and biology.

There are many kinds of complexity. Mendelian genetics appears complex to some. Atoms appear complex to people who have acquired fears of mathematics. But the problems raised here present us with a different kind of complexity, one that often is summarized as "the whole is more than the sum of its parts." A cup of coffee is more than the sum of the properties of cream, sugar, coffee, and water. Mixed together these ingredients combine to give rise to a fifth set of properties, a set that cannot be derived by summing the separate properties. Considering an even simpler example, chemists who know very much about the properties of the soft metal sodium (Na) and the inert gas chlorine (Cl) are unable to predict from the separate properties of these two elements all of the properties of sodium chloride (NaCl), common table salt. Even salt appears to be more than the sum of its parts.

If we cannot predict the structure and function of DNA from the properties of carbon, oxygen, nitrogen, phosphorus, and hydrogen, why should we expect to predict the behavior of a child from the little we know about human history and the child's history, biology, and psychology? This is not to ignore either our current knowledge or the creativity and the insights that have spawned that knowledge. The point is to prevent our intuitions from impersonating facts and our abstractions from becoming real things. (See An Aside)

There is more to say about defining behavior and about designing experiments to determine how behavior emerges during development. But enough has been said to start us thinking critically about behavior, especially as a complex of disparate behaviors each said to result from the action of a specific gene.

A Central Form of Behavior Called Intelligence

If you were to read the full text of the history of scientific approaches to the study of human intelligence (see *The Mismeasure of Man*, by Stephen Jay Gould), you might become convinced that the scientists who are searching for it are simply chasing rainbows. But there are striking differences among them. Some put forth biological arguments that attempt to justify the segregation of people into types—African from European, rich from poor, Christian from non-Christian (Herrnstein, 1971). Others argue in support of equal access to self esteem (Lewontin, Rose, and Kamin, 1984), not to be confused with biological sameness. Some scientists claim to have evidence for the existence of intelligence genes and use it to defend the first kind of argument. Others advance the second kind of argument by disclaiming such evidence, rejecting the definition of the phenotype (intelligence) and suggesting that the polemic is more political than scientific.

In terms of what is known about population genetics, the major human populations carry essentially the same constellations of genes. It seems rather clear now that there is more genetic variation within each of the major populations than there is between them (Figure 5.3), which is to say that the differences between populations are quantitative, not qualitative.

This fact illustrates one of the glaring differences between the statistical description of differences between two breeding populations and the argument that the two populations are different races that exhibit qualitatively different biological, mental, or moral characteristics. In the language of genetics it can be

An Aside

Many who have studied human intelligence and tried to devise accurate measures of it have defined intelligence as the ability to think abstractly. What is usually meant by this phrase is the ability to convert a real, physical problem into a mathematical expression. More generally, though, the ability to think abstractly means the ability to create mental images of real things. The mental images must retain a recognizable identity with and at the same time be clearly different from their real counterparts. Abstract painting and music are examples, but those of us who are not artists or musicians may have trouble identifying the real counterparts of the abstractions we see and hear. In science the relationships between abstract images (often called models) and their real counterparts are more graphic, but even so most nonscientists do not grasp these relationships without a lot of practice.

There is a reverse twist to relationships between real things and abstractions of them. Consider the following: within many modern industrial societies there are divisions of labor, some of which are determined by skills, some by age, some by ethnicity, and some by sex. If we study the divisions of labor during the Victorian age (the latter half of the 19th century), we find that women were seldom picked to build railroads or steamships, or to mine coal. Indeed, women were discouraged from studying mathematics. It is likely that many men of that period, when thinking of women (that is, creating abstractions of them) probably endowed them with few if any mathematical skills: "Women cannot think abstractly." Women were encouraged to become house workers, secretaries, nurses, and other kinds of service workers; that is, these social practices grew from and reinforced the abstraction by designating the kinds of societal activities that women were able and therefore *ought* to do.

Years later some women began to question divisions of labor along sex lines. A few women argued that women should study mathematics if they are so inclined. The opposition simply reiterated the abstraction: "Women can't think abstractly." Indeed, the opposition devised mathematics exams to prove their point. And sure enough, females of all ages scored lower than males on a variety of mathematics tests. In short, the abstraction was made real through social practice. Abstractions made real through social practice are called **reified abstractions**. In other words, in our minds we create abstractions of real things and with our societal practices and policies we are able reify some of them. That we are capable of abstracting the real, and of converting abstractions into real things, sets our "mind's eye" apart from the "mind's eye" of all other species. But there is a price to pay for these remarkable abilities. One of the greatest sources of friction within human societies is over proprietorship of the knowledge and of the methodology necessary for distinguishing between abstractions and reality.

Luckily we possess an equally admirable characteristic, the desire to know the closeness of fit between our images of reality and reality. It has been suggested by philosophers and other scholars that the desire to know how close our abstractions of reality match reality had something to do with the development of a scientific attitude, which by one definition is an attitude that propels the search for better and better images of reality.

Probably no other abstraction in the history of psychology and biology has been reified in more different ways by social practices than has our abstraction of intelligence. Intelligence is not a material thing, like an arm or a leg; rather, it is an abstraction drawn from different forms of behavior. Publicly intelligence may signify the ability to get high marks on standardized tests; it may symbolize a complex neurobiological mechanism, e.g., a battery of chemical reactions set in motion by an as yet undisclosed battery of genes. In the vernacular we may say that persons who are good at mathematics are smart, but what about the mathematician who smokes two packages of cigarettes a day? We may say that the student-athlete with a 3.9 grade point average is smart, but is she smart if she becomes addicted to anabolic steroids? It turns out that there are as many uses of the word intelligence as there are for words like love, beautiful, and good.

We can never acquire abstract images of reality that precisely match reality. For one thing, reality is always changing. This fact is a never-ending source of humor among intellectuals. Oxford University, for example, has enshrined on one of its halls the sage advice, "Do not adjust your mind; the fault is in reality."

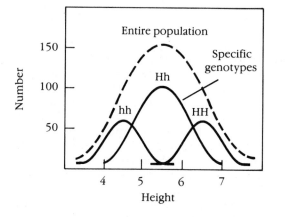

This hypothetical population is made up of many different genotypes. The three genotypes shown have different but overlapping ranges of phenotype, all of which contribute to the population's range of phenotypes.
The phenotypic variation shown here was observed in Environment 1.

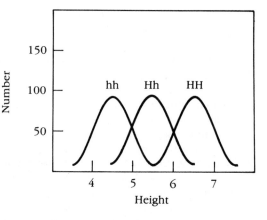

The same genotypes within the same population, but in a different environment, Environment 2.

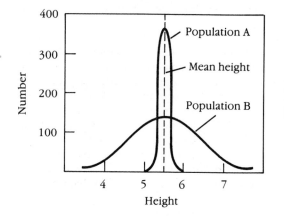

These two populations have the same mean (5.5) but very different distributions of variation (from 5 to 6 for A and from 3.5 to 7.5 for B). That is, the populations are indistinguishable as judged by their means, but many individuals in B are shorter and many are taller than any of the individuals in A.

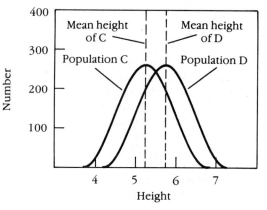

These two populations show identical distributions of variation but different means. Judged by height, the majority of people in both populations might belong to either population. That is, only a few individuals in C are shorter that the shortest in D, and only a few in D are taller that the tallest in C. It is possible then to distinguish the two populations by their means, but not to distinguish the majority of individuals by their individual heights.

Figure 5-3 Variation exists within and quantitative differences exist between related populations.

said that population A differs from population B by allele frequencies at 12 gene loci, and that the allele frequency differences can be specified within narrow limits of error (e.g., the ratio of A to *a* is 0.7 to 0.3). The argument that such populations are different races often takes a very different form; the terms superior and inferior may be used, as well as words such as ugly, beautiful, criminal, law abiding, sexually obsessed, sexually constrained, intelligent, and unintelligent.

This argument may go even further, claiming that these subjective mental and moral characteristics are biologically based, in the same way that eye and skin color are encoded by genes. The Nobel laureate William Shockley asserted that "nature has color-coded groups of individuals so that statistically reliable predictions of their adaptability to intellectually rewarding and effective lives can easily be made and profitably be used by the pragmatic man in the street" (Shockley, 1971). Shockley failed to define intelligence and he failed show how "intelligence genes" can be identified other than by their cohabitation with pigment genes. Not all Nobel laureates use the prestige of The Prize to win others to their private prejudices, but Shockley did. He was not a geneticist but an electrical engineer. He was awarded the Nobel Prize for co-inventing the transistor.

What is Intelligence?

In 1903, Alfred Binet of Paris published the first paper in history describing how to distinguish, among school children who were not learning, those who were able to learn from those who were unable. In other words Binet devised the first diagnostic test for mental retardation. Without comment about how well the test worked, Binet's was a unique and creative contribution to psychology.

Binet selected sets of questions for each chronological age group of children between 6 and 16 years. Each set of questions was selected after **standardizing** each test item upon large numbers of children of specified age groups. For example, the questions standardized for 8-year-olds were selected such that the average 7-year-old answered fewer than 50% of the items correctly, the average 8-year-old answered 50% correctly, and the average 9-year-old answered more than 50% of the items correctly (Figure 5.4). Binet selected test items such that within large populations of children, correct answers and **chronological age (CA)** were positively correlated. Scores on these tests were said to measure **mental age (MA)**.

In 1912 the concept of intelligence quotient, or IQ, was developed by the German psychologist L. William Stern. IQ was defined as the quotient of MA and CA (MA divided by CA) multiplied by 100. This concept of intelligence was incorporated into the famous **Stanford-Binet IQ tests**, developed by Lewis Terman of Stanford University. The structure of these tests included a set of six test items for each age group. The conversion of test results into mental ages was based on the rule that a correct response to a test item equals two months of mental age. For example, a 10-year-old child who provided a correct response to all of the items up to and including the 8-year-old set (8 years of mental age), four of the 9-year-old set (8 months of mental age), three of the 10-year-old set (6 months of mental age), and 2 of the 11-year-old set (4 months of mental age), was said to have a mental age of 9.5 years (8 years + 8 months + 6 months + 4 months = 9.5 years). A mental age of 9.5 years divided by the chronological age of 10 years and multiplied by 100 = 95, the child's IQ. An 8-year-old child with the same mental age was said to have an IQ of 117 (9.5/8 = 1.17 x 100 = 117). The key assumptions here, for which there still is no experimental support, are that IQ test scores measure mental age and that normal intelligence requires that mental age and chronological age remain in sync throughout childhood.

By use of age-adjusted sets of questions, the mean IQ for a population is set at 100, except in Garrison Keillor's Lake Wobegon, where all of the children are

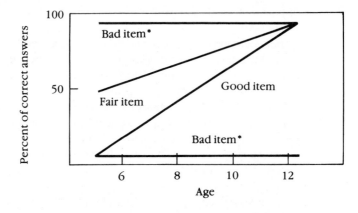

*Bad items are those that all age children answer correctly (What is your name?) or incorrectly (During which dynasty were the Egyptian pyramids constructed?)

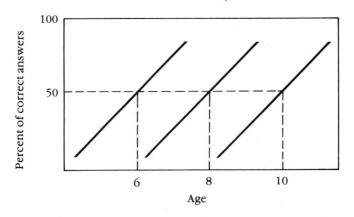

Each line represents one test item selected for one age group.

Figure 5-4 The selection of test items for IQ tests is based on responses to them by hundreds of children; good items are those that show high correlations between the percentage of correct answers and the age of the children.

above average (Figure 5.5). The bell shaped curve that describes the distribution of the population's variation shows that as IQ scores rise or fall from the mean, they do so symmetrically; i.e., there are as many people in the population with an IQ of 110 as with an IQ of 90, and the same for IQ scores of 120 and 80. The majority of children have IQ scores near the mean, while the numbers of children with higher and lower scores become rarer with increasing distance from the mean. A bell shaped curve of a population's phenotypic variation is a classic description of quantitative variation.

In the United States the Stanford-Binet IQ test became a standard by which millions of school children were ranked for "native intelligence." It hasn't been the statistical treatment of IQ scores that lies at the root of the acrimony about IQ, intelligence, and genes; rather, it is the meaning of intelligence and whether IQ tests measure it. Here we find that the instrument used to measure the phenotype at the same time defines the phenotype (intelligence). In the absence of a test score intelligence is very difficult to define. Since Binet's contribution to measuring mental functions, intelligence has been regarded as the ability to make good judgements, to comprehend and reason, or simply the ability to think abstractly. In 1969 Jensen said that intelligence is what intelligence tests measure. In their reference to the book, *What is Intelligence?*, edited by R. J. Sternberg and D. K. Detterman, Nelson and Jurmain state that "we find in the book viewpoints of twenty-four of the foremost

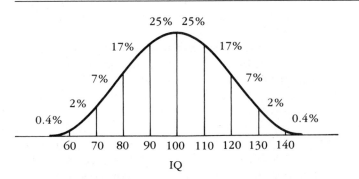

A bell-shaped curve showing the distribution of IQ scores and the percentage of people in Euro-American populations with IQ scores in each 10-IQ-point seqment of the distribution range.

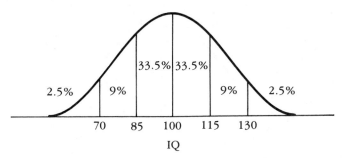

The same distribution range divided into 15-IQ-point segments. Fifteen IQ points is one standard deviation (SD) and is based on the convention that 67% of the population is less than one SD from the mean; 95% are within 2 SDs of the mean.

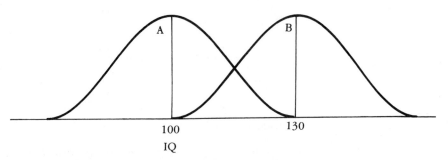

Population A is the Euro-American population in the United States. Population B is the population of children in Lake Wobegon, MN., according to Garrison Keillor.

Figure 5-5 The mean IQ score is 100, as set by the means for standardizing test items; in Lake Wobegon, Minnesota, "All of the children are above average."

experts in the field, and two dozen definitions of intelligence" (See also Gould, 1981).

One of the main criticisms of IQ testing and of equating IQ scores with intelligence is of the belief that intelligence is a discrete and unitary phenotypic character, as are height and weight. There may be many kinds of intelligence; the tests may reflect the cultural biases of those who design them; the children who take them may not be educated equally in test taking skills beforehand, and so forth. Is an IQ test score a rank of intelligence as 5'6" is a rank of height (Figure 5.6)?

Many **psychometricians** have defined intelligence in the language of moral philosophy rather than in the language of biology, or they loosely define intelligence as the ability to think, which is to say that everyone is intelligent. The definition by Jensen is silly. If intelligence is what an intelligence test measures, then how will we know if the test is working as well as it might? Unfortunately, these kinds of definition influence interpretations of the test results. Consider a former IQ test item: Your mother sends you to the store for a loaf of bread. You get to the store and the store is closed. What do you do next? Usually children who live in cities answer, "I would go to another store." Usually children from rural areas answer, "I would go home." Both answers are based upon experience since there are many stores in cities but only one or few in rural areas.

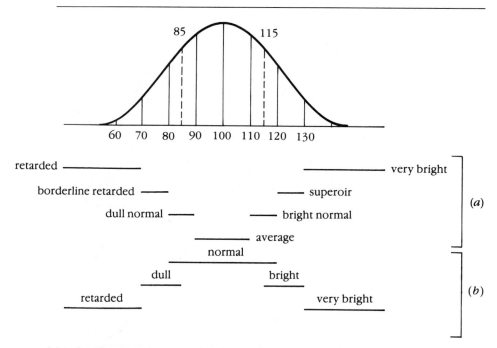

(a) A classification scheme used in the Wechsler Adult Intelligence Scale.

(b) A general classification scheme that summarizes the main features of many schemes. The question raised is whether these statistical categories of IQ correspond to a rank of intelligence, as 5'6" corresponds to a rank of height.

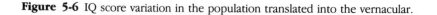

Figure 5-6 IQ score variation in the population translated into the vernacular.

Nevertheless, the second answer is wrong, and would be interpreted by the "standardized" IQ test to be an indicator of low IQ, i.e., low intelligence.

The Minnesota Study of Twins Reared Apart

If genetically identical individuals are phenotypically different, the phenotypic differences cannot be attributed to genetic differences between them. Can the similarities between genetically identical individuals though, be ascribed wholly to their identical genotypes?

From the largest study ever made of monozygotic (MZ) twins reared apart, Thomas J. Bouchard and his colleagues at the University of Minnesota conclude that about 70% of the IQ score variation observed within the greater population would vanish if all of the people in it were genetically identical, that is, genetic variation is responsible for 70% of the population's IQ score variation. According to the formula introduced in Chapter 4, $V_g = 0.7$; therefore, $V_e = 1 - 0.7 = 0.3$ (Figure 5.7). (Recall that V_g and V_e are the genetic and environmental contributions to the population's total phenotypic variation, V_p.)

In a summary table the authors compare MZ twins reared apart (MZA) with MZ twins reared together (MZT) for nine classes of biological and psychological variables. For example, the correlations between genetic similarity and finger-print ridge count similarity were nearly perfect: 0.97 for MZAs and 0.96 for MZTs. For height the correlations were 0.86 for MZAs and 0.93 for MZTs. The tests for brainwave variation showed correlation coefficients close to 0.80 for both MZAs and MZTs. The next highest correlations were found to be those of mental ability. Five different kinds of test were administered and the correlations ranged between 0.69 and 0.78 for MZAs and between 0.76 and 0.88 for MZTs. For other measures of mental ability and other personality variables, the correlations were generally below 0.5 (Figure 5.8a).

In a section of the paper entitled "Why Are MZA Twins So Similar?" the authors discuss three possibilities: (1) General intelligence or IQ is strongly affected by genetic factors, (2) the institutions and practices of modern Western society do not greatly constrain the development of individual differences in psychological traits, and (3) MZA twins are so similar in psychological traits because their identical genomes make it probable that their effective environments are similar (Bouchard, et al., 1990).

In their discussion of general intelligence being strongly affected by genetic factors, IQ test scores were converted into general intelligence without explanation, as were correlation coefficients (of 0.7) into estimates of heritability (of 0.7). Two secondary points were made: first, since none of the twins studied were "reared in poverty or by illiterate parents and none were retarded," the heritability estimate of 0.7 should not be transposed to the disadvantaged sectors of society. Second, the authors quoted other studies which show that "the average IQ test score has significantly increased in recent years." Their only response to this was to suggest that "This increase may be limited to that part of the population with low IQs" (Bouchard, et al.). A very different kind of explanation is given by Stephen J. Ceci in the September 1991 issue of "Developmental Psychology" wherein he presents evidence that if children remain in school their IQ scores rise, and that if they don't, as during summers or when skipping a year, their scores drop. (This pattern describes my tennis game: if I play regularly, my game improves, and if I don't, it doesn't.)

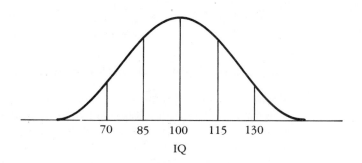

All of the area under the bell-shaped curve = V_p; the experimental task is
to fractionate this area into components called V_g and V_e, such that $V_p = V_g + V_e$.

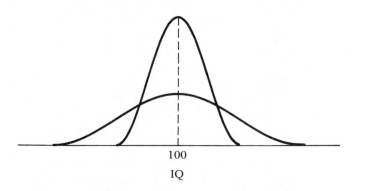

If V_g = 70%, and if V_g is removed from the normal population, the population
will have 70% less variation, only that which can be attributed to V_e.
Thirty percent of the variation will remain, but the distribution of this variation
may be very different from what is expected.

In human populations V_g is determined as a correlation coefficient between
closeness of genetic relatedness and IQ. In some plant populations V_g is
determined as the difference between the V_ps of isogenic and heterogenic
populations. For example, $V_p = V_g + V_e$ in heterogenic populations,
whereas $V_p = V_e$ in isogenic populations; therefore $V_g + V_e - V_e = V_g$.

Figure 5-7 An illustration of the equation, $V_p = V_g + V_e$

In summary, then, the discussion of the high heritability of general
intelligence (1) equates general intelligence with IQ, which is conjecture; (2) fails
to explain what heritability means in terms of cause-effect relationships between
genes and phenotype; and (3) leaves the reader to wonder why heritability is so
important if changes of environment so easily bring about large changes of
phenotype. (Recall that the average heights of several populations have
increased with changing diets; now we learn that average IQ scores have
increased greatly, for reasons unknown.).

Bouchard's argument that "the institutions and practices of modern Western
society do not greatly constrain the development of individual differences in
psychological traits," begins with this sentence: "The heritability of a psychological trait reveals as much about the culture as it does about human nature."
This certainly isn't the case for Mendelian and molecular genes, which tell us

(a)

Variables	MZAs		MZTs	
	R	**pairs**	**R**	**pairs**
Fingerprint ridge count	0.97	54	0.96	274
Height	0.86	56	0.93	274
Brainwave variables	0.80	35	0.81	42
Mental ability				
Test 1	0.69	48	0.88	40
Test 2	0.64	48	0.88	40
Test 3	0.71	48	0.79	40
Test 4	0.78	42	0.76	37
Test 5	0.78	43	—	—
Mean of 18 California Psychological Inventory Scales	0.48	38	0.49	99

Adapted from T.J. Bouchard, Jr., D.T. Lykken, M. McGue, N.L. Segal, and A. Tellegen, "Sources of Human Psychological Differences: The Minnesota Study of Twins Reared Apart," *Science* 250(1990):223–228.

(b)

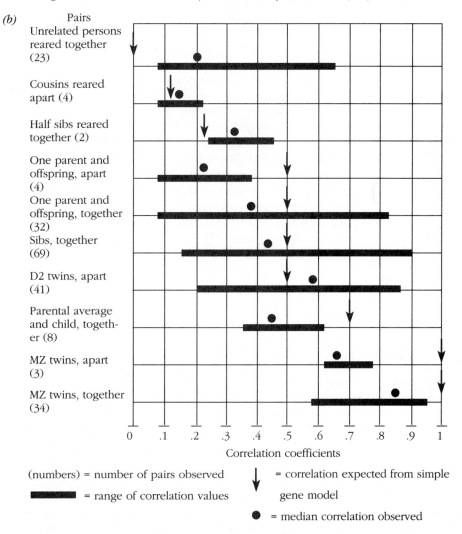

Figure 5-8 Correlation coefficients between genotype and phenotype similarity for *(a)* monozygotic twins reared apart (MZAs) and together (MZTs) and *(b)* Correlation coefficients between genetic relatedness and IQ among many paired subjects.

Adapted from T.J. Bouchard and M. McGue, "Familial Studies of Intelligence: A Review," *Science* 212(1981):1055–1059.

nothing at all about culture and probably very little about human nature. What is heritability? What the authors try to explain is that if the culture is homogeneous, and if every member of the culture has an opportunity to partake of culture's advantages, V_e will contribute little to V_p, that is, most of V_p will be determined by V_g. V_g, then, becomes a bigger fraction of V_p. We are not told whether V_p goes up or down when cultures become more homogeneous, and neither are we told about how genomes and environments interact and integrate into one another's domains, being simultaneously transformed by the processes. Thus we still have not learned what intelligence is, nor have we learned how it emerges throughout the development of individuals.

The third reason proposed to account for the similarity between MZAs is that identical genomes seek out similar environments, meaning that even though MZ twins are reared apart they will, in effect, seek environments so similar as to cancel out the effects of V_e. This is to say that whether reared together or apart, MZ twins will experience more homogeneous developmental environments than will dizygotic (DZ) twins or non twin siblings. The experimental evidence needed to verify this conjecture is impossible to obtain by use of human subjects.

The Minnesota twin study has added nothing new to the many attempts that have been made to demonstrate cause-effect relations between genes and intelligence. In fact, it presents very similar correlation coefficients and very similar arguments that correlation coefficients are a reasonably good method for sighting genes. However, unlike some of the earlier studies, the Minnesota study does not get into the hornet's nest of "race" differences in intelligence. Therefore the question of whether IQ tests are biased against some ethnic groups does not come to the surface in the way it did when Arthur Jensen published his now infamous paper in the Harvard Educational Review in January 1969 entitled, "How much can we boost IQ and scholastic achievement?" Jensen claimed that the mean IQ of African Americans is 15 points below that of Euro-Americans.

An Aside

Arthur Jensen's conclusion that 80% of the population variation in IQ scores means that 80% of an individual's intelligence is determined by intelligence genes was followed by wave after wave of criticism but also by more claims that intelligence, like bad teeth, is the outcome of genes. However, the controversy was fanned by the claim that human races are characterized by gene pools carrying different proportions of high and low intelligence genes, a conclusion that cannot be reached by estimates of heritability drawn from only one population. Critics exposed the cultural biases of the tests that allegedly measure intelligence; they exposed the differential social opportunities provided for children in different social classes and ethnic groups to learn test-taking skills; they exposed what was called the racist ideology that inevitably led to such conclusions in the first place; and of course they criticized the scientific error of equating estimates of heritability with Mendelian genes. Fewer of these kinds of inflammatory mistakes are being made by psychometricians today, but of course the debate still rages about whether heritability estimates can be made of human, quantitative, phenotypic variation. The character of these debates has not been changed by the more recent studies, which is to say that Jensen's and Bouchard's heritability estimates are of the same general quality.

How are "Intelligence" Genes Sighted?

This discussion of alleged intelligence genes speaks not only to the Minnesota twin studies but to all studies that use heritability estimates for sighting intelligence genes. Heritability is the portion of a characteristic's variation that can be attributed to genetic differences (V_g) among individuals within a population. The portion of the variation due to environmental differences (V_e), to a component called interactions between genes and environments (V_i), and to other components, accounts for the remainder of the phenotypic variation (V_p). In shorthand, $V_g + V_e + V_i = V_p$, the same formulation of components of variance used in the Bouchard paper discussed above. The experimental task of the psychometrician is to tease apart these components of variance and to discover the relative contribution of each to a population's IQ score variation.

The first step toward making a case for cause-effect relationships between genes and intelligence is the recognition of correlations between genotypic and phenotypic similarity. For example, correlation coefficients are higher for siblings than for unrelated people, and higher still for MZ twins (Figure 5.8b).

Correlations between genotypic and phenotypic variation are discovered in a variety of ways. Consider the following: a **sample** of people representative of the general population is taken by selecting pairs of individuals at random. The IQ scores of the members of each pair are determined and the difference (phenotypic variance) between the two scores is recorded. Pair one may differ by 5 points, pair two by 12 points, and so on. The pair differences are added and divided by the number of pairs to give a mean difference. Suppose the mean difference is 12 points.

Next we will measure pairs of individuals who share 50% of their genes in common, brother-brother, brother-sister, sister-sister, and parent-offspring pairs. Again we calculate the mean difference among the pairs. In the event that we could locate them, we would repeat the procedure with MZ twins. Let us suppose mean differences of 8 and 4 points for these two groups. As shown in Figure 5.8b, a positive correlation exists between genetic and phenotypic variation. (Compare lines 1, 6, and 10 in Figure 5.8b.)

While the pairs of MZ twins are not phenotypically identical, in this hypothetical population the IQ score difference between the members of a twin pair is on average 67% less than the difference between the members of a pair of unrelated persons (a mean difference of 4 compared to a mean difference of 12). In other words, *in the absence of genetic variation there is only one third as much phenotypic variation.*

With data like these, statistical techniques have been used to convert correlations into estimates of heritability. In the Minnesota twin study discussed above, the correlation coefficients were used directly as estimates of heritability. The correlation between test scores and genotypes is high for both MZT twins (0.85, Figure 5.8b) and MZA twins (close to 0.7). From correlation coefficients that range between 0.6 and 0.8 it is claimed that intelligence genes have been sighted.

What is wrong with this conclusion? First, it ignores environmental correlations among individuals with common genotypes, especially identical twins. It ignores the environmental differences between populations, such as those between the populations from which the test subjects are picked and the populations within which low IQ individuals are said to live. These biases inflate heritability estimates. Second, the conclusion ignores the norms of reaction of the genotypes in the population over the full range of environments within which intelligence develops. Third, heritability estimates do not isolate specific gene-phenotype combinations as do Mendelian and molecular genetic studies. Rather, heritability is indicative of rough correlations between genetic and phenotypic similarity within environments that cannot be controlled. The argument for the existence of intelligence genes is weak.

Environmental Influences Upon IQ

That IQ scores miss their intended mark of measuring intelligence is a commonly held view among psychologists and geneticists, but some psychologists and a few geneticists disagree. Many nonprofessionals have been pleased to learn that the experts disagree, but in the main, mothers and fathers tend toward despair upon learning that their child has a low IQ test score. Even parents who feel that IQ scores are not reliable indexes of the worth of their children know that IQ is used as a determinant of societal success and failure. These same illusions sometimes mask the results of studies which show clearly that IQ scores are not fixed, as are mid-digital hair and eye color. Evidence abounds that IQ scores of children are correlated with opportunity and with the incentive to learn the skills required

---

Perspective 5.1

If the phenotypic variance observed within a population is partitioned into genetic and environmental components such that $V_p = V_g + V_e$, the experimental task is to isolate one or the other of the two components. With many plant and a few animal species it is possible to obtain rather large, **isogenic** populations (i.e., of genetically identical individuals) and analogous populations of genetically **heterogenic** individuals. If both kinds of population are exposed to a wide variety of environments, V_e can be measured directly within the isogenic populations (all of the variation within the isogenic populations will be attributed to environmental factors), and the combined contributions of $V_e + V_g$ can be measured within heterogenic populations. The difference between these two kinds of population variance is an estimate of the total genetic variance, V_g.

This strategy for estimating the genetic component of V_p is of no use to geneticists who study human populations since there are no isogenic populations of humans. Even if there were, it would be impossible to measure their variance within many environmental conditions, or to compare that variance with the variance observed within heterogenic populations living in the same sets of environments.

Even if such an estimate of the genetic component of variance could be made, it still would not tell us what we want to know, namely, which genes shall we call intelligence genes? There are, in fact, two types of heritability: one, called broad heritability, H^2, cannot be used to identify specific gene-phenotype relationships; the second, called narrow heritability, h^2, can be used to identify the contributions of specific genes to phenotypic variation. H^2 and h^2 are very different constructs; each has a specific meaning and a specific usage in genetics, but they are often confused. (The heritability estimates of IQ score variation reported by Jensen, H^2, is now a classic example of the misuse of heritability.)

H^2 is simply the ratio of the genetic variance to the total phenotypic variance ($H^2 = V_g/V_p$). In the studies of nonhuman populations the genetic variance is the combined influence of all the alleles in the population that act directly to influence the phenotypic trait. This means that estimates of H^2 have meaning only within specific populations in which the alleles that influence behavior have been isolated in this way; it also means that comparisons between genetically different populations are meaningless. Even within specified populations *H^2 does not describe a property or characteristic of the popu-*

to succeed on IQ tests. Children from low socioeconomic classes have been adopted into families in higher socioeconomic classes, and in every case studied, the adopted children's IQ scores are on a par with those of the children born into the high socioeconomic, foster families and higher than those of children in the classes from which they came.

The fact that IQ scores can be raised by opportunity and incentive does not tell us what IQ tests measure, nor does it clear up the question whether genes influence scores. The fact simply means that high scores and the skills necessary to make high scores are needed for entrance into high socioeconomic classes (one can be born into an upper class without qualifications but one cannot enter without qualifications). This is worthy of a parent's concern, but the fact need not mystify relationships between scores and intelligence and between intelligence and intelligence genes.

In one study of adopted children (Scarr, 1984), 130 African American and children of ethnic mixed matings who had been adopted by advantaged Euro-American foster parents showed a mean IQ score of 106 (6 points above the national "white" mean, 21 points above the national African American mean (according to Jensen), and 12 points higher than the mean of the children living in the neighborhoods from which the adopted children came). Scarr concluded that "the major findings of the study support the view that the social

lation, properties that vary with changes in the environmental conditions. Broad heritability is of little use to geneticists; it does not reveal how much of a phenotype's variation is due to nature and how much to nurture.

Narrow heritability is used in plant and animal breeding programs when it becomes important to evaluate the effectiveness of artificial selection as a way to shift phenotypic means, as described earlier for increasing milk production in dairy cows. But still, the idea is based on the model of many genes contributing to the phenotypic variance. However, in the case of h^2 it is necessary to know the additive effects of each causative allele within the genomes. This is called **additive genetic variance** (Figure 5.9). h^2 is the ratio of additive genetic variance to total phenotypic variance. It is impossible to estimate the influence of h^2 upon phenotypic variation in human populations.

Even if this were possible in a select, genetically homogeneous population living within a narrow range of environments, *the values of h^2 would have no meaning outside that population and outside that environment.* Why? Because, as first discussed in Chapter 1, we know that specific genotypes develop very different phenotypes in different environments. In other words, a genotype's norm of reaction must be understood before it can be estimated how much influence genes exert upon behavior and performance.

In all major human populations there are many different genotypes and many local environments. If we consider nothing more than the phenotypic variation that arises from interactions between genotypes and environments (Figure 5.10), it would be, for practical purposes, impossible to compute them. (If there are p genotypes and q environments, there would be (pq)!/p!q! kinds of interaction. For example, if p = 5 and q = 5, very low numbers, then pq = 25, and pq! = 25 x 24 x 23 x 22 x 21 .... x 1; p! = 5 x 4 x 3 x 2 x 1, and q! = 5 x 4 x 3 x 2 x 1; and so on.)

The state of the art of identifying genetic correlates of behavior is rudimentary. Human behavior is exceedingly complex; it varies continuously over wide ranges within all populations, and it varies during individual development, which is some indication of a genotype's norm of reaction; that is, one individual (one genotype) may exhibit a wide range of behaviors during her or his life-span. (For a review of the concepts of behavior genetics, see Hirsch, McGuire, and Vetta.)

Parents   red kernels   ×   white kernels
Genotype $R_1R_1R_2R_2R_3R_3$ $r_1r_1r_2r_2r_3r_3$
Gametes $R_1R_2R_3$ $r_1r_2r_3$
F_1 $R_1r_1R_2r_2R_3r_3$ intermediate color
F_1 gametes $R_1R_2R_3$ $R_1R_2r_3$ $R_1r_2R_3$ $R_1r_2r_3$
                 $r_1R_2R_3$ $r_1R_2r_3$ $r_1r_2R_3$ $r_1r_2r_3$

$F_1 \times F_1$

	$R_1R_2R_3$	$R_1R_2r_3$	$R_1r_2R_3$	$R_1r_2r_3$	$r_1R_2R_3$	$r_1R_2r_3$	$r_1r_2R_3$	$r_1r_2r_3$
$R_1R_2R_3$	6	5	5	4	5	4	4	3
$R_1R_2r_3$	5	4	4	3	4	3	3	2
$R_1r_2R_3$	5	4	4	3	4	3	3	2
$R_1r_2r_3$	4	3	3	2	3	2	2	1
$r_1R_2R_3$	5	4	4	3	4	3	3	2
$r_1R_2r_3$	4	3	3	2	3	2	2	1
$r_1r_2R_3$	4	3	3	2	3	2	2	1
$r_1r_2r_3$	3	2	2	1	2	1	1	0

The numbers in each square give the number of dominant alleles in that genotype, and the number of dominant alleles can be estimated accurately by the color of the kernels, i.e., 6 = red, 5 = lighter red, and so on to 0 ($r_1r_1r_2r_2r_3r_3$) = white. Each dominant allele adds an increment of color to the wheat kernels.

h^2 = additive genetic variance ÷ V_p

Figure 5-9 Three genes influence kernel color in wheat. The recessive alleles add nothing to color, but the dominant alleles each add an increment of color.

environment plays a dominant role in determining the average IQ level of children and that both social and genetic variables contribute to individual variation among them." The first part of the conclusion is supported by evidence, but where is the evidence for the genetic variables? The grand conclusion will remain ambiguous as long as evidence for the genetic variables remains elusive.

An earlier observation of this type was of 63 children who were adopted into families with high socioeconomic standing. The mean IQ of the mothers of the children was 85, and the mean IQ of the children was 106, 21 points higher. There is a greater spread between these two means than between the national African American and Euro-American means (15 points) reported by Jensen. Again the evidence favoring the view that African American and Euro-American IQ mean differences are due to socioeconomic differences rests on a better foundation of facts than does the evidence supporting the claim that these mean differences are due to unequal portions of intelligence genes.

Another finding that came from a study of a large, middle-class Euro-American population showed IQ differences based on four "types" of children: whether they were single birth, twin birth, triplet birth, or twin birth but only one survivor. The mean score of 48,913 single-birth children was 100.1. The mean score of 2,164 twins was 95.7; of the 33 triplets, 91.6; and of 148 children born as members of a twin pair but reared as singles, 98.8. The variation among these mean scores was not a function of economic or ethnic variation, and there is no obvious explanation of the results by way of genetic variation. It doesn't stretch the imagination, though, to posit that singles experience more child-parent bonding than child-child bonding, that twins will be somewhat intermediate and that triplets experience the most child-child bonding. It could be that child-adult interactions do more to increase the performance on IQ tests than child-child interactions.

The issue of genetic versus environmental influences upon intelligence will not be resolved soon. Not only does the concept of intelligence symbolize an exceedingly complex aspect of what it means to be human, but we still do not know what "intelligent" people have more of than "unintelligent" people. If we do not know what intelligence is it is unscientific to claim that we can measure it. But does this mean that genes have nothing to do with intelligence? Emphatically not! Consider the following; most people have two kidneys, a few have one, and even fewer have three. The variation in number of kidneys is due to environmental factors (surgery); genetic variation contributes nothing to the observed phenotypic variation. Now, does zero heritability of variation in kidney number mean that genes do not contribute to the development of kidneys or

---

a) Maze learning in mice: two inbred strains of mice were selected for fast and slow learning, called maze bright and maze dull, in environment B. Both strains were then tested in environments A and C, called open and restricted. The results are shown in the following graph:

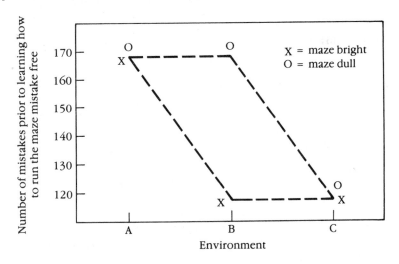

Within environment B the mice were selected (over many generations) for fast and slow maze running. Within B the mice bred true. Within a restricted environment (A) the bright mice were as dull as the dull mice, which were no duller than they were in B. In the open environment (C) the bright mice did no better than they had done in B, but the dull mice were as bright.

b) In the maze-learning study, there are 2 genotypes (p = 2) and 3 environments (q = 3): (pq)!/p!q! = 720/12 = 60 possible kinds of interaction

For 5 genotypes (p = 5) and 5 environments (q = 5), (pq)!/p!q! = 25!/5! × 5! = 14,592,350,223,313,890,816,000,000 possible kinds of interaction

Figure 5-10 Gene-environment interaction.

to the number of kidneys per person? Absolutely not. Zero heritability simply means that none of the observed variation in kidney number is caused by genetic differences among those whose numbers are different.

Heritability does not identify specific genes that encode proteins or that can be identified by inheritance patterns of contrasting phenotypic traits. For example, good vision is a factor that influences the early learning of children. Children with good vision may not learn to read well, but the acquisition of reading skills is slowed by bad vision. A child with poor vision may develop a negative attitude toward reading before corrective lenses are supplied. Poor reading skills inevitably lead to low IQ scores. A child with a low score initiated by poor vision will in fact be genetically different from his or her sibling with 20:20 vision and a higher IQ score. Consider the following sequence of cause-effect events: gene → non round eye shape → poor vision → no interest in books → poor reading skills → low IQ score → high heritability. But note: the high heritability at the *end* of the cascade identifies a gene(s) that influences eye shape, not intelligence.

Is There More to Intelligence Than a Test Score?

According to Harvard's R. J. Herrnstein, there is more to intelligence than a test score. Here is his view of relationships among genes, intelligence, prestige, and social standing:

> If differences in mental abilities are inherited, and
> If success requires those abilities, and
> If earnings and prestige depend on success,
> Then social standing will be based to some extent upon
>   inherited differences among people.

In other words, good genes encode high intelligence, which leads to high earnings and prestige, which in turn leads to high social standing. Whatever else can be said about Herrnstein's syllogism, it gets high marks for simplicity.

While the discussion of intelligence thus far has considered it to be an aspect of individual rather than group or societal behavior, this classification scheme presents a false picture of the controversy that has dogged studies of intelligence for more than a century. If we understood better what intelligence is and how it emerges during development, we could see more clearly the correlations between what have been called intelligence and societal values, legal systems, manners, and motives. We might also find that the so-called intelligence differences between societies boil down to little more than differences in customs.

Compare Herrnstein's syllogism with the following facts. A child with an IQ score in the top 10% (above 120) is 50 times more likely to be in the top 10% income group than is a child with an IQ score in the bottom 10% (below 80). (This is evidence used by Herrnstein to defend his syllogism.) However, if the top 10% IQ and the bottom 10% IQ children are from the same income group, the difference is only 2 times more likely for the high IQ children. Now, a child in the top 10% income group with an IQ of 100 is 25 times more likely to be in the top 10% income group than is a child with the same IQ from the bottom 10% income group. Herrnstein's syllogism suggests that "He is rich because he is smart," but it may be that "she is smart because she is rich."

The societal implications of high and low intelligence transcend the claim that IQ tests measure intelligence. The social uses of intelligence range from criteria of "race" and social class to credentials for admission into colleges, universities, the stock exchange, and social clubs. Indeed, the concept of intelligence has so mystified many of us that we have learned to partition people into "smart" and "dumb" much like the old western movies earmarked "good" and "bad" people with white and black hats.

The assumptions that underlie the grand generalization are as follows: First, intelligence is a unit phenotypic character, as is height. That is, it is not a socially constructed behavior, as are good and bad manners. Second, intelligence can be measured as can height (not with yard sticks but with IQ tests that can be converted into increments of intelligence). Third, genetic variation is the main contributor to IQ score variation, and therefore social relations are constructed upon an **IQ-genetic meritocracy**, which implies that social relations are correspondingly immune to change through social planning. Thus individual and societal behavior become intertwined, which brings the first discussion in this chapter together with the second.

Genes and Human Societies

No doubt the most vulgar misuses of the explanatory power of genes occurred under the leadership of the Third Reich, between 1933 and 1945. The entire medical program of the Nazi Party became an integral part of the "Final Solution" to the "Jewish Problem," and thereupon medicine became as brutal as were the gas chambers and firing squads. During that period of history the gene was employed to explain the differences between the "master race" and the "degenerate races," differences that were exaggerated beyond recognition in order to gain public support for the now infamous genocidal social policies. The social program for putting "master genes" into the service of the Nazi Party was called "race hygiene." The science of genetics was conscripted to prevent the "Aryan gene pool" from becoming "polluted with degenerate genes."

It would be dead wrong to blame a mad dictator, or a cloak-and-dagger fraternity of mad men for that scheme. True, a mad dictator and his mad servants did commit heinous crimes against humanity, but they did not invent the ideology within which it becomes possible to transform crimes against humanity into "sanitized solutions" to social problems.

Social Darwinism

The most prominent biologist of the 19th century was Charles Darwin. The only geneticist to later become prominent was Gregor Mendel, but that didn't happen to him until long after his death. There were many prominent psychologists, including Wilhelm Wundt, the "father" of experimental psychology; Edward B. Titchener, credited with transforming experimental psychology into **structuralism**, which fractures experiences into parts by way of **introspection**; William James, who rejected structuralism for **functionalism**, arguing that we process experiences continuously and are continuously being modified by them; Sigmund Freud, the founder of psycho-analysis, and others.

An Aside

The concept of sanitizing solutions to social problems is seldom discussed in public schools and universities, not because teachers are unaware of them but because they know that almost all political solutions to social problems are controversial. Most proposed solutions to societal problems are backed by one or another political objective, or they turn out to be hopeless compromises. During the Third Reich, the Nazi Party did not debate its opposition, they annihilated it. To convince the army and the police to defend the Party with passion, the Party explained its policies in a sanitized language. The explanations relied on euphemisms such as "The Party is taking the bull by the horns to preserve our race before it becomes swamped by the degenerate races." The language used was somewhat like that of a physician who proposes the surgical removal of a tumor before it swamps the patient.

The experience of the Third Reich, however revolting, illustrates rather graphically how the characteristics of a society can be rewritten as a "script" for the gene. Germany had suffered defeat in World War I and experienced economic depression for years after that war. The future of the country seemed bleak. In a very deliberate political move the Nazi Party decided to gain power by putting the blame for the social disaster on the Jews. With racism as its slogan the Party acquired control of the state in January 1933, less than 15 years after the Armistice.

Almost all modern political leaders including Adolph Hitler have paid lip service to democracy. They claim democracy is their ideal and goal. Hitler, however, was quick to add that democracy could not be achieved if there were degenerate types within the ranks who took it upon themselves to sabotage democracy. He argued that in order to achieve democracy those types must be identified and destroyed. Thus Hitler's final solution to the economic woes said to be created by Jews was to annihilate them.

Hitler's scientific advisors presented the Nazi Party with much the same kinds of information that William Shockley presented to California Governor Ronald Reagan in 1971 when he promised the Governor that California could be shed of its need for prisons, mental hospitals, and welfare programs simply by sterilizing all people with IQ scores below 100. (At that time the vast majority of the people Shockley had in mind were of Mexican and African origin; recall that Shockley claimed to be able to correctly identify genes for low intelligence simply by looking at skin color). Incidentally, Hitler's scientific advisors advanced their prestige within the Nazi Party by inventing the fiction that Jews not only carried genes for economic disaster but genes for communism as well.

We now know that while Hitler paid lip service to democracy his real goal was to acquire absolute power over Germany and, yes, over the human species. Why did so many people throughout Europe and America support his quest for absolute power? Were they convinced that Jews are a genetic threat to democracy? Or were they simply afraid of democracy? The history of the holocaust tells us that these kinds of questions were suppressed in the minds of many people, in large part by the doctrine that *biology determines destiny* (Muller-Hill). In Nazi Germany the Party had only to "whip up" latent fears that the Jews were genetically dangerous. It was then a logical next step for the Party to convince the fearful to carry out orders to exterminate the Jews, which many did, proudly, patriotically, and resourcefully.

Nazi Germany was not a freak historical happening; the German people during the Third Reich were not the only people in history to willingly participate in the execution of innocents. Indeed, the ideas of race hygiene, of improving "the human stock" by breeding and selection programs, and of sterilizing the socially "unfit" were imported by the Nazi Party from Great Britain and the United States. Some who studied eugenics and **social darwinism** planned eugenics programs to improve the quality of the species and thereby to steer the course of human history long before the German Nationalist Socialist Party (Nazi) was formed in 1923 (See Chase, Kevles, Muller-Hill, Lewontin, et al.)

Sociology also acquired prominence in the 19th century, in particular because of the efforts of Herbert Spencer. Spencer wrote his first papers on what came to be known as social darwinism before Darwin had written *The Origin of Species*. Spencer attributed social evolution to forces of natural selection, as did Darwin, but he adopted a very different model of evolutionary change, understandable in that Spencer was a sociologist, not a biologist. The forces of natural selection in Spencer's view propelled the development and emergence of nation-states, societies, and human cultures, not species.

It would be difficult, if not impossible, to exaggerate the influence that Spencer had upon the thinking of European and American intellectuals and politicians. He was said by some to be the greatest thinker ever, and by others to be at least as brilliant as Isaac Newton. He was invited by the political, industrial, and intellectual elite from all across America to speak, write, endorse, and to "praise success." The ideas of social darwinism crept into every sphere of Western thought, including economics, international relations, physics, biology, and, of course, sociology. One of the most comprehensive books written on the subject of social Darwinism is by Richard Hofstadter, in which he shows the ambitious attempt made by Spencer to scientize sociology:

> The aim of Spencer's synthesis was to join in one coherent structure the latest findings of physics and biology. While the idea of natural selection had been taking form in the mind of Darwin, the work of a series of investigators in thermodynamics had also yielded an illuminating generalization. Joule, Mayer, Helmholtz, Kelvin, and others had been exploring the relations between heat and energy, and had brought forth the principle of the conservation of energy. . . The concept won general acceptance along with natural selection, and the convergence of the two discoveries upon the nineteenth-century mind was chiefly responsible for the enormous growth in the prestige of the natural sciences. . . . Among the new thinkers, Spencer most resembled the eighteenth-century philosophers in his attempt to apply the implications of science to social thought and action. (Hofstadter, p 36)

Spencer adopted a view of evolution somewhat like Darwin's, but in Spencer's scheme evolution has a different end-point for individuals than it has for societies. Individuals are born and die; the biologically successful ones last only long enough to procreate. But societies arise over long periods of time and don't die; rather, they eventually get better, precisely because fit individuals leave more offspring than do unfit individuals. Thus societies, not individuals, become the beneficiaries of natural selection, and become stable, kind, inviting, and beneficent. Spencer coined the phrase "survival of the fittest" as a slogan symbolizing the relationships between the struggle for existence among individuals and the ever-increasing quality of the societies to which the survivors belong. Hunger, famine, and other hardships were said to be beneficial to societies, since hardships weed out the unfit. (Spencer's and Darwin's definitions of "fit" were as different as their respective personalities.)

Spencer's main thesis was that humans are perfectible, that evolution is purposeful, and that the laws of nature exist to ensure human perfection. Progress may be slow, he believed, but progress is progress is natural law! This is why Spencer so hated state power. He objected to any and all forms of state power on the grounds that it interferes with the processes of perfection. For example, nature puts each individual on trial and each will live and reproduce or die childless. Yet for reasons that do not fit the theory, societies evolved the tendency to feed the poor, cure the ill, and house the homeless—all abominations against nature because such tendencies allow the unfit to live to reproduce their kind, and eventually to destroy societies from within as a cancer destroys an individual.

In America Spencer's image among the rich and powerful was akin to that of a road-show, television evangelist. Industrial tycoons explained the rise of their empires as "the workings of the law of the survival of the fittest." James J. Hill, John D. Rockefeller, Andrew Carnegie, and others fondled Spencer's ideology as if it were gold, or at least an absolution of sin. To touch the cloth of Spencer's cloak was a transcendental experience. Carnegie stated that Spencer had freed him from theology and a belief in the supernatural by leading him to the truth of evolution, which Carnegie interpreted to mean "All is well since all grows better" (Hofstadter, p 45). Then he adds "If Spencer's abiding impact on American thought seems impalpable to later generations, it is perhaps only because it has been so thoroughly absorbed. His language has become a standard feature of the folklore of individualism, 'You can't make the world all planned and soft.' 'The strongest and the best survive—that's the law of nature—always has been and always will be.' " (Hofstadter, p 50)

Sociobiology: A New Paradigm?

The vestiges of social darwinism that influence current evolutionary thought are debatable. The debate is long and complex but there are scientific and philosophical problems that even after 150 years of polemic won't go away. For example, a major feature of the social darwinist view is that cultural evolution demands the existence of *inherited cultural behavior*. Yet the processes of cultural inheritance remain as intractable as they were in 1850. Evolutionary theories that attempt to "Darwinize" cultural evolution *imply the existence of genes,* but the genes remain unknown.

Nevertheless it is interesting to examine the long history of biology in the context of what Thomas Kuhn calls "scientific revolutions." In a book entitled *The Structure of Scientific Revolutions*, Kuhn describes the history of science as a succession of **paradigms**, or frameworks of thought and experiment. For example, Galileo aided the development of scientific inquiry when he began to systematize physical laws into mathematical formulations; the old paradigm, i.e., the old way of looking at and describing the world gave rise to a new paradigm which eventually replaced the older one. No one takes seriously anymore the pre-Galilean model of the solar system, with the earth at its center. Darwin introduced a new way of thinking about the history of biology, another new paradigm. Freud created an entirely different paradigm from which to look for explanations for why we behave as we do. Newton, Einstein, and Mendel did not change the world, only our way of looking at it, and in some cases they left us with a new methodology for changing the way we look at it.

Today there is widespread debate about whether sociobiology is a new paradigm, one that will eventually replace what is called the **neo-Darwinian synthesis**, the view of evolution introduced in Chapter 4. The term sociobiology was not invented by the man whose name is most often associated with the creation of the new paradigm, Edward O. Wilson of Harvard University, but it was Wilson who called sociobiology "the new synthesis." *Since paradigms are new syntheses,* according to Kuhn, Wilson prepared from the beginning to introduce sociobiology as a new way of thinking about our evolutionary past.

Wilson is recognized as one of the world's leading experts in the field of insect social behavior. In his many papers and books describing his studies of the social behavior of ants, he concludes that the behavior of ants and other social insects is tightly controlled by their genotypes and influenced little, if any, by learning. In his 1975 book *Sociobiology: The New Synthesis*, Wilson reviewed the social behavior of many species of animal from which he concluded that social behavior becomes less tightly bound to genotypes as animal complexity increases (e.g., from invertebrates to vertebrates, within the vertebrates from amphibians to reptiles, fish, and birds to mammals, to primates, and finally to humans). In fact, Wilson makes an analogy between the tightness of genotypic control over behavior and the length of a leash, claiming that chimp and human behavior is tied to a relatively long leash and that the behavior of the social insects is held in check by a short leash. In other words, the behavior of the social insects forms a rather clear window through which the expression of "behavioral genes" can be observed. The windows through which biologists observe complex animal behavior become increasingly opaque with increased biological complexity.

But whether leashes are long or short the bottom line is that the social behavior of all animal species is said to be held in check by a leash, that is, *there is a genetic base upon which all animal social behavior, including human societal behavior rests.* This idea seems trivial if we consider that most animals with lungs walk or fly, that most animals with gills swim, and that most animals with wings fly—different behaviors that correspond to different morphologies and hence to species genetic differences. But Wilson goes further, identifying the scientific problem as that of discovering the lengths and the contents of species-behavior leashes, which in formal terms means discovering the genetic components of behavior variation within species, and the evolutionary links among different kinds of behavior.

Wilson defines sociobiology as "the systematic study of the biological basis of all social behavior," and therefore "a branch of evolutionary biology and particularly of modern population biology." Its grandiose ambition is to explain social and societal behavior as outcomes of gene action. In order to understand sociobiology and to assess it, however, the grand synthesis must be divided into smaller segments. (This has been attempted by Bock, Kitcher, and others. Only a few of the segments are discussed here.)

In overview, sociobiologists appear to view *the evolution of social behavior* in Darwinian terms. It is said that genes influence the emergence of behavior and explain behavior variation; therefore the "behavior genes" that contribute to the reproductive success of the individuals carrying them will tend to increase within gene pools. New mutations that lead to more successful behaviors will eventually replace their alleles. In short, *the neo-Darwinian synthesis and the Mendelian gene* both are necessary to the sociobiological explanation of the evolution of social and societal behavior.

As was mentioned in the last chapter, a great deal of evidence used to support both the theory of evolution and the fact that species are genetically related (remember that breeding experiments across species boundaries are impossible) comes from comparative studies of amino acid sequences of related proteins and of base pair sequences of genes that encode related proteins. The sequence evidence that chimps, gorillas, humans, and many species of monkey are genetically related is overwhelmingly convincing. Now the question is whether behavioral similarities among related species will serve, as do protein and DNA similarities, as a support base for biological evolution. Are there enough similarities among primate behaviors to convince us that behavior, like hemoglobin, can be used to construct evolutionary trees?

Beginning in the 1920s, Mendelian and population genetics put new meaning into Darwin's paradigm of evolution. Darwinism was given a "face lift" that some have classified as a new paradigm. Darwinism's new name became the neo-Darwinian synthesis. Many evolutionists today consider themselves to be neo-Darwinians.

Sociobiologists insist that the neo-Darwinian synthesis is inadequate in that it fails to explain the emergence of species-specific behavior. Prior to sociobiology animal behavior was divided into human behavior, the domain of psychology, and other-animal behavior, called **ethology**. Ethologists did not study animal behavior from the perspective of genetics, but rather from that of neurobiology and physiology, in essence the animal counterpoint of earlier studies of human psychology. It was Wilson's thesis that all of these disparate fields of behavior and biology belonged within the same *new synthesis,* for which he chose the name sociobiology.

Not only does the neo-Darwinian synthesis sidestep the subject of behavior, it also fails to explain a category of behavior called altruism, defined by sociobiologists as behavior that reduces the reproductive success of the altruist but enhances that of the benefactors. Darwinism, social darwinism, and the neo-Darwinian synthesis all included competition, struggle, and the survival of individuals who won struggles. Indeed, this fact introduces a contradiction. While these theories do not fully explain human behavior, competition is one form of human behavior. In fact, it is the quintessential form of animal behavior in that, according to these theories, competition is and always has been the individual's response to the forces of natural selection. Informally, many evolutionists feel that competition is the only means for making evolutionary headway. In this context the neo-Darwinian synthesis explains competitive behavior. But the early theories do not explain altruistic behavior. Some sociobiologists have stated that sociobiology's explanation of altruistic behavior is evidence enough for its status as a new paradigm of evolution. What is that explanation?

In the vernacular, altruism means "devotion or concern for others," and is described as unselfish behavior that testifies to that devotion and concern. Sociobiologists have added to that meaning that altruistic behavior actually puts the reproductive success of the altruist at risk. Therefore they ask, "Why should an animal behave altruistically if such behavior reduces its fitness?" "How does a gene that encodes altruistic behavior compete successfully against its allele in the gene pool?" We know that mutations are rare and that only one individual among thousands will carry a new mutation. The death of that individual means death for the new mutation. How, then, do the forces of natural selection allow such a mutation, dangerous to the individual but beneficial to a population of altruistic individuals, to survive?

Consider an example. In a prairie dog colony in Western South Dakota, the boundaries of a territory are known by all of its inhabitants. Yet stationed around the boundaries of the territory are a few prairie dogs that spend their time watching for intruders, in particular dangerous ones like hawks and coyotes. When it appears to a "watch dog" that the members of the colony are in danger, it will emit a characteristic barking signal that sends all of its colony mates underground (Figure 5.11).

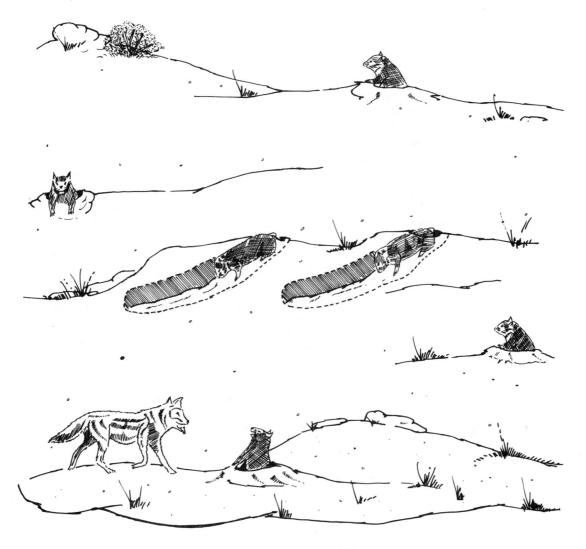

Figure 5-11 Prairie dog territories are protected by altruistic behavior.

The job of the "watch dog" is risky. Its location is closer to that of the intruder than are the locations of its colony mates. The watch dog, then, is the most likely member of the colony to become the intruder's next meal, and if this happens its gene(s) encoding altruism will be removed from the gene pool. It is one problem to determine why a prairie dog would "accept such a job," it is another to determine how its genes for altruism gained entrance into the gene pool in the first place. According to sociobiological theory the prairie dog accepts the job because it has no choice in the matter; it has inherited a gene for altruism that programs it to protect its colony mates.

The question of how such a gene becomes established within a gene pool is answered less easily, but in a way that distinguishes sociobiological from neo-Darwinian theory. It comes down to expanding the meaning of fitness. Darwinian fitness is measured not merely by the numbers of offspring a mate-pair produce but by the numbers of progeny produced by those offspring and so on down through the generations. An individual is fit only if his or her offspring, grandchildren, and great-grandchildren are fit.

Within the paradigm of the neo-Darwinians, fitness is a measure of the number of an individual's genes that are transmitted during successive generations, and a gene's fitness is measured in terms of its contribution to the reproductive success of the individuals carrying it. Sociobiology brought into clearer focus the fact that an individual's genes (or copies of them) can be transmitted to successive gene pools by the individual's relatives — sisters, brothers, cousins, and, of course, progeny. This view draws attention away from the individual to ask only whether his or her genes have been transmitted. Thus individual fitness is a function of the fitness of relatives, or kin.

We are not talking here about the kinds of altruism expected of good girl scouts who help old men cross the street; we are, instead, limiting the discussion to genetically programmed altruism. The prairie dogs within a colony are closely related, meaning that many within a colony are capable of passing to the next generation some of the watchdog's genes in the event that the watchdog becomes an intruder's meal. If the watchdog sacrifices its genes that two or more of its sibs will live to reproduce (or eight or more of its first cousins), then from the gene's point of view "all is well." Certainly nothing is lost, and there is always the chance that something will be gained, i.e., that many of the relatives saved by the warning call will pass on the genes that encode altruistic behavior.

There are two problems here: (1) we still do not know how the gene for altruism became established in the gene pool, that is, how many members of the colony came to inherit it, and (2) it would appear that what is important here is the gene, not the prairie dogs who carry the gene.

Is Problem One Solved by Inclusive Fitness?

If behavior is considered in terms of one individual acting and another individual being acted upon, then in terms of selfish, cooperative, and altruistic behavior, it becomes possible to make predictions about the fates of the genes that encode these different forms of behavior, as did the biologist W. D. Hamilton more than 10 years before Wilson popularized sociobiology. Selfish behavior increases the fitness of the actor and may decrease the fitness of the actee. Cooperative behavior increases the fitness of both, and altruistic behavior decreases the fitness of the actor while increasing the fitness of the actee.

If genes encode these types of behavior, then natural selection will favor genes that induce individuals to enhance the reproductive success of their genetic relatives. This mechanism for ensuring the survival of genes is called **kin selection**; upon adding kin selection to all of the other forces of natural selection, the result is called the **inclusive fitness theory**. In other words, sociobiology is advanced over the neo-Darwinian synthesis by the additional explanation of how genes are preserved within gene pools.

One of the architects of the neo-Darwinian synthesis, J. B. S. Haldane, anticipated the inclusive fitness theory with the suggestion that he would sacrifice his own life for the lives of two siblings or eight first cousins, pointing out that two siblings or eight first cousins are as likely to transmit his genes into the next gene pool as he is. He was in a pub at the time of the statement, but sober enough to recognize that there is no record of such an act in human history. Even if there were a dozen or so examples of such an act, the main problem is still left unexplained, namely, we are taking on faith that there are genes for altruism and that such genes increase within gene pools even though the individuals who carry them are less fit. Kin selection may explain how a gene that puts their carriers at risk may survive after many individuals have become carriers, but the theory does not explain how the first mutation survived so that many individuals would become carriers. Why would a family of selfish individuals allow a single altruistic offspring to survive? If an altruistic offspring took risks to ensure the reproductive success of its selfish kin, which allele would survive? Since altruism is not a neutral behavior genetic drift is not a convincing answer to these questions.

Wilson's primary studies have been with ants, and ants do exhibit an interesting form of social behavior that correlates with an interesting genome difference within kinships. First, within social ant colonies the "workers" are sterile females. Their work is considered to be altruistic in that it can never enhance their own reproductive success. Indeed, all of their work tends to increase the reproductive success of the "queen," called a gene machine by some sociobiologists. Without question the queens lay lots of eggs. The males do no work at all except to wait in the wings for the opportunity to mate with a queen.

The genetic relationships among the social ants are different from those described earlier for peas and humans. The females are diploid and the males are haploid (Figure 5.12). As the queen lays her eggs she is able to fertilize some of them with sperm stored after copulation, and she is able to lay some eggs without their being fertilized. The fertilized eggs develop into females, half of whose genes are maternal and half paternal. Now, only half of the mother's genes are inherited by each daughter while all of the father's genes are inherited by each daughter. Thus each daughter is related to her mother as we are, by 50% of their genes, but the daughters are related to each other by 100% of their father's genes, which is to say that their relationship to one another and to their sister, the queen, is 75%. The theory of kin selection predicts that the degree of altruistic behavior of the females to one another and to the queen should be greater than it is within families that share only 50% of their genes in common. It was from the behavior of the social insects that exhibit this haploid-diploid mode of reproduction that the kin selection theory was developed. But as yet no one has discovered the evolutionary antecedents of this mode of reproduction or of this form of altruistic behavior. It is a leap of faith to draw conclusions about human societal behavior from ant social behavior; even the vast majority of insect species do not behave like social ants and there aren't many vertebrate species known to behave like the social insects.

Another twist to the story of altruistic behavior is called **reciprocal altruism**, the idea that "I will help you if you will help me." The actor does something helpful for the actee in the expectation that the actee will reciprocate the favor. Again, behaviors have been described that fit such a model, but the genes alleged to program such behaviors have not been tracked down, at least not by methods that satisfy geneticists. One example of such behavior often used to illustrate the point is that of vampire bats. The food of vampire bats is animal blood. As the bats search for animals during the night, some will be successful and others will return home "empty." The successful bats have been found to regurgitate blood for their unsuccessful cavemates. The following night the fortunes of the bats may be reversed, and the recipient of the night before will become the donor. It isn't clear why vampire bats behave in this way; not all species of bat do; not all species of mammal do. Blood hemoglobin of mammals is a better indicator

(*a*) Workers (diploid females) work to bring food to a colony's communal hive,
then they work to raise the queen's offspring.

(*b*) Consider only two pairs of chromosomes:

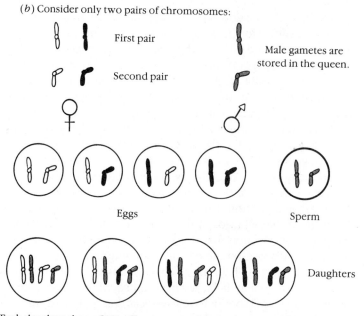

Each daughter shares 50% of her genes with her mother, and daughters share
100% of their father's genes with one another. On average they share 75% of
their genes with one another and with their sister, the queen.

Figure 5-12 Social insects exhibit a peculiar form of altruistic behavior that correlates with genetic relatedness.

of species relationships than is the reciprocal sharing of captured blood by
vampire bats.

The neo-Darwinian synthesis is better at explaining the genetic basis of
speciation than is sociobiology at explaining the genetic basis of social behavior.
This is even more obvious in the case of human societal behavior than it is with

social insect and vampire bat behavior. But before discussing human societal behavior it is necessary to pick up the thread of the second problem mentioned above, that of the gene replacing the organism as the unit of selection.

Problem Two: Is the Gene the Unit of Selection?

The idea that genes program prairie dogs to ensure the survival of prairie dog genes is a new twist to Darwinian evolution. Recall Samuel Butler's adage, "a chicken is the egg's way of making another egg." From Darwin's perspective the forces of natural selection act upon individuals, and individuals compete with one other for survival. The successful competitors will contribute genes to the next gene pool. Clearly, with Darwin's scheme it is difficult to explain how altruistic behavior survives in a world "red in tooth and claw," that is, a world in which every (prairie) dog is out for itself. However, the shift of the focal point of natural selection from individual to gene allows us to feel less hurt at the loss of individuals and to suppose that there are strategies by which genes ensure their arrival into succeeding gene pools without depending upon the competitive edge of the individuals who carry them.

Consider kin selection again. The idea is that within kinships (extended families), natural selection favors genes that are preserved by way of altruistic behavior among relatives. Unwittingly, altruistic behavior becomes an agent of natural selection by acting to ensure the survival of the genes that encode it. The altruistic behavior expressed by one individual to another may appear to be motivated by the actor, but it is not; the actors inevitably die; their genes live on through their kin. But the gene, according to this view, is the ultimate biological vehicle. It is the only concern of the forces of natural selection. Individuals are but genetic strategies designed by genes for the survival of genes. We are, then, our genes' way of making more genes.

But there is a contradiction in this scenario. For the genes that encode altruistic behavior to survive they depend upon the behavior of individuals within kinships; indeed, the entire kinship appears to be the the unit of selection, rather than the gene or the individual. To say that genes "want" to survive and that altruism is one "strategy" devised by them to ensure their survival may simplify the processes of evolution a bit too much. Dawkins argues that genes are selfish, that they have no cares in the world except to survive, and that they will do anything to survive. He does not mean that genes think about these things, only that genes within kinships are preserved by the vicissitudes of kinships, not by thought processes of genes. It is the characteristics of kinships that make it appear as if genes design their own survival. Yes, a honeybee will commit suicide by stinging an intruder; a blackbird will be eaten by a hawk as it warns its flockmates to flee; a successful vampire bat may share its cache of blood with an unsuccessful cavemate. But questions remain: Are these behaviors simply survival strategies designed by selfish genes? Can such behaviors be used, as amino acid sequences of proteins are used, to construct evolutionary family trees? The evidence isn't in yet.

Arguments about the units of selection will continue for some time, as will arguments about the details of evolutionary processes. This is no embarrassment. The history, characterization, and causes of behavior will remain unknown for years to come. Curious people keep striving for better and better explanations; a few strive for new paradigms. It is exciting when they succeed and it is no cause for embarrassment when they fail. The point isn't that gene selectionists are good guys or bad guys (the prominent ones are guys, however). The point is to join in the fun; cheer for a winner if you choose and razz an opponent if it makes the action more meaningful. But if our understanding of biology improves through time we must learn how to jettison wrong explanations and to examine new ones. If science becomes too much like professional hockey, it will cease to be science.

Does Sociobiological Theory Explain Human Societal Behavior?

If the social behavior of ants and chimpanzees is difficult to fathom, the societal behavior of humans is even more so. Indeed, the word *societal* is used in the place of the word *social* for describing human as opposed to nonhuman behavior. Human societal organizations are more complex, no doubt because so much of human behavior is based on historical precedent, learned behavior, legal systems, religious beliefs, habit, and so forth. Yet despite the complexity of human behavior some sociobiologists unabashedly assign genes to nearly every quirk of behavior observed.

Philip Kitcher, in his book *Vaulting Ambition*, calls much of human sociobiology **pop-sociobiology**, arguing that the steps it takes to explain behavior are first to observe it, call it a phenotype, assign to it a causative gene and then tell a story about how such a behavior led to reproductive success among our ancestors. This pathway toward explanation, Kitcher elaborates, lies outside the boundaries of science. The explanations of behavioral adaptation derived therefrom fall into two very different kinds, depending upon whether they are considered to be competitive or altruistic. Elaborate scenarios are drawn up to explain competitive behavior, especially behaviors that involve males competing for females. Altruistic behavior, on the other hand is said to be preserved by kin selection. This is simply a statement that genes harmful to individuals but good for kinships are preserved within "family gene pools" by kin selection for the good of the kinship.

For example, kin selection is invoked to explain altruism and homosexuality, both of which are said to reduce the fitness of individuals and to enhance the fitness of kinships. (There is little evidence that same sex preference is determined by genes, but, like altruism, the neo-Darwinian synthesis fails to explain its existence. Wilson has drawn up a scenario explaining how such a behavior can exist within human populations and at the same time lead to the reproductive failure of the individuals who behave that way.) Good old-fashioned aggression, territoriality, and male dominance over females, however, is explained with good old-fashioned neo-Darwinism and social Darwinism.

The last chapter of Wilson's *Sociobiology* introduces the subject of human behavior. Wilson made an attempt there to distinguish between individual and universal behavior. For example, he refers to the heritability estimates of introversion and extroversion, neuroticism, depression, schizophrenia, and so forth, establishing that individuals within populations differ. He describes these differences in much the same way that IQ score differences are described.

Universal behaviors, in contrast, are more conservative. Indeed, all members of a species may exhibit the same behavior and therefore may be genetically identical with respect to those behaviors as they are for having skeletons, bones, veins, and arteries. He predicts that conservative behaviors should be found in related species, in other primates with whom we have shared our recent evolutionary history. Wilson calls these universal, conservative behaviors "aggressive dominance systems," such as males dominating over females or one male dominating over all others within a kinship or close-knit social group.

Wilson draws parallels between hunter-gatherer societies and modern industrial societies. He says that in both the women and children stay at home while the men forage for food or for money to buy food. The women and children maintain social ties within kinships by means of conversation and shared cooking facilities. That the same patterns of social behavior exist within ancient and modern societies, then, is presented as evidence that such behavior has a genetic base. In an article printed in the *New York Times Magazine* (October 12, 1975) Wilson wrote, "Even with identical education and equal access to all professions, men are likely to continue to play a disproportionate role in political life, business, and science" because of historic wisdom encoded within our chromosomes.

Some sociobiologists have expanded upon issues like those of male dominance and female passivity, suggesting in effect that since women produce only a few eggs, natural selection has programmed women to be choosy about who fertilizes them, the result being that in matters of sexuality, women are coy, restrained, and highly gifted in *sales resistance*. On the other hand, males, having gametes to spare, are less concerned about which eggs are fertilized by them and in fact may dream about fertilizing every egg in the community; males therefore have developed skills in *salesmanship*. There is, then, a genetic base for macho and coy sex behavior. Pop-sociobiologists may incite more kinky behavior than they explain.

Exotic explanations of human behavior and vehement criticisms of these explanations have followed in the wake of Wilson's *Sociobiology* in 1975, but the real controversies are pretty much confined to human sociobiology. Biologists who describe insect, bird, fish, and bat behavior, and who do not make analogies between the behaviors they observe and human behavior are little bothered by the critics. Nevertheless, the issues drawn up by Wilson and other sociobiologists will not go away. They will be debated for years to come.

For example, human aggression manifests itself differently than does aggression of other species. Conscious decisions to wage wars, to travel on highways, and to create differential access to food, and medical support systems, lead to more intraspecies deaths, proportionate to population size, than do the combined aggressive acts of all the other primate species. How do genes influence these conscious decisions?

Human territoriality also is different, even within the species. Many groups of people, even in very recent years, have shown no evidence of understanding the meaning of private property or the ownership of land. These include the people who lived in the Hawaiian Islands prior to the intrusions of Europeans, native Americans prior to their loss of America; Inuit, Yupik and Yuit peoples living in the Northernmost latitudes (Canada, Alaska and Siberia); and many groups of African peoples. If genes for territoriality were of survival value early on during primate evolution and if territorial behavior is to be a clue to genetic relatedness, how shall we explain these facts?

Many sociobiologists who claim that genes determine human societal behavior have responded to the critics by professing a strong belief in civil and human rights. While racism and sexism are programmed by genes, they say, we still ought to strive for less racism and less sexism in our societies. But there are two dangers, they warn. One is that a society that devotes its energy to fighting against its *biological disposition* will risk losing to societies that don't, in the inevitable competition among nations for world real estate. The second problem is that if we are programmed by genes to behave as we do, it should be impossible to shift the focus of natural selection from genes to individuals — how might a human brain designed by genes learn to subvert the strategies of the genes that encoded it?

In some ways pop-sociobiology emulates social darwinism. It considers the social status quo to be a biological given, as human geneticists regard eye color as a given; then it speculates the pathways by which evolution produced the result. It is now agreed that social darwinism was and is pseudo-science; it has been suggested by some critics that the same is true of pop-sociobiology.

Similarities and Differences Revisited

Scientists, historians, economists, writers, and artists have discovered and reported on important aspects of the human condition. Opinions differ, facts are interpreted differently, self-interests underlie the alleged common good, and confusion always seems to reign supreme. But we would feel the same way about the Egyptian pyramids, the Roman aqueducts, the Empire State Building, and the Hoover dam if we had witnessed only about one hour of their construction, and if we had no knowledge of how they were planned or how the finished product was to turn out.

A great deal has been learned about biology, human behavior, and human societies, but we are witnesses to only a tiny fraction of the processes that have led to the events that we witness. We know little or nothing about the planning stages or how the original planning strategies became transformed into habits of thought that still cannot be broken; we will never live long enough to see the finished product. Yes, it is confusing, but perhaps not as confusing as it first appears.

Kenneth Boch, a student of history and of sociology, presents us with a very different view of the study of human cultures, in particular the studies of differences within and between human cultures. For example, the approach one takes toward the study of human cultures will vary depending upon the initial views of what cultures are. At one extreme, a starting premise may be that "culture is the expression of a universal human nature." From such a position one begins to look for evidence in support of the belief that human cultures will be more similar to one another than different. The search would be for universals which are characteristic of all cultures. At the other extreme, each human culture might be viewed as the outcome of a single history. Therefore it would be expected that each culture is unique, and the search would be conducted with the expectation of discovering differences among cultures. Admixtures of universal expressions of a human nature and single histories can be modeled if for some reason in some locations some aspects of human nature appear not to be expressed, and out of which developed an apparently different culture. For example, differences between cultures have been studied as outcomes of differences between races, between physical environments, even between contrasting adaptations to different environments.

Bock concludes from this that:

> The kind of approach to cultural differences one chooses is a serious matter, both practically and theoretically. A search for universals through identification of similarities can take the form of prescribing the limits of culture, and so the

boundaries of possible action. And a focus on similarities obviously tends to obscure the theoretical problem of cultural differences. Insistence on the absolute individuality of each culture can, on the other hand, create romantic visions of cultural destinies flowing from culture souls. In this case the methodological implication is that cultures are not comparable and that scientific knowledge about culture histories is not, therefore, possible. The alternatives lying between these extremes involve in some measure the difficulties of each.

Before leaving the subject of whether genes influence the development and character of human cultures, let us return to a statement made in Chapter 4 to review what we know about genes and their influence upon phenotypes. Interesting phenotypes such as individual behavior are not known to be determined by Mendelian and/or molecular genes. Behavior, while interesting and important, cannot be reduced by inheritance patterns and transcription products to the actions of specific genes. On the other hand, the genes we know so much about and that have been studied in detail usually are found to influence phenotypic traits that are comparatively less interesting, or, if they are interesting they have been neglected because of a greater interest in the genes. However, geneticists continue to discover more genes, and to learn more about how gene expression influences phenotype. Yet far less is known about how phenotypes emerge than is known about how genes express themselves, but geneticists are beginning to understand how phenotypes emerge during development. It seems certain that much will be learned about development during the next decade, but progress toward an understanding of cultural evolution promises to be much slower.

In this chapter the focus has been upon studies of human intelligence and human societal behavior, "phenotypes" that have invited genetic explanations. No one doubts that genes will be included in the future explanations, but the roles that genes have played in the emergence of intelligence and evolution of human cultures can only be guessed at today.

At this moment in history we might be well advised to treat all speculations about how genes evolved to control individual and societal behavior as "just so" stories, as Richard Lewontin calls them. One colleague said to another, "you have to smile through times like these." The other colleague countered, "How can I smile when my tax dollars are paying the salary of a man who is studying the inverse relationship between IQ and penis size?" (The man referred to is J. Philippe Rushton of The University of Western Ontario. See reference. Rushton's pseudoscientific suggestion would, however, lead us to predict that women have much higher IQs than men.)

Prejudices do not arbitrate the meaning of facts and theories in the long run, but in the short run they may. One thing we can do to shorten the life span of prejudices is to put them on the table, talk about them, and search for alternative explanations.

A Philosophical Problem

Isaac Newton made a good case that ocean tides, falling bodies, and the orbiting of planets in our solar system are all explicable by the theory of gravitation. Darwin made a good case that all extant species of animals, plants, and microorganisms arose from pre-existing species. Molecular geneticists have made a good case that genetic information is stored in DNA, and is much the same in all living forms. These theories are about aspects of the physical world, and so are of no use for predicting the vicissitudes of the stock market, the winner of the next election, or what to do about the production, sale, and widespread use of addicting drugs.

There are economic theories, but, though complex, their predictive powers aren't good yet, certainly not in comparison with atomic theory or the theory of the gene. In the field of psychology there are theories of human behavior associated with Freud, Pavlov, Skinner, Piaget, and Adler. But like economic theories, theories of psychology are not yet reliable in making predictions, as is the theory of the gene.

The theory of gravity is of no help to economists. It is being debated whether the theory of evolution can be of any use at all to psychologists or whether it is merely misleading. Gravity and economics are about different kinds of phenomena, and knowledge of them is built from facts obtained by very different methods of study. Planetary motion and falling bodies are not known to influence any of the phenomena economists theorize about. It takes a leap of faith to draw cause-effect relationships between planetary motion and falling stock markets, as you know if you have dabbled in astrology.

There is some reason to say the same about relationships that have been forged among evolution, genetics, psychology and sociology. While genes appear to influence human behavior more than do the orbits of planets, evidence of causal relationships is weak. The apparent connections are fortified by faith and ambition. Indeed, not one gene's activity has been shown to lead to the development of normal behavior, except, to say trivially that all 100,000 human genes may be necessary to normal behavior.

Mutant genes that have been shown to lead to abnormal behavior in fact lead to mutant enzymes (e.g., PKU), mutant physiology (e.g., cretinism), or mutant morphology (e.g., excessive height or short stature, the absence of hands, feet, fingers, etc.). To date we know only that these mutant genes lead to mutant biology, which in turn leads to mutant behavior. From the vantage point of biology, behavior is elusive; yet it is common to hear it said that this or that gene has been shown to cause this or that behavior. Behavior is the focus of psychological study, but psychologists have yet to construct a theory of behavior that can be used as physical theories are used. Human behavior hasn't been reduced to mathematical, physical, chemical, and genetic parameters. Therefore skepticism is justified upon hearing that intelligence genes have been sighted. Claims to have united theories of biology and theories of behavior through cause-effect phenomena and by way of the predictive power expected of a unified theory must be examined closely and critically.

- Are there alternative viewpoints? Yes, but these too must be subjected to critical analyses. Indeed, the development of behavioral patterns within the human species involves a continuous interaction among individuals and their environment. The ways in which individuals interact with their environments are unknown, and are suspected of being too complicated for our current analytical skills. Biological determinists tend to ignore human history, discounting the role it plays in shaping human behavior. To the extent that a knowledge of human history is necessary to an understanding of human behavior, we would then have to disentangle influences of the whole of human history from the influences of each of our developmental histories, and then both of these from the influences of our biological particularities. Otherwise our explanations of ourselves will remain forever vague. As our understanding of ourselves improves, atoms, genes, and chromosomes will be factored into the stories we tell to explain ourselves, but probably only as participants, not as determinants.

There is some thought being given to the view that organisms do not adapt to their environments. Rather they become parts of their environments, changing them, sometimes destroying them, but always transforming them. As environments are transformed, so are the organisms, since in the process they become vital parts of their environments.

In other words, we may eventually begin asking, "How do cultures — behavior, intelligence, adults — emerge?" and stop asking "Which genes cause culture, intelligence, and particular kinds of behavior?" The logic of determinism was powerful before we learned a bit about the world we live in. Now we know that the properties of things and of phenomena are shaped by interactions, integration, and emergent processes. We know that organisms interpenetrate their environments, and that environments interpenetrate the organisms that inhabit them, and that both become parts of a whole, each transforming the other, thereby the character of the whole is continuously changed.

This view of nature will insist that biological systems cannot be understood outside the context of their general and individual histories. Cultures will be understood not in terms of their parts but in terms of their histories, just as species are coming to be understood in terms of their evolutionary histories. This view does not negate the discrete character of Mendelian genes; rather it integrates this discrete character with processes of emergence that makes sense only within an historical context. This may be one of the most important changes now taking place within the sciences, and in particular within the biological sciences.

Summary

The theory of evolution, beginning with Darwin's explanation of speciation and including all of the implications of Mendel's explanation of inheritance patterns, is a biological theory. But biological organisms exhibit individual behavior, and groups of organisms exhibit social behavior. The question arises whether these forms of behavior are as easily explained by evolution theory as is animal biology.

This question is controversial. At the level of individual behavior many scientists have tried to demonstrate that genes are causal agents, not in the trivial sense that animals with lungs extract oxygen from air and that animals with gills extract oxygen from water, but in the sense that individual behavior includes many specific and discrete forms of behavior, each of which can be traced to one or a few genes. Since most forms of behavior vary quantitatively within populations, the demonstrations of genetic causes have, for the most part, taken the form of "isolating" the genetic component of population variance.

For example, it has been claimed that at least 70% of the IQ score variation in Euro-American populations is due to genetic variation. These claims have been challenged because the methods used to estimate heritability within human populations are inadequate. With some plant and animal species it is possible to compare isogenic and heterogenic population variation within many kinds of environment, and thereby to estimate the genetic component of population variance. This cannot be done with human populations.

The problem is compounded by the fact that we do not know whether IQ scores are a reliable measure of intelligence. Since we do not know what intelligence is, it is risky to use it as a criterion of race or of social worth.

The use of a biological theory to explain cultural phenomena, and in particular cultural evolution, is questionable. Before making a final conclusion about the role of genes in shaping societal behavior, we ought to know more about societal behavior, cultures, and cultural histories. For example, genes are composed of atoms, but atomic theory is of little use in predicting what genes do and how they do what they do. The risks of using gene theory to explain human behavior include the temptation to define behavior according to the characteristics of genes, before understanding behavior.

However, some sociobiologists have proposed that the study of cultural evolution, human behavior, anthropology, and even economics be subsumed into the theory of evolution. In other words, all animal behavior is believed to have the same causal relations to genotypes as have all animal metabolism and morphology, albeit some behaviors are more modifiable by non-genetic forces.

Since it is impossible to understand the fullness of biological evolution in the absence of a knowledge of species histories, it would be risky to subsume cultural evolution into biological evolution theory. Even if we knew those histories we still would need to know how cultures modify human behavior and how behavior modifies cultures — that is, how both are modified by interaction, integration, competition, and altruism.

Study Problems

1. Outline the relationships among theory, hypothesis, experiment, and fact. Illustrate, using this outline, that Darwin's theory of evolution is a biological rather than a psychological or a sociological theory.

2. Explain why it is easier to predict the morphological sex type of the adult that emerges from a 44XX zygote than it is to predict the sexual behavior of the same adult. (Recall that a scientific theory is judged, in part, by its predictive power.)

3. Argue for or against the proposition that the **P** gene, that encodes the amino acid sequence of phenylalanine hydroxylase, is a behavior gene. (Recall that persons with genotype pp, are phenylketonuric.)

4. From what you have learned about relationships between phenotypes and genotypes, do you think it will become possible in the near future to identify the phenotype called intelligence, and from the genomes of intelligent persons to isolate, with restriction enzymes, one or a few genes that encode intelligence? Explain.

5. Did Alfred Binet design tests that distinguished between intelligent and unintelligent children? Explain.

6. Explain how the IQ of a child is determined.

7. Among all of the definitions of intelligence with which you are familiar, which, in your opinion, is the best?

8. Are mean IQ score differences between populations reliable criteria for separating those populations into races? Explain.

9. If it is observed that the heritability of IQ score variation is high within the Euro-American population, does this prove that genes cause intelligence? Explain.

10. Do you think that human societies today would, or would not, benefit from the efforts of ethicists to evaluate the research projects of scientists who study individual and societal human behavior? Explain.

11. Can different kinds of behavior be used to show evolutionary relationships among related species of animals, as different kinds of proteins, bones, and muscles are used? Explain.

12. This problem is based upon the concept of gene-environment interaction upon phenotype. Indeed, it is based upon actual results of a research project with mice, but the animals and their behavior have been changed to protect the innocent. You are the director of a large research laboratory studying dolphin behavior. Since you are famous you are away from your laboratory most of the time, and in your absence one of your graduate students yielded to the temptation to express her scientific creativity. She bred two lines of dolphins, one with high, and one with low ability in probability. In one of the two large faculty swimming pools, she selected a pure line all of whose members learned to play poker after 10 practice games. In the other pool she selected a pure line whose members required 50 practice games, just to learn "five-card draw." Further selection failed to extend the extremes of brightness or of dullness. There came to be a problem of space, so half of the bright dolphins were transferred to the small and crowded student swimming pool, and half of the dull dolphins were transferred to a beautiful bay, near Progreso, Mexico. To the graduate student's surprise, the bright dolphins in the small student pool became as dull as the dull dolphins, and the dull dolphins near Progreso became as bright as the bright dolphins. The graduate student then demonstrated her own brightness; she transferred dull dolphins to the student pool and bright dolphins to Progreso, to discover that the dull dolphins are equally dull in the faculty and student pools, and that bright dolphins are equally bright in Progreso and in the faculty pool. The student telegraphed her results to you just as you were boarding a flight home. You have two hours to prepare your response to her data. Arrange your response in the form of a graph, and then in prose explain the meaning of the graph.

13. Explain the difference between Darwin's use of the word fitness, and Hamilton's use of the phrase inclusive fitness.

14. By use of the inclusive fitness theory, predict whose children men would most jealously protect in (1) monogamous societies, and in (2) promiscuous societies. Explain. Do you know of any actual data that can be used to support this theory? Explain.

15. Working within contemporary concepts of morally right and wrong, how would you try to determine the influence of genes upon human, altruistic behavior? Upon the laws against murder in all contemporary, industrial societies? Upon the relaxation of laws against murder during wartime?

Suggested Reading

1. Banfield, E. C., 1970. *The Unheavenly City*. Boston: Little Brown & Co.
2. Boch, K., 1980. *Human Nature and History: A Response to Sociobiology*. Columbia University Press, New York. An easy to read expression of the importance of human history to the study of human societal behavior.

3. Bouchard, T. J., Jr., D. T. Lykken, M. McGue, N. L. Segal, and A. Tellegen, 1990. "Sources of Human Psychological Differences: The Minnesota Study of Twins Reared Apart." *Science,* Vol 250, No. 4978, pp 223-228.

4. Chase, A., 1977. *The Legacy of Malthus*. New York: Knopf. This is a detailed history of eugenics, from Britain, to the U. S., to Germany, and back to the U. S. again.

5. Davenport, C. W., 1911. *Heredity in Relation to Eugenics*. New York: Henry Holt & Co. Davenport was probably the most avid eugenicist among those who have been classified as a professional geneticist. He had immense influence both in the public sector and within the scientific community.

6. Dawkins, R., 1976. *The Selfish Gene*. New York: Oxford University Press. A second edition of this book was published in 1990 that includes many responses, pro and con, to his first edition.

7. Galton, F. G., 1869. *Hereditary Genius*. New York: Macmillan.

8. Gould, S. J., 1981. *The Mismeasure of Man*. New York: W. W. Norton. Gould offers a comprehensive history of the attempts that have been made to measure human intelligence. He shows us the mistaken premises, mistaken methodology, and the statistical manipulations that have been devised to hide the mistakes, from the time that head size was used to measure intelligence to the era of IQ testing.

9. Haldane, J. B. S., 1938. *Heredity and Politics*. New York: W. W. Norton & Co. Haldane, a prolific and famous 20th century biologist, exposed the major problems of using genes to explain individual and societal behavior; his writings before World War II seem contemporary, except that he added more flare and humor that is usual today.

10. Herrnstein, R. J., 1971. *IQ in the Meritocracy*. Boston: Little Brown. Herrnstein is one of the most popular proponents of the view that the social class structure of society is genetic in origin, for example, that genes determine IQ, and that IQ determines social standing. This book was an expansion of the article, "IQ," which appeared in "Atlantic," September, 1971.

11. Hirsch, J., T. R. McGuire, A. Vetta, 1980. "Concepts of Behavior Genetics and Misapplications to Humans," in *The Evolution of Human Social Behavior*, Edited by Joan S. Lockard, Elsevier, New York. Even though this paper is a bit more technical than the text, its explanations of the misapplications of genetics to human behavior are very readable.

12. Hirsch, J., 1981. "To unfrock the charlatans." *SAGE Race Relations Abstracts*, 6(2), 1-65. Sage, London.

13. Hofstadter, R. 1959. *Social Darwinism in American thought*. New York: Houghton Mifflin Co. This book is a comprehensive coverage of the ideas of social Darwinism and Herbert Spencer.

14. Jensen, A. R., 1969. "How much can we boost IQ and Scholastic Achievement?" *Harvard Educational Review,* 39, 1-123. This article ignited the issue that intelligence is a valid criterion of "race" differences. It is as clear as any article published since in defense of the existence of "intelligence genes."

15. Kevles, D. J., 1986. *In the Name of Eugenics: Genetics and the uses of human heredity*. New York: Penguin Books.

16. Kitcher, P., 1987. *Vaulting Ambition: sociobiology and the quest for human nature*. Cambridge: The MIT Press.

17. Kuhn, T., 1962. *The Structure of Scientific Revolutions*. Chicago: University of Chicago Press.

18. Lewontin, R. C., S. P. Rose, and L. J. Kamin, 1984. *Not in Our Genes*. New York: Pantheon Books.

19. Rushton, J. P., 1988. "Race Differences in Behavior: A review and evolutionary analysis." *Personal Individual Differences,* 9, 1025-1033.

20. Scarr, S. 1984. *Race, Social Class, and Individual Differences in I. Q.* Hillsdale, New Jersey: Lawrence Erlbaum Associates. This book covers the topics listed in its title. It is controversial both in the topics picked for discussion and in the interpretation

of data, but it is well written and an effort was made by the author to be clear about both sides of every issue.

21. Shockley, W., 1972. "Dysgenics, Geneticity and Raceology." *Phi Delta Kappan,* Vol. 53, No. 5, p 305.

22. Slater, P. J. B., 1983. "The Development of Individual Behavior." *Animal Behavior,* edited by T. R. Halliday and P. J. B. Slater. Oxford: Blackwell Scientifiic Publications. This is a good place for those who are curious about animal behavior to start learning about concepts of behavior, and the difficulties of translating these concepts into experimental designs.

23. Spencer, H., 1969. *The Man vs. The State*. New York: Penguin.

24. Wilson, E. O., 1975. *Sociobiology: The New Synthesis*. Cambridge, MA: Harvard University Press.

Answers To Study Problems

Chapter 1

1. a) Explained by contrasting forms of genes:
   attached vs unattached ear lobes
   presence vs absence of mid-digital hair
   ability vs inability to roll tongue
   brown vs blue eyes
   sickle cell anemia vs non-sickle cell anemia
   b) Explained by contrasting experiences:
   belief vs disbelief in family region
   support vs no support for family political tradition
   science vs poetry
   monolingual vs polylingual
   c) Experiential modification: skin color via UV; a gene that leads to susceptibility to alcoholism via not drinking alcohol
   Medical modification: eye sight via corrective lenses; hearing via hearing aid; PKU via phenylalanineless diet

2. A norm of reaction is the range of phenotypic variation expressed by a single genotype. Since it is impossible to produce many copies of individual human genotypes, and to "test" identical genotypes (e.g., MZ twins) in many different environments, it is impossible to measure norms of reaction of human genotypes.

3. Many non-scientists perceive science as providing the means for discovering exact answers to questions about the physical world. Seldom is it perceived that science also produces abstract models of reality. Mendel's genes were not *real*; his factors were simply abstractions modeled to explain *real* segregation patterns of contrasting phenotypic traits.

4. If both parents are genotype MDHmdh, both will have mid-digital hair, but if their child inherits the *mdh* allele from both parents, that child will lack mid-digital hair. If either parent is MDHMDH genotype, none of the children is expected to lack mid-digital hair, and if both parents are mdhmdh all of their children will lack mid-digital hair.

5. mdhmdh x mdhmdh matings can never produce progeny with an MDH allele, that is with mid-digital hair, because both parents produce only mdh gametes. All eggs of mdhmdh females, and all sperm of mdhmdh males are mdh, which is to say that all progeny from such matings will be mdhmdh.

6.

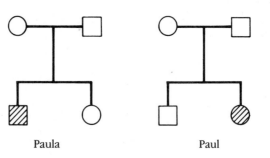

Paula   Paul

a) All of the parents, Paula's and Paul's, are heterozygous, Pp
b) Paula's PKU brother and Paul's PKU sister are homozygous, pp
c) The probability that Paula is Pp is 2/3; the probability that Paul is Pp is 2/3;
d) The probability that both of them are Pp is 2/3 × 2/3 = 4/9.
e) If both are Pp, the probability that their first child will be pp is 1/4; therefore, the probability of Paula's first child being PKU is 4/9 × 1/4 = 4/36 = 1/9.
f) The probability that their first child will be PKU is 1/9; that their first child will be a boy is 1/2; that their first child will be a PKU boy is 1/2 × 1/9 = 1/18.

7. a) The genotypes of I-1, I-2, I-3 and I-4 are Pphdhd; I-5 is PPhdhd, and I-6 is PPHdhd.
   b) The probability that II-3 carries the p allele is 2/3; the Hd allele, 0.
   c) The probability that II-6 carries the Hd allele is zero.
   d) The probability that III-6 carries the Hd allele is 1/2, and nearly zero for the p allele.

8. The genotype pp has the widest norm of reaction. PKU babies reared on normal infant diets become severely mentally retarded, but if PKU babies are fed a phenylalanineless diet they develop into normal children and adults. So far $\beta^S\beta^S$ individuals exhibit similar phenotypes in all environments within which they have developed from babies to children to adults.

9. A little knowledge about genotypes may be used to avoid a lot of human suffering. While genotype of one's mating partner may not be the most important issue facing a couple preparing to have a child, it also is not the least important.

10. a) The smooth-yellow seeds will be one of four genotypes, SSYY, SSYy, SsYY or SsYy.
    b) The four backcrosses, then will be
        SSYY × ssyy → all SsYy progeny
        SSYy × ssyy → 1/2 SsYy, and 1/2 Ssyy progeny
        SsYY × ssyy → 1/2 SsYy, and 1/2 ssYy progeny
        SsYy × ssyy → 1/4 SsYy, 1/4 Ssyy, 1/4 ssYy and 1/4 ssyy progeny
    The outcomes of self fertilization will be
        SSYY × SSYY → all SSYY progeny
        SSYy × SSYy → 1/4 SSYY, 1/2 SSYy and 1/4 SSyy progeny
        SsYY × SsYY → 1/4 SSYY, 1/2 SsYY and 1/4 ssYY progeny
        SsYy × SsYy → 9/16 S-Y-, 3/16 S-yy, 3/16 ssY-, and 1/16 ssyy progeny
    c) If S and Y are linked 10 map units apart, the results of one of the backcrosses will be
        SsYy × ssyy → 45% SsYy
                       5% Ssyy
                       5% ssYy, and
                       45% ssyy progeny

d) Smooth Green = SSyy or Ssyy
    Backcross: SSyy × ssyy → Ssyy
                  Ssyy × ssyy → 1/2 Ssyy : 1/2 ssyy
      Selfing: SSyy × SSyy → SSyy
                  Ssyy × Ssyy → 1/4 SSyy : 1/2 Ssyy : 1/4 ssyy
Wrinkled Yellow = ssYY or ssYy
    Backcross: ssYY × ssyy → ssYy
                  ssYy × ssyy → 1/2 ssYy : 1/2 ssyy
      Selfing: ssYY × ssYY → ssYY
                  ssYy × ssYy → 1/4 ssYY : 1/2 ssYy : 1/4 ssyy
Wrinkled Green = ssyy
    Backcross: ssyy × ssyy → ssyy
    Selfing: ssyy × ssyy → ssyy

11. A fertilized egg cell, or zygote, is a diploid cell. After it divides by mitosis (Figure 1.11a), it is transformed into a two-cell zygote, and both cells are diploid; in addition, the two cells have identical chromosomes and identical genes. If these two cells come apart, they will develop into identical embryo-fetus-child-adults.

12. No two eggs and no two sperm produced by the same individuals will be alike; sex cells are the outcomes of meiotic cell divisions (Figures 1.11a and 1.11b) during which homologous chromosomes recombine and alleles segregate into different cells. Therefore, no two zygotes of same parents will be genetically identical.

13. Population size = 1,000,000
    a) $1/10 \times 1/10 = 1/100 \; \beta^A\beta^S \times \beta^A\beta^S$ matings.
    b) From such matings, 1/4 of the children are expected to be $\beta^S\beta^S$, therefore $1/100 \times 1/4 = 1/400$ babies is expected to be SCA.
       If $\beta^A = p$, and $\beta^S = q$, then $\beta^A\beta^A = p^2$, $\beta^A\beta^S = 2pq$, and $\beta^S\beta^S = q^2$.
       If $\beta^S\beta^S = 1/400 = q^2$, then $q = 1/20$, and $p = 19/20$;
       Thus, $p^2 = 361/400$; $2pq = 38/400$; and $q^2 = 1/400$.
    c) $\beta^A = 19/20$ and $\beta^S = 1/20$, or, of 1,000,000 people, 10% (100,000) are $\beta^A\beta^S$ and 900,000 are $\beta^A\beta^A$, roughly 1,900,000 β^A and 100,000 β^S alleles.
    d) The β^S allele will decrease in frequency until an equilibrium is reached between the mutation rate from β^A to β^S, and the deaths of $\beta^S\beta^S$ individuals.

14. The number of Europeans living on the island of Mauritius is about 13,000, therefore the number of Hd genes is about 26,000. Twenty six people have had or have the disease, and 25 are at risk. The 26 people with the disease carry 26 *Hd* alleles and the 25 at risk carry at least 10 and as many as 16, which means, for the population, 36 to 42 *Hd* alleles per 26,000 genes, roughly 1 *Hd* per 650 *hd* alleles. On the shores of Lake Maracaibo, of the 2,600 descendents (5,200 Hd genes) of the founder, about 100 have the disease and about 1,000 are at risk. That is, of the 5,200 genes, 600 are in the *Hd* form, a ratio of 1 *Hd* to 86 *hd* alleles. In the general population the ratio is 1 : 40,000.

15. Sunlight darkens skin color: the dd genotype in rodents dilutes the colors initiated by at least three other genes.

16.

	Phenotypes	Genotypes	
BbCc × BbCc →	9/16 B-C- (black)	1/16 BBCC, 1/8 BBCc, 1/8 BbCC, and 1/14 BbCc	
	3/16 bbC- (brown)	1/16 bbCC, 1/8 bbCc	
	4/16 B-cc and bbcc (albino)	1/16 BBcc, 1/8 Bbcc, and 1/16 bbcc	

Chapter 2

1. a) The phrase, "information is stored in DNA," is a metaphor signifying that the primary structure of DNA determines the primary structure of RNA, and RNA the primary structure of proteins. It also signifies that parent DNA molecules encode progeny DNA molecules.

    b) Information is stored in DNA in the form of base pair sequences, a unique sequence for each gene and hence each protein. DNA is double stranded but usually only one of the two strands within each gene encodes mRNA and protein primary structure.

    c) Yes. Both strands encode their complements during DNA synthesis.

2. Mendelian genes are abstractions assigned to contrasting phenotypic traits if, and only if, those phenotypic traits are inherited in ratios that accord with the laws of chance. As discovered later, Mendelian genes obey the same laws of inheritance as are observed for homologous chromosomes; therefore it was postulated that Mendelian genes reside on or are integral to chromosomes. Later it was shown that specific segments of DNA can be followed through meiosis in gametes, and ultimately into progeny organisms. Today it is routine to follow specific genes (DNA) from parents to progeny such that both the laws of Mendel (the inheritance of contrasting phenotypic traits) and the laws of chemistry (the two alleles of a gene can be shown to be chemically different) are obeyed by the same gene.

3. Begin with any contrasting phenotypic trait, e.g., non-SCA vs SCA, or non-PKU vs PKU. Follow the inheritance pattern of these traits through two or three generations, to show that the inheritance patterns obey the laws of Mendel. Then discover the contrasting forms of a single protein and prove that each contrasting form of phenotype is always associated with one of the contrasting forms of the protein, e.g., non-SCA with Hb-A and SCA with Hb-S. Then locate the gene that encodes the protein, e.g., the β gene that encodes the β protein, and compare the β gene carried by non-SCA persons with the β gene carried by SCA persons. If the contrasting forms of the β gene are chemically different, and if these differences are always correlated with the differences between Hb-A and Hb-S, then the gene can be defined as a segment of the chromosomal DNA that encodes a single protein. The skills needed to understand probability problems are not found to obey the laws of Mendel, and no protein has been discovered whose contrasting forms are correlated with the ability and the inability of people to develop such skills.

4. The DNA of a human genome includes some 100,000 genes; the DNA of a broad bean chromosome may include hundreds; probably thousands of genes. The genomes of bacteria include several hundred genes, and of viruses between 5 and 6 and 100 genes. The experiments that provided evidence that the genetic material is DNA could not have explained what genes are because the experiments utilized whole genomes, not isolated genes.

5. Genetic information that is transmitted from one generation to another is always in the form of DNA. For example, the DNA that resides in a primary oocyte (distributed among 46 chromosomes) is transmitted via meiosis to secondary oocytes and then, usually, to three polar bodies and one egg cell (each with 23 chromosomes). Egg cell DNA is not transmitted to a succeeding generation, however, until the egg is fertilized. A fertilized egg cell (zygote) usually develops into an individual. Genetic information that is transmitted within body cells is first transcribed into RNA and then translated into proteins. The work of cells, tissues, and organs is performed by these genetically informed proteins.

6.

Metaphase	Anaphase	Replication	Metaphase

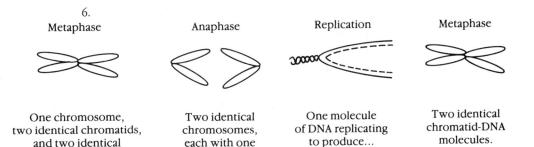

One chromosome, two identical chromatids, and two identical DNA molecules.

Two identical chromosomes, each with one molecule of DNA.

One molecule of DNA replicating to produce...

Two identical chromatid-DNA molecules.

7. The hemoglobin molecules of non-SCA persons (Hb-A) are made up of two β and two α proteins of known primary structure (the amino acid sequences of these two kinds of protein are known). The hemoglobin molecules of SCA persons (Hb-S) also are made up of two β and two α proteins. The α proteins of Hb-A and Hb-S are identical, but the β proteins differ in primary structure. Hb-A functions normally, does not stack, and does not alter the shapes of red blood cells. Hb-S does not function normally. In the absence of oxygen, the hemoglobin molecules stack, and the stacking changes the shapes of the red blood cells; the irregular shaped red cells often plug up capillaries, causing blood clots. The only difference between Hb-A and Hb-S β proteins is found at the 6th amino acid position, where Hb-A carries glycine and Hb-S carries valine. This amino acid difference between the two proteins accords with a single base pair difference between the β alleles, where CTC of β^A encodes gly and CAC of β^S encodes val.

8. The first 18 pairs of β$_A$ are: G T G C A T T T A A C T C C T G A G
    C A C G T A A A T T G A G G A C T C
The first 6 codons of mRNA are: G U G C A U U U A A C U C C U G A G
The first 6 amino acids of Hb-A are: val   his   leu   thr   pro   glu
a) To encode glu.his.leu.thr.pro.glu, the first codon could become GAG
b) To encode val.his.leu.thr.pro.val, the last codon could become GUG
c) To encode val.his.phe.thr.pro.glu, the third codon could become UUU
d) To encode val.his.leu.thr.his.glu, the fifth codon could become CAU
For these codon changes to occur the sense strand of the β$_A$ gene would have had to mutate to, a) C T C G T A A A T T G A G G A C T C
    b) C A T G T A A A T T G A G G A C A C
    c) C A T G T A A A A T G A G G A C T C
    d) C A T G T A A A T T G A G T A C T C

9.

	ncs DNA								
ncs DNA	TTT	TTT	TTT	GTG	**GTG**	ATG	CAT	ACT	TAA
cs	AAA	AAA	AAA	**CAC**	CAC	TAC	**GTA**	TGA	ATT
mRNA	**UUU**	UUU	UUU	GUG	GUG	AUG	CAU	ACU	**UAA**
tRNA	AAA	**AAA**	AAA	CAC	CAC	UAC	GUA	**UGA**	-----
protein	phe	phe	**phe**	val	val	**met**	his	thr	STOP

10.   pro ⌐
       GGA ⌋

       Growing
       protein
       ↓
       met   ← tRNA
       try   UAC ← anticodon
    his ⌐
    GUA
              phe ⌐ lys ⌐   ACC
              AAA   UUU

              CCUCAUUUUAAAUGGAUG   mRNA
gene (DNA) { GGAGTAAAATTTACCTAC   coding strand
             CCTCATTTTAAATGGATG   non coding strand

11. The *P* allele encodes functional phenylalanine hydroxylase. The *p* allele does not encode functional phenylalanine hydroxylase. In the absence of functional pheylalanine hydroxylase, the amino acid phenylalanine accumulates in the blood, is carried to the brain where it inhibits enzymes necessary to proper brain function. If PKU babies are provided a phenylalanineless diet, they develop normally. The genes are inherited. The enzymes are synthesized according to the information carried by these genes. The active enzyme converts phenylalanine to tyrosine, fast enough to ensure that phenylalanine does not accumulate in the blood. Normal health emerges. In the presence of two *p* alleles, excess phenylalanine, and abnormal health emerges.

12. If E is an essential metabolite, and if enzyme 4 is not synthesized in sufficient quantity to maintain normal health, then diet supplements of E may improve health. At the same time, metabolite D may accumulate and act as an inhibitor of one or more essential enzymes, or D may be transformed into another metabolite that acts as an enzyme inhibitor.

13. If the chromosomes of pp or $\beta^S\beta^S$ zygotes, destined to develop into PKU or SCA individuals, were implanted with a few P or β^A alleles, it might be possible to cure these diseases. With both diseases the genes and the enzymes necessary for health are known and the genes can be isolated in preparation for implantation. In the cases of alcoholism and schizophrenia the relationships between genes, enzymes, and diseases are not at all clear; indeed, very different relationships may exist.

14. The genomes of lytic viruses do not become integrated into host genomes, whereas genomes of some latent viruses do. So far it has been impossible to selectively target viral genomes within host genomes by means of a "magic bullet."

15. Avoid exchanging blood with persons who are infected with HIV. The virus is not airborne or waterborne, but resides within white cells of the blood. Whole white cells must be transmitted from carrier to non-carrier.

Chapter 3

1. A son of the mother's sister who bled to death following circumcision inherited the *h* allele from his mother. Since the two women are sisters the probability that the mother is a carrier is 1/2; if she is a carrier the probability of each of her sons inheriting the *h* allele is 1/2. Whereas, the son of the mother's brother who bled to death following circumcision inherited the *h* allele from the brother's wife, not from the brother.

2. a) about 3,000,000 — 2 million carried by 1 million females, and 1 million carried by 1 million males.
   b) 50,000 of the 1 million X chromosomes carried by the males carry the *d* allele, i.e., about one in 20. 1/20th of 3 million = 150,000.
   c) If 1/20 males carry the *d* allele, then $1/20 \times 1/20 = 1/400$ females must be dd.
   d) The probability of a dd × d- mating is $1/400 \times 1/20 = 1/8,000$
      The frequency of Dd females is $2 \times 19/20 \times 1/20 = 38/400 = 19/200$, therefore
      The probability of a Dd × d- mating is $19/200 \times 1/20 = 19/4,000$.
      The probability of a Dd × D mating is $19/200 \times 19/20 = 361/4,000$.

3. Boys never inherit the *d* allele from their fathers because boys never inherit an X chromosome from their fathers.

4. Given that the daughter had been socialized for 15 years to behave as a female, I would encourage her to continue her life as a female. Socialization is at least as important an influence upon behavior as is biology.

5. The models for developmental changes and cell differentiation are based upon too few data, that is, the actual mechanisms of change are unknown. However, based upon what is known, inactive genes become active at different times in the lives of cells and embryos. The triggering of activity may be due to changes in the gene's environment; these changes may be due to the actions of other genes. The sequence of gene-on-gene-off events is understood in the broad sense, but the actual circumstances of these events are not understood.

6. The male hormone, testosterone, is an important developmental trigger toward maleness; testosterone is necessary for masculinization. Therefore an XY embryo may develop toward femaleness if a) the receptor sites for testosterone are missing, or b) if one or more of the genes necessary to the synthesis of testosterone are mutant. The SRY gene is necessary to male development in that testes are not formed in srysry embryos.

7. XY males are heterozygous for sex chromosomes. Persons heterozygous at any gene locus will produce two kinds of gamete in equal numbers, e.g., Aa persons produce A and a gametes in equal numbers; for the same reason XY persons produce X and Y gametes in equal numbers.

8. Many males living in maximum security prisons exhibit anti-social behavior; also, all persons who know or suspect that they are objects of a "scientific" study will tend to behave differently than they otherwise might. If many males of all chromosomal composition were randomized and coded (so that neither they nor the investigators knew their chromosome composition) and allowed to engage in their normal lives without being disturbed, it might be possible to compare XY, XXY, and XYY chromosome types with respect to one or more aspects of behavior.

9. Within our society there exist cleancut social sex roles. Most nursing is done by women. Most surgery by men. Most child rearing by women. Most "bread winning" by men. Most women want to have children. Most men want to earn a wage. It looks natural and it is said to be natural. Used in this way the word natural means "caused by genes." However, we a) can't distinguish between social habits and biology by looking only at the phenotypes, and b) no one has isolated a gene for social behavior.

10. Skin color is inherited and can be altered greatly by ultraviolet light. Inherited eye shapes may alter vision greatly, but most bad vision can be corrected with eye glasses. PKU and many inborn errors of metabolism can be corrected by adding or subtracting metabolites to and from the diet. Susceptibility to various diseases can be offset with antibiotics.

11. Once males had acquired social hegemony they have wanted to keep it. Once in power they created explanations for why women are not, some of which were that women are not bright enough to hold and wield power, women are easily derailed from emotional stability by cyclic hormone production, women's creative energies are consumed by reproduction, etc. Once the social practice had been established wherein women managed homes and men earned wages, men then prepared themselves for politicking, doctoring, lawyering, and scientizing; women have not had these social opportunities throughout most of history. Until women have and feel that they have societal equality it will be difficult if not impossible to determine whether genes are determinants of societal sex roles.

Chapter 4

1. The Creationist: The earth was created less than 7,000 years ago. Species that exist today were created as such; species are immutable and do not evolve. The evidence for this view comes from the Bible, a literal explanation of the his-

tory of the earth and all forms of life thereon. The Biblical scriptures are literal truth.

The Evolutionist: All contemporary species evolved from pre-existing species, and these from the first forms of life which date back to about 3.6 billion years. All species gene pools are in flux; new variations arise from mutations; natural selection, drift, and gene flow continuously change allele frequencies within breeding populations. There is no purpose or direction to evolution; many of the driving forces of evolution are chance events. We know what we know about evolution through studies of the real world, whether fossil remains of past events or contemporary events and processes, living and non-living, not from authority or faith.

2. $A = 0.9$, $a = 0.1$. If $p = A$, and if $q = a$, and if $p + q = 1$, then the genotypic frequencies must be $(p + q)^2$, i.e., $p^2 + 2pq + q^2$. Thus, the frequency of the three genotypes will be AA = 0.81, Aa = 0.18, and aa = 0.01. That is, 18% of the population is heterozygous.

3. Zero. None. If $A = 0.5$ and $a = 0.5$, a new population might begin with 50% AA and 50% aa individuals, with 100% Aa individuals, or with 25% AA, 50% Aa and 25% aa individuals. In all cases the genotypic distribution will become 25% AA:50% Aa:25% aa. This distribution of genotypes will never change by means of sexual reproduction or genetic recombination, that is, in the absence of mutation, selection, chance, and/or gene flow.

4. New mutations may enter, remain in, or leave a gene pool without interference from natural selection if they are neutral and if chance falls their way. Mutations may be neutral with respect to redundancy of the code or with respect to non-conserved segments of protein primary structure. Neutral mutations do not rule out the influence of natural selection; many mutations are not neutral and may enhance or diminish the reproductive capabilities of individuals who inherit them.

5. Since mutations are rare events they exert little influence upon allele frequencies within gene pools. Within the context of evolution, the important thing about mutations is that they are the source of genetic variation within all populations.

6. Selection acts against the a allele by acting against individuals of aa genotypes. Selection cannot "see" the a allele in heterozygous individuals as evidenced in this case by the fact that Aa individuals have as many offspring as AA individuals.

7. *If aa = 1/10,000, then a = 1/100. Thus, A = 99/100. Then Aa = 2 × 99/100 × 1/100. *If aa = 1/100, then a = 1/10. Thus, A = 9/10. Then Aa = 2 × 9/10 × 1/10. Genetic drift, founder's effect, or different selection pressures in different environments; in some past environments Aa individuals may have been more reproductively successful than AA individuals.

8. Species are populations within which individuals cannot reproduce with individuals of any other species. Species are reproductively closed populations. So-called races are populations within species that are fertile with all other within-species populations, which leads to the definition of breeding populations. Breeding populations are defined not as populations surrounded by reproductive barriers, but as populations providing access to potential mates.

9. Breeding populations and ethnic populations have in common the feature that most matings occur between within-population individuals. However, breeding populations are described in terms of genetic composition (allele frequencies) whereas ethnic populations are described in terms of culture (food, music, dance, clothes, and so on).

10. In order to make meaningful estimates of heritability of continuously varying phenotypes it is necessary to be able to a) compare the phenotypic variation observed within populations of genetically identical individuals and populations of genetically diverse individuals, or b) perform selection experiments that allow one to measure the extent to which the population mean can be changed, up or down. Neither of these two kinds of program can be engineered with human populations.

11. MZ twins are genetically identical, therefore phenotypic differences between members of a twin pair cannot be ascribed to genetic differences between them. However, at most the genotype can be exposed to only two environments. In real life, most MZ twins enjoy very similar environments; only a few are adopted into different environments. But even here the "different" environments do not represent the full range of environmental differences experienced by the members of our species. Therefore the phenotypic differences observed between members of a twin pair can tell us that genes do not fix phenotypes absolutely, but they cannot tell us the norms of reaction of genotypes.

Chapter 5

1. A theory is the result of an attempt to unite all that is known about a phenomenon into a single generalization. Facts are verifiable small bits of information, usually the outcomes of research and careful observation. From theories hypotheses are generated (if that is so, then this must be so), but hypotheses are good only if they can be tested by experiment or careful observation. If the hypothesis is shown to be correct, new facts will have emerged from which the theory will have been strengthened. It is a fact that human chromosomes are composed of DNA and proteins; it is theory that all chromosomes of all plants and animals are composed of DNA and proteins. From the generalization we may deduce (hypothesize) that our pet dog's chromosomes are DNA and protein; a simple experiment proves the hypothesis correct; the theory is strengthened. Another hypothesis may suggest that the chromosomes of the bacteria in our gut are DNA and protein; an experiment tells us that the bacterial chromosomes are DNA only, with no protein. What does this do to our theory? It turns out that the theory is not correct for bacteria and viruses, and therefore requires modification. The relationships among theory, hypothesis, experiment, and fact are always in flux; they get better, but they are never perfect.

2. Not all 44XX females are biologically identical, but the vast majority fit the criteria used to define the biological female. The same is true for 44XY males. Sex chromosomes are easy to identify, and so are the morphological criteria of sex. Sexuality is a cultural phenomenon; its manifestations differ within and between cultures; sex chromosomes cannot be used to predict the sexuality of adults; cultural criteria are better predictors, but not at all good. This suggests that very little is known about sexuality.

3. PKU individuals (genotype pp) develop normal mental abilities and normal behavior if provided a phenylalanineless diet during their first decade of life. In other words they develop normally if their blood phenylalanine levels are low. Without diet control blood levels of phenylalanine are high because pp individuals do not possess normal phenylalanine hydroxylase, an enzyme that converts phenylalanine to tyrosine. High concentrations of phenylalanine inhibit the actions of other enzymes, some of which are necessary to normal brain development. Brains that do not develop normally often are correlated with abnormal behavior. But, the *P* allele of the P gene encodes normal enzyme, and with normal enzyme low levels of blood

phenylalanine are observed. But, low levels of blood phenylalanine do not tell us much about normal behavior, with which it is positively correlated.

4. Many people argue that intelligence is a social construct, not a biological thing, like an arm or a neuron. If intelligence is a social construct it is highly likely that some individuals will develop socialized intelligence and that others won't, and it is highly likely that such differences will be correlated with biological differences (tall males are more likely to fulfill the expectations of National Basketball League owners than short males or, for that matter, tall females; and there may be genetic differences between tall and short males; but if the baskets were lowered a foot and if only males shorter than 6 feet were allowed to play, the quality of play would be as good, and maybe better). But it will be difficult to find DNA sequences that encode the socialized concept of intelligence, which is different in different cultures, and which changes from time to time within a given culture.

5. No. Binet designed tests to identify the mentally retarded children who differed from other children who were not learning by lacking the capacity to learn subjects taught in Paris schools at the time. Binet's use of the concept of mental age (MA) did not include the concept of intelligence quotient (IQ). MA was simply a referent from which to identify children who could not learn.

6. IQ is determined by dividing mental age by chronological age and multiplying the quotient by 100. Chronological age is the actual age of a child, and mental age is determined by a test score (tests that now are called intelligence tests).

7. Since it is unclear what intelligence is, it is wrong to give it a definition. Personally I like the way Robert Graves deals with the subject in his poem "In Broken Images."

> He is quick, thinking in clear images;
> I am slow, thinking in broken images.
> He becomes dull, trusting to his clear images;
> I become sharp, mistrusting my broken images.
> Trusting his images, he assumes their relevance;
> Mistrusting my images, I question their relevance.
> Assuming their relevance, he assumes the fact;
> Questioning their relevance, I question the fact.
> When the fact fails him, he questions his senses;
> When the fact fails me, I approve my senses.
> He continues quick and dull in his clear images;
> I continue slow and sharp in my broken images.
> He in a new confusion of his understanding;
> I in a new understanding of my confusion.

Like beauty, intelligence may be in the mind of the beholder.

8. No. The tests may be biased with respect to the cultures of the two populations; the tests may not measure but a small fraction of total information characteristic of any culture; the fraction of mental ability measured may be more a function of socialization that of genetic determination. In the last analysis we don't know what intelligence is or what races are.

9. No. In the first place it is impossible to make reliable estimates of heritability within human populations, and in the second place if the estimates that have been made do in fact show a correlation between genotypic and phenotypic similarity, the genotypes that influence performance on so-called intelligence tests could just as easily influence biological phenomena that secondarily alter test performance, e.g., sight, hearing, metabolism, hormonal output and so forth.

10. Based on the ways that biological misinformation has been used against people in the past, and on similar values that influence human relations now, ethical

considerations could help to avoid some of the gross tragedies of the past. It always seems to help to have persons with a broader perspective than is needed to perform experiments oversee the societal outcomes of the usage of the knowledge gained through those experiments. Not that there exists an absolute ethical standard, but by paying attention to ethics better standards may be developed.

11. No. While the skeletons and muscles of the primates illustrate remarkable similarities among them, behavior does not. Some species of monkey are very playful and agile; most species of ape are less playful and much less agile; only members of the human species write books and sing in operas. There is no known biological relationship among the many species-specific behaviors.

12. First notice that only in one environment are the two lines of dolphins different in learning speed. In favorable environments dull and bright dolphins are bright; in unfavorable environments dull and bright dolphins are dull. That is, we see two lines of dolphins in three environments, but only in one environment is the genetic difference between them reflected by behavior difference. The data are shown in graph forms as follows:

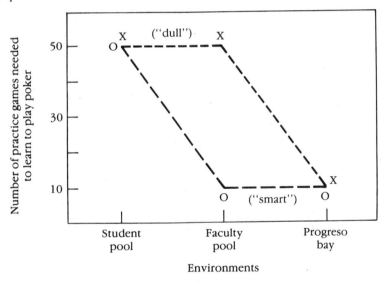

13. Darwin's usage of the word *fitness* included the reproductive success of a couple, of their children and their children's children, and so on. Hamilton's usage of the phrase *inclusive fitness* applies the same concept of reproductive success to kinships.

14. In a monogamous society a man would tend to mind and protect his own children, that is, the children resulting from matings with his wife. In a promiscuous society a man could never know if his wife's children were sired by him, so he would tend to mind and protect his sister's children. There are no data to back up this conjecture, and in our current society, neither entirely monogamous nor entirely promiscuous, we see little evidence that men without children, for whatever reason, go to any great lengths to help their sisters raise children.

15. The necessary experiments cannot be done. If in the future it becomes possible to make better estimates of the genetic influences upon societal phenomenon, the task of identifying a genetic influence will be eased, but only if the societal stimuli to steal, rape, and murder are removed.

Glossary

Acquired Immune Deficiency Syndrome, AIDS A disease of the immune system believed to be initiated by HIV and which, to date, progresses until death. However, the deaths of people with AIDS are due mainly to secondary infections, since the probability of contracting secondary infections increases as the immune system weakens.

Active Site The position, or site, on an enzyme or on any functional protein that recognizes the molecules that will be bound to and/or modified by that protein.

Adaptation Adaptations in the broad sense are equilibria among organisms and their environments. Both organism and environment are changed by the equilibrium. In the narrow sense, an adaptation refers to a single change in an organism that increases its chances of survival within a specific environment. The gene that causes such a phenotype "adapts" into the gene pool, and the phenotype caused by that gene is adapted into the biology of that species.

Additive Genetic Variance Phenotypic variance that is influenced by many genes, but that can be partitioned into small increments of variance each of which can be attributed to each participating gene. That is, the experiments designed to measure the additive effects of genes upon continuous phenotypic variation permit detection of the substitution of one allele for another upon examination of phenotypes.

Adenine One of the five bases and one of the two purine bases found in DNA and RNA. Adenine always pairs with thymine in double-helical DNA molecules and with uracil in DNA-RNA complexes. One of the four letters in the genetic alphabet.

Aggressive Behavior In general, behavior that seems necessary to reaching a stated goal, or that seems necessary to avenge a perceived wrong. What looks like aggressive behavior may not be perceived as such by the actor, but aggression that is provoked by pain or by frustration usually is perceived to be aggressive by the actor. Usually other people are the victims of such behavior. However, the many forms of aggression cannot be lumped into a single unit of behavior.

AIDS dementia complex, ADC A complication of AIDS characterized first by changes of behavior and second by deterioration of the central nervous system. It seems certain today that HIV attaches to and enters brain nerve cells, and that these cells are destroyed as T4 cells are destroyed. HIV may enter the brain by way of macrophages, through the bloodstream.

AIDS Related Complex, ARC About 90% of individuals who test HIV-positive show none of the symptoms of AIDS. About 1% are diagnosed with AIDS. The remaining 9% show some but not all of the symptoms of AIDS. Early medical descriptions of pre-AIDS symptoms included the phrase AIDS-related complex.

Alcaptonuria An "inborn error of metabolism" characterized (a) by an accumulation of homogentisic acid in the urine, which turns the urine black when it is exposed to oxygen and (b) by an inheritance pattern characteristic of autosomal recessive genes.

Allele Sex cells carry one copy of each chromosome and of each gene. Zygotes carry two copies of each. The two copies of each gene are called alleles, which may be identical (AA and aa) or different (Aa). Alleles occupy analogous positions (loci) on homologous chromosomes.

Altruistic Behavior In sociobiology, behavior that aids the reproductive success of other to the expense of self. In the vernacular, kind deeds and acts toward others.

Amino acid The "building blocks" of proteins. During protein synthesis, amino acids are joined together by the formation of peptide bonds (Figure 2.34); proteins are sometimes called polypeptides. Each amino acid carries a basic, an amino, and an acidic COO^- group (Figure 2.26), but the peptide bonds that unite amino acids into proteins destroy them, and their basic and acidic properties, except for the amino group on the amino acid at one end of the protein and the COO^- group at the other end.

Amniocentesis A medical procedure for removing fluid from the amniotic sac within which fetuses develop. Cells of the fetus become suspended within the amniotic fluid, and these can be removed, then cultured and inspected for a variety of genetic alterations (Table 2.2 and Figure 2.34).

Antibody Protein molecules (immunoglobulins) synthesized in response to specific antigens (non-self proteins). Mature antibodies assume specific three-dimensional shapes that allow them to bind tenaciously with the antigens that provoked their synthesis.

Anticondon Three base segments of tRNA molecules that attach to codons located on mRNA transcripts during protein synthesis. Anticodons are located on the side of tRNA molecules opposite the site of amino acid attachment.

Antigen Any substance that triggers the formation of antibodies. Most antigens are proteins.

Asexual reproduction Reproduction without the union of egg and sperm (i.e., without a genetic contribution from two individuals to a new individual). A bacterial cell will divide by fission into two identical cells; yeast cells by the budding of a progeny cell from a parent cell; strawberry plants and sequoia trees by underground "runners" (rhizomes) from which roots develop downward and stems upward.

Avirulent Refers to a strain of bacteria that does not cause disease and is not harmful to its host. As used here, avirulent refers to a strain, or mutant form, of an otherwise virulent strain of *Pneumococcus.*

Bacteriophage The name given to viruses that infect bacteria; also called phage.

Base Pair The two strands of each double-helical DNA molecule are joined by A-T and G-C pairing. Adenine, Thymine, Guanine, and Cytosine are nitrogen-based molecules, and this is abbreviated, base; therefore, the A-T and G-C pairs are called base pairs (bp for short). The information encoded by a DNA molecule is in the form of a specific sequence of these base pairs.

Behaviorism A school of psychology based on the view that an organism's behavior is an objective measure of one or more forms of its mental activity, particularly in the case of stimulus-response behavior. This leads to the view that learning is the most important aspect of an organism's development.

Biological Determinism Explanations of individual and societal behavior as outcomes of gene action. Biology is destiny.

Blastocyst An early stage (reached about five days after fertilization) of embryogenesis characterized by a nearly hollow ball of 100 to 150 cells with an inner cell mass, the amnion, that will form the embryo proper. The outer layer of cells is called the chorion and will become, among other things, the outer layers of skin, nails, hair, etc.

Breeding population A within-species population recognized by a high frequency of within-population matings. Members of a breeding population are more likely to mate with members of their own than with members of different breeding populations.

Breed True Parents that breed true have children genetically identical to themselves; there is no variation among the offspring.

CA Chronological age, the time between one's birth and the present.

Cancer General name given to uncontrolled cell growth. Uncontrolled cell growth may give rise to a tumor or to "floating" cancer cells, as in lymphoma and leukemia.

Carcinogen Any agent, whether chemical or a portion of the electromagnetic spectrum, that induces a genetic or chemical change within a cell that transforms it into a cancer cell.

Carrier A person heterozygous at a gene locus, the recessive allele of which initiates a genetic disease if homozygous.

Catalyze The process of greatly speeding the rates of chemical reactions. Substances or molecules that speed up chemical reactions, without being changed in the process, are called catalysts. Biological catalysts are enzymes.

Cause-effect relationship The phrase cause-effect refers to the observed change(s) associated with the action(s) of one thing on another (others). A moving billiard ball will cause another to assume motion if the first hits the second. In genetics the issue is the action(s) of gene(s) upon phenotype.

Central Dogma After it was discovered that genetic information is transmitted from its source, DNA, through RNA to proteins, the discovery was said to be universal (i.e., genetic information is the same within all species, and therefore it is transmitted in the same way in all species). Enthusiasts referred to these "universal" properties as the central dogma of genetics.

Chance Events that occur without known causes or forces and without design. It is a matter of chance whether a tossed coin lands head or tail, whether the *A* or the *a* allele will be included within the next egg cell produced by a heterozygous woman, etc.

Chorion villus sampling A procedure by which cells of developing embryos, called villi cells, can be aspirated (a form of biopsy) from the embryo. As with any biopsy, the isolated cells can be cultured and examined for gene and/or chromosome alterations.

Chromatid Prior to the segregation of homologous chromosomes, during the first stage of meiosis, each chromosome is a double structure. Each half of each chromosome is called a chromatid. The two chromatids of a chromosome are identical. During the second stage of meiosis the chromatids of each chromosome may be different and each will separate and move to opposite poles of the cell, at which time each becomes a mature chromosome.

Chromosome Long, rod-shaped structures found in the nuclei of cells. The chemical composition of chromosomes, discussed in Chapter 2, helps to explain the linear arrays of genes along their lengths.

Chromosome map Genes appear in a linear order, from one end to the other, along each chromosome. A chromosome map shows this order. There are several methods for constructing chromosome maps, one being that of observing recombination between linked genes.

Clone An individual cell or organism or a population of individuals derived from a parent cell or organism by any one of several forms of asexual reproduction.

Clotting Factor VIII A protein that functions about midway in the cascade of chemical reactions that lead to the synthesis of fibrin, a protein necessary to the blood clotting process. If clotting factor VIII is missing, the blood does not clot, which is the case for persons with hemophilia A.

Codon A genetic "word." Three bases of mRNA to which anticodons of tRNA attach during protein synthesis. Each codon calls for one amino acid, but since there are 61 codons and only 20 amino acids, some amino acids are invited to mRNA molecules by more than one codon.

Complementary The two strands of a DNA molecule are complementary; each is a complement of the other in that the base sequence of one is determined by the base sequence of the other. Such complements are neither exact copies nor mirror images of one another; indeed, complements are oriented in opposite directions and at each homologous base pair attachment site the complementary strands present a different base.

Components of variance Variance refers to phenotypic differences among individuals within a specified population. Components of variance refer to the forces that contribute to that variation. For example, different genotypes may account for variation of eye color, and different diets for some of the variation of body size. Genotypes and diets are component of variance.

Concordant Individual objects of a type are characterized by their size, shape, and articulation of parts. There is a consistent pattern that best describes the unity of objects in forming a type or class (e.g., human females and human males). Females whose individual size, shape, and articulation of parts fits the general pattern are said to be concordant with respect to all of the criteria used to form the pattern; the same is true for males. See discordant.

Conservative replication   Prior to the Meselson-Stahl and the Taylor experiments, it could not be ruled out that during DNA replication both strands of a parent DNA molecule might be conserved and that both strands of a progeny molecule might be new. The phrase conservative replication was reserved for this possibility. See semiconservative replication.

Contrasting Forms of Phenotype   See phenotypic trait.

Correlation   Related variation of two variables during the same course of time. For example, among children, as height increases, weight increases; two variables that increase together are positively correlated.

Creationist   Fundamentalist Christian whose view of human origins arises from a literal interpretation of the Old Testament, and who actively opposes the teaching of evolution in public schools.

Cross   When the geneticist makes a forced mating between two plants or two animals to study the phenotypic outcomes and inheritance patterns of certain alleles, the forced mating is called a cross (e.g., a cross between a green and yellow plant or between a white and a black mouse).

Cross-fertilization   The union of gametes produced by different parents, independently of whether the parents are hermaphroditic.

Culture   All of the ways of living developed by a group of people that are passed from one generation to the next, including everything from table manners to music, art, literature, science and history. Legal systems codify certain ancient religious beliefs, but in no case do we have a complete historical record of any culture.

Cytogenetics   The study of the internal structures of cells, but most often the word is used to describe the internal structures of

cells that participate in the transmission of genetic information.

Cytosine   One of the five bases and one of the three pyrimidine bases found in nucleic acids. The other pyrimidines are thymine (found only in DNA) and uracil (found only in RNA); the two purine bases are adenine and guanine.

Data   Facts and figures obtained from experiments or observations that are used in making generalizations or conclusions about the objects or subjects being studied.

Deoxyribonucleic acid, DNA   Large molecules found in all genomes (except for a few viruses) that store, replicate, and transcribe genetic information. The molecules of which genes are composed. DNA is double stranded; the strands are complementary; both strands serve as templates during DNA synthesis; one strand, the sense, or coding, strand, serves as a template during RNA synthesis.

Deutan red-green color-impaired   A form of red green colorvision impairment characterized by seeing red and green as red. The green-sensitive opsin is either missing or mutant. Therefore both red and green light is registered by the red-sensitive opsin.

Development   The processes of growth and cellular differentiation that ultimately transform fertilized eggs into embryos, embryos into fetuses, and so on into old age and finally death, at which time the processes stop.

Developmental Genetics   A growing branch of genetics that seeks for genetic causes of development, that is, of cell growth and of cell differentiation.

Differentiation   The complex processes that oversee the emergence of specialized cells from stem cells. For example, during the time of embryonic development, all of the major organs are developing, which means that heart muscle cells; leg muscle cells; eye, ear, toenail, and hair cells are becoming different from each other and from the original zygote cell, all the while remaining genetically identical.

Digestion   A general name given to the processes of breakdown of relatively complex biochemical substances into relatively simple ones, e.g., of starch into sugars, and of sugars into CO_2, H_2O, and energy, of proteins into amino acids, and so on). Digestion is the opposite of synthesis.

Dihybrid In general a hybrid is the offspring of parents that are genetically different. Specifically a monohybrid is heterozygous at one gene locus, and a dihybrid is heterozygous at two gene loci.

Diploid Two sets of chromosomes; an organism or cell carrying two sets of chromosomes.

Discordant See concordant. Objects with size, shape, or articulation of parts of more than one type. For example, persons with femalelike bodies and malelike testes are discordant with respect to the criteria used to distinguish between female and male sexes.

Dispersive replication See conservative replication. Before it became known that DNA is synthesized semiconservatively, this phrase was reserved for the possibility that DNA synthesis is preceded by the partial digestion of parent DNA molecules and that synthesis used old DNA segments salvaged from the digested molecules to form progeny DNA molecules.

Distribution pattern An arbitrary description, or depiction, of classes of phenotypic variation. See histogram. Variation of human height can be depicted in classes of 1 cm, 1 inch, or 4 inches, or as a bell-shaped curve. The point is to show the population's distribution of heights in terms of a range and the frequency of persons in each height class.

Dizygotic twins Children born at the same time, of the same mother, that resulted from the fertilization of two eggs.

Dominant One of two words that describe a phenotypic hierarchy among alleles. Given the genotypes, AA, Aa, and aa, there will be two or three possible phenotypes. If there are two phenotypes, that is, if Aa is like AA, the *A* allele is said to be dominant to the *a* allele.

Down's syndrome A syndrome characterized by an extra 21st chromosome (i.e., trisomy 21), and markedly different physical and behavioral features from those expected of 44XX and 44XY norms.

Egg Cell Female gamete; cell that when fertilized by a sperm cell becomes a zygote; haploid, female sex cell.

Ellis-van Creveld Syndrome A rare genetic condition characterized by short stature and six fingers on each hand. This syndrome was introduced into the Amish population (of Lancaster County, PA) in 1744 by Samuel King or his wife. Today there are nearly as many individuals with the syndrome among the Amish as there are in the rest of the world.

Embryo A beginning stage of development. In the case of humans, the stage between the first cell division of a zygote and the emergence of a fetus, roughly the first two months after fertilization.

Encode To make or to send a coded message. The genetic code is stored within DNA molecules in most species of plants, animals, and microorganisms, and genetic codes are relayed to RNA molecules and then to protein molecules. In other words, DNA encodes RNA, which in turn encodes proteins.

Endonuclease A class of enzymes characterized by their ability to digest (cut) nucleic acid molecules, DNA and RNA.

Environmental component of variance The proportion of total phenotypic variance that can be attributed to environmental differences among the population's members. The ratio of environmental variance to the total phenotypic variance.

Enzyme All enzymes are proteins, and their general function is to increase the rates of chemical reactions within cells, tissues, and organs, that is, enzymes are biological catalysts.

Epidemic Within the field of medicine, an epidemic refers to rapid spread or increase in the incidence of a disease if and when the incidence surpasses what is called the "normal" incidence.

Ethnic A common cultural, religious, or social background that unites a group of people. The Lakota Native Americans, for example, share an ethnic experience that sets them apart from other ethnic groups.

Ethology A branch of biology that concerns itself with non-human animal behavior, in particular their behavior in natural environments and as an extension of their physiology.

Eugenics The name coined by Francis Galton to describe societal programs designed to improve the quality of the "human stock" by selection and breeding. His program was designed after plant and animal improvement programs, which were designed to make bigger, better, and higher yielding varieties and stocks.

Exon A segment of genes, that is, of transcribing units of DNA whose base sequences are represented in mature transcripts—mRNA molecules—and in the amino acid sequences of the proteins encoded by genes. See intron.

Explanation In philosophy, explanation is revealing the essence of the object in question. In science it is showing that the object obeys or fits into a certain law or theory. In both science and philosophy, explanations are restrained or expanded by knowledge of the object or phenomenon in question. Explanations may include cause-effect relations as well.

F$_1$ First filial generation. Mendelian genetics experiments usually begin with parents that are homozygous for different alleles of the same gene. The heterozygous progeny resulting from a mating between such parents are called F$_1$ progeny.

F$_2$ Second filial generation. The progeny of F$_1$ matings.

Factor What are referred to today as genes Mendel called factors. He proposed that factors are passed to progeny through sex cells and that contrasting forms of factors account for contrasting forms of phenotypic traits. Factors were renamed genes in 1906.

Family Pedigree A history of all members of an extended family that records all members with and without certain phenotypic traits, whether eye color or other morphological differences among members of the family, or genetic diseases. See pedigree chart.

Fertilization Union of two sex cells, egg and sperm, resulting in the formation of a zygote.

Fitness, Darwinian A term that Darwin used for reproductive success. Individuals who contribute genes to succeeding gene pools are **fit**; those who do not are **unfit**. But the spectrum of differences between fit and unfit shows that fitness is a quantitative measure of reproductive success.

Founder effect A phrase given to a known source, an individual or a small population, of a specific allele within the gene pool of the population founded by the source. For example, Samuel King and his wife are founders of the Ellis-van Creveld condition within the Amish population in Lancaster County, PA.

Functionalism William James is credited with shifting the emphasis from the structure to the function of conscious experience, as a means for understanding human behavior.

Gametes Haploid (one set of chromosomes) sex cells. Among animals male sex cells are called sperm, and female sex cells are called eggs. Male sex cells of plants are called pollen grains. The union of male and female sex cells produce zygotes (with two sets of chromosomes).

Gametogenesis The genesis of gametes. The general name given to the processes by which gametes are formed, in both sexes, and in all species that form gametes.

Gender identity The sex one "feels"himself or herself to be, which usually is not but may be discordant with biological sex. For example, transsexuals are discordant with respect to biological sex and gender identity; other people, knowingly or not, may be less strikingly discordant.

Gene Mendelian genes are defined as units of inheritance recognized by their influence upon phenotype. Their locations at specific sites on specific chromosomes are determined by frequencies of recombination with other genes. Mendelian genes are mutable; indeed, a gene cannot be discovered unless it exists in two or more allelic forms. Molecular genes are defined in terms of the information they encode. Commonly a gene is known as a segment of DNA that encodes the amino acid sequence of a protein, or the base sequence of an RNA molecule.

Gene flow Migrations into and out of breeding populations are, genetically gene flows into and out of populations. Usually migration describes the movements of organisms, but we could as well refer to gene migration, or gene flow.

Gene Pool The sum of genes carried within all of the genomes of all individuals within a species or a breeding population with the potential of being included in the next gene pool.

Genetic Code The information carried within mRNA molecules in the form of three-letter segments called codons. The sequence of the letters, A, U, G, and C, three at a time, are code words for amino acids. See Figure 2.31, the dictionary showing the coding relations between all codons and all amino acids.

Genetic Coding The processes of transferring genetic information from DNA to RNA and from RNA to protein.

Genetic drift Differences of allele frequencies between parent and progeny gene pools may occur by chance in that the matings that give rise to progeny gene pools do not select perfect, representative samples of alleles from the parent gene pools. These chance deviations are said to arise from gene drift.

Genetic variability Populations seldom are made up of genetically identical individuals. The extent to which genetic differences characterize individuals within a population is the extent of genetic variation therein.

Genome The full complement of genes and chromosomes carried by an individual.

Genotype The sum of genes within a cell or within the cells of an organism. In practice the word genotype is used to describe only the one or the few genes being studied at the time (e.g., the gene associated with eye color or Huntington's disease, as the genotypes BB and Bb lead to brown eyes and the genotype bb to blue eyes).

Geographical isolation Breeding populations within species can be kept from interbreeding if they are isolated by any one of a number of geographical barriers (e.g., a mountain range, a river, or large body of water). Any geographical barrier that prohibits matings between two populations.

Germ Line cells Cells that give rise to sex cells, that is, cells that carry the chromo-somes and genes that will be passed from one generation to the next. In humans the germ line cells are carried by ovaries and testes; the sex cells are eggs and sperm.

Growth Either an increase in cell number or both an increase in cell number and in cell size, in particular during the processes of embryonic, fetal, and child development.

Guanine One of the two purine bases, and one of the five bases found in all nucleic acids; the other purine base is adenine, and the three pyrimidine bases are cytosine, thymine, and uracil.

H-Y Antigen The H-Y antigen is encoded by a gene on the long arm of the Y chromosome. When it was discovered it was thought to be the testis determining factor, but recent experiments show that it is not. It is a uniquely male antigen, and its function is not known but is now thought to be important to spermatogenesis.

Haploid One set of chromosomes. A cell carrying one set of chromosomes.

Hardy-Weinberg equilibrium The observation and mathematical expression of the fact that allele and genotype frequencies will remain constant within breeding populations from generation to generation unless acted upon by one or more forces of change (e.g., natural selection, migration, drift, and mutation). In other words, reproduction and genetic recombination do not disturb allele and genotype frequencies.

Hemoglobin A protein found in red blood cells whose main function is to transport oxygen from the lungs to body cells and carbon dioxide from body cells back to the lungs (See Chapter 2).

Hemolytic anemia A form of anemia caused by the deaths of red blood cells as they burst from internal pressure. Weakness of red blood cell membranes accompanies the loss of the enzyme glucose-6-phosphate dehydrogenase.

Hemophilia A An X-linked recessive condition characterized by the inability of the blood to clot.

Heredity The study of the transmission of traits from parents to offspring, that is, the study of why offspring resemble their parents.

Heritability As used here, the ratio of the genetic variance to the total phenotypic variance, that is, the proportion of phenotypic variance that can be attributed to genetic variance.

Hermaphrodite An individual with both female and male reproductive organs, as a pea plant or earthworm. Originally an organism possessing opposite sex qualities, originating in Greek mythology: the son of Hermes and Aphrodite was named Hermaphroditos.

Heterogenic As used here, an intra-species population of organisms that are genetically different from one another.

Heterozygous The state of carrying different alleles at one or more gene loci (e.g., AaBb).

Histogram A diagram showing the distribution pattern of classes of quantitative variation. Each class is represented by a vertical column, and the height of the column indicates the frequency of individuals from the population in that class.

Histones Proteins found on chromosomes that bind tightly to DNA. There are five classes of histones that are found in all plant and animal chromosomes; their primary function is to "spool" long DNA molecules into highly coiled structures as DNA molecules become condensed into chromosomes.

Homologous chromosomes Each of the 23 chromosomes of a human egg cell has a counterpart among the 23 chromosomes of a human sperm cell. Zygotes, then, carry 23 pairs of chromosomes. The members of each pair are structurally identical, that is, homologous. One member of each pair is the homologue of the other.

Homologue Each chromosome of a homologous chromosome pair is a homologue of the other. One of a pair of homologous chromosomes.

Homozygous The state of carrying identical alleles at one or more gene loci (e.g., AABB or aabb).

Human Immunodeficiency Virus, HIV A virus discovered in 1983 that is believed to be the causative agent of AIDS. The virus attacks certain white blood cells, and, upon killing them, weakens immune systems. People with weakened immune systems become susceptible to bacterial and viral infections. Most people with AIDS die of secondary infections.

Human Leukocyte Antigen, HLA Antigens located on the surface of many kinds of blood cell, encoded by four major genes located on chromosome 6. Since there are many alleles of these four genes (e.g., 23 for HLA-A, 47 for HLA-B, 8 in the HLA-C gene, and 23 in HLA-D), there are thousands of combinations within the human population. However, each person inherits four HLA alleles from each parent, making a total of 8; thus each individual will carry between 4 (if homozygous at all four gene loci) and 8 (if heterozygous at all four loci) HLA antigen types.

Human Society All the ways of living of a group of people that are expressed within a prescribed time frame; a time segment of a culture. Human societal behavior is distinguished from animal social behavior because it changes rapidly through time by eliminating old and adding new practices; it is more easily modified by learned behavior.

Huntington's disease A genetic disease initiated by a rare gene, known as an autosomal dominant mutation. Persons who inherit the causative gene usually do not exhibit symptoms of the disease until during the fourth decade of life, the first symptoms of which signal a degeneration of the nervous system that continues until death.

Hybrid An individual whose parents have different genotypes. Such an individual is heterozygous at one or more gene loci.

Identical Twins See monozygotic twins.

Immunology The study or science of the immune system—the cells, the genes, and the proteins that function to determine self-antigens from nonself antigens and processes by which immune systems destroy the latter.

Inborn Errors of Metabolism Metabolism is a word that symbolizes all of the chemical reactions that take place in the cells of our bodies. Many of these reactions are absolutely essential to health and life. Each chemical reaction is dependent upon a functioning gene; thus errors of metabolism are often due to nonfunctioning genes (i.e., inborn errors).

Inbreeding The tendency for persons with like genotypes to mate. Geneticists working with organisms like *Drosophila* and mice will produce large populations of nearly genetically identical individuals by making brother-sister matings for many generations.

Inclusive Fitness Darwinian fitness, i.e., of individuals and their offspring, plus the fitness afforded by kin selection, the passage of one's genes into subsequent gene pools by relatives other than offspring.

Information, genetic In general, information is knowledge gained, stored, transmitted, and received. The sequence of base pairs within the DNA segments called genes is called genetic information. Genetic information is stored in the form of the chemical structure of genes and is replicated and passed to progeny generations of cells and organisms. Genetic information is transformed into biological work through proteins, whose primary structures are determined by it.

Inheritance Patterns The frequency of appearance in families, in successive generations, of contrasting phenotypic traits. The contrasting traits caused by contrasting alleles of one gene appear in accordance to the laws of chance and in accordance to the segregation of homologous chromosomes during meiosis.

Innate Behavior Inborn behavior, or absolutely genetically determined behavior.

Intron Segments of DNA within genes that are transcribed into pre-mRNA molecules but whose information is excised from pre-mRNA molecules as they are processed into mature mRNA transcripts. See exon.

Introspection The subject's interpretation of his/her own behavior, a technique of observation introduced by structural psychologists.

IQ-Genetic Meritocracy An explanation of the class structure of society based on the idea that genes determine IQ and IQ determines individual positions within the class hierarchy, i.e., merit is determined by genes.

Isogenic A population of genetically identical, or nearly identical individuals.

Isotope Chemical elements exist in two or more forms as judged by atomic weight and nuclear properties. Different forms of the same element are called isotopes. Isotopes of the same element have the same chemical properties but different atomic weights.

Kaposi's sarcoma A type of cancer associated with HIV infection and AIDS. Kaposi's sarcomas spread without metastasis, and they form darkly colored spots just under the skin layer.

Kin-selection The transmission of one's genes into the next gene pool by relatives other than one's offspring, a process that extends the concept of fitness to inclusive fitness.

Klinefelter's syndrome Individuals with 47 chromosomes, 44XXY, who appear as males at birth but who often develop femalelike secondary sex characteristics.

Labeling The process of adding "marker chemicals" (e.g., isotopes) to normal biological molecules (e.g., proteins, nucleic acids); a technique that permits the scientist to identify "marked biological molecules" by the properties of the isotope used (e.g., radioactivity).

Latent infection Infection by a class of viruses whose genomes become integrated within host genomes and whose effects on host cells during the time of integration may be nil. However, in time, the integrated, viral DNA will transcribe its information into RNA transcripts, which at once become mRNAs that direct the synthesis of new viral proteins, and new viral genomes that will be encased within the viral coat proteins. After the new viral particles are assembled, the host cells are killed and the particles are released.

Laws of Chance Refers to the probability of things happening if and when we do not understand what causes them to happen. For example, we do not know what causes a coin to land head after it has been flipped into the air, but we do know that the probability that it will is 1/2.

Learned Behavior Behavior that has developed from experiences and from knowledge actively learned.

Linked genes Genes located on the same chromosome are said to be linked. The linkage between linked genes is broken by recombination between homologous chromatids in frequencies roughly proportional to the distance between the genes.

Locus The site of a gene along the length of a chromosome (i.e., the gene locus).

Lymphocyte, T and B T and B lymphocytes develop from lymphoid stem cells. B lymphocytes produce antibodies after being triggered by T4 cells. There are two types of T lymphocytes, T4 and T8. T4 lymphocytes are the first to respond to foreign antibodies, and this response is signaled to other cells in the immune system. T8 cells bind directly with invading organisms and are called killer cells. T8 cells also emit "all-clear," and "back-to-normal" signals after the "invasion" has been thwarted.

Lymphoid stem cell One of two types of secondary stem cells that arise from the primary stem cells that give rise to all blood cells; lymphoid stem cells eventually differentiate into T and B lymphocytes.

Lytic infection Infection caused by a class of viruses that, upon entry into a host cell, stops the host cell from responding to its own genome and switches it to respond to the genetic information encoded within the viral genome. The viral genome thereupon is replicated and transcribed and its coat proteins are synthesized quickly. New virus particles are assembled, and finally the host cell is lysed and the new virus particles are released to attack neighboring cells.

MA Mental age, in psychology an IQ test score.

Mean The arithmetic average. For example, the mean height of a population is calculated by summing the heights of all the individuals and dividing that figure by the number of individuals measured.

Medical-research community Medical refers to the entire health care delivery system (physicians, nurses, hospitals, supplies, and equipment), and research refers to the scientists and scientific discoveries used by the health care delivery system.

Meiosis One of two types of cell division (mitosis is the other). During meiosis the chromosome number is reduced by one half, as when diploid germ line cells give rise to haploid sex cells, or gametes. In humans, diploid germ line cells carry 46 chromosomes and gametes carry 23. See Perspective 1.1

Messenger RNA, mRNA An mRNA molecule is a transcript of a gene that encodes the amino acid sequence of a protein. mRNA transcripts translate genetic messages into protein primary structures during protein synthesis.

Methodology A form of logic that follows the usage of certain procedures and methods to gain further knowledge within a specialized branch of science.

Mitochondrial DNA Most cells are inhabited by organelles called mitochondria; these organelles possess a tiny, circular chromosome, composed of DNA, that resemble the chromosomes of bacteria.

Mitosis One of two types of cell division (meiosis is the other). Mitosis is sometimes called equational division because the progeny cells of a mitotic event are genetically identical to the parent cell. Prior to cell division, every chromosome in the cell will have undergone a doubling, such that every chromosome will consist of two chromatids. Mitosis ensures that each progeny cell will receive one of the two chromatids of each chromosome.

Mode The most frequent class within a distribution pattern.

Monohybrid See dihybrid. A monohybrid is heterozygous at one gene locus.

Monozygotic twins Two children born at the same time, of the same mother, resulting from the fertilization of one egg. Following the first division of the zygote into a two-celled embryo, the two cells become separated, and each develops into an individual. MZ twins, therefore, are genetically identical.

Müllerian duct Five-week-old embryos possess a pair of hermaphroditic gonads. To each are attached two ducts, one müllerian and the other wolffian. As the embryo develops, the hermaphroditic gonads develop into ovaries within XX embryos and into testes within XY embryos. As the process proceeds, müllerian ducts become fallopian tubes within XX embryos and disintegrate in XY embryos.

Multigene family A group of genes with a common origin, that is, with a common ancestor gene. Genes are sometimes duplicated within genomes by unequal crossing over, after which a mutation in one is harmless because the normal function is performed by the other. Mutations that occur within one of a pair of such genes may lead it to encode a different function. This process of duplication and mutation may lead to multigene families of 6 to 12 genes, each with different but related functions.

Mutagen Any agent, whether chemical or a portion of the electromagnetic spectrum, that induces gene mutations.

Mutate, Mutation To mutate refers to the action characterized by a change from one gene form into another; mutation refers to the end result (i.e., a change in a gene results in a mutation, or a mutant gene).

Myeloid stem cell One of two types of secondary stem cells that arises from the primary stem cells that give rise to all blood cells. Myeloid stem cells, in turn, further differentiate into red blood cells, platelet cells, and a few types of white blood cells.

Nanometer One billionth of a meter, one thousandth of a micron, and about one 25 millionth of an inch.

Natural Selection The sum of natural circumstances that act upon individuals within species and breeding populations to determine their reproductive success or fitness.

Neutral mutations Mutations said to be "blind" to the forces of natural selection (i.e., mutations that remain in, or are lost from gene pools by chance alone).

Nature (human, biological, social) Often the phrase human nature refers to the characteristics of being human, but just as often the phrase implies that all human characteristics are biological.

Neo-Darwinian Synthesis The concept of biological evolution introduced by Darwin and modified by Mendelian genetics during the 1920's and 1930's. Of the three principles united in theory by Darwin, variation, heredity, and selection, it was heredity about which Darwin knew little; the new science of Mendelian genetics supplied the missing information, and with it Darwin's theory became the neo-Darwinian synthesis.

Neutrophil A class of white blood cells characterized by its direct attacks upon all invading microorganisms, in particular bacteria and viruses.

Nondisjunction The failure of homologous chromosomes to disjoin, or segregate, during meiosis. Following nondisjunction, one secondary gametocyte will carry both homologues and the other neither.

Norm of Reaction The simplest example of an NoR is the variation observed among a population of genetically identical individuals, divided so to experience a wide variety of environments, for example, the variation in height expressed at different elevations by plant 1 (Figures 1.3 and 1.4). A complete norm of reaction is the full range of phenotypes expressed by a single genotype over a wide range of environments.

Nonsense Sequences of DNA base pairs or RNA bases that do not encode functional proteins. Meaningless sequences.

Objectivity Analyses without regard to personal feelings or opinions; in some cases analyses of phenomena other than personal feelings or opinions; the attempt to see things as they really are, as opposed to things as our feelings dictate them to be.

Oogenesis The sum of cell divisions and cell differentiation that occurs between the time of primary oocyte formation, from oogonial cells to the formation of an egg cell.

Oogonium A mitotically active, germ-line cell that will give rise to two primary oogonial cells, each of which will give rise to primary oocytes, cells in which the first step of the first meiotic division has taken place.

Opsin A pigment protein located on photoreceptor cells in the retina of the eye. The three color opsins that absorb red, green, and blue lightwaves are located on cone shaped cells, and rod-opsin, which mediates sight in dim light, is located on rod shaped cells.

Ought An expression of one's relationship to moral duty or belief system (one ought to be kind to one's neighbors). In philosophy, is versus ought signifies the difference between that which is discovered by way of objective inquiry and that which is dictated by morals, beliefs, etc.

Oxidize The process by which molecules combine with oxygen atoms or lose hydrogen atoms, both of which often lead to the same chemical state.

Paradigm A set of forms or relationships among known particulars that limit the ways we can think about the whole; e.g., without genes it is difficult to explain differences and similarities among people related by descent. With genes, sets of relationships appear obvious. A way of thinking about groups of facts and interpreting new facts; a way of thinking that can be changed by the discovery of new relationships among facts.

Particles, virus We refer to individual members of higher plant and animal species as individuals (persons in the case of humans), but individual members of viral species are known as viral particles or simply particles.

Pedigree chart A genealogical record of a family. A pedigree chart includes as many of the phenotypic characteristics within a family as are of interest to the study, whether eye color, one or more of many diseases, etc.

Persistent infection Some viral, lytic-like infections do not kill all of the target cells. Either some cells of the target area do not become infected or the virus fails to shut off the host cells response to its own genetic information, whereupon the host cell "reads" from its own genome as well as from the viral genome. In either case, the host target area remains partially functional.

Phenotype Any feature or characteristic of an organism or any group of characteristics (e.g., of metabolism, physiology, or morphology). The word phenotype may refer to all of the characteristics of an organism or to one phenotypic trait.

Phenotypic Trait One aspect of phenotype (e.g., eye color, ear shape, distribution of body hair, etc.). Mendel noticed the importance of studying traits that appear in two or more contrasting forms; in humans such contrasting forms of phenotype are blue and brown eyes, attached and unattached ear lobes, the presence or absence of mid-digital hair.

Phenotypic variation The observed phenotypic differences among the members of a species, population, or family. Mendelian genetic studies usually focus upon one or a few of the differences (see phenotypic trait), but within populations there are many such differences.

Phenylketonuria PKU An "inborn error of metabolism" characterized by the accumulation of the amino acid phenylalanine in the blood and urine. Untreated, PKU children become severely mentally retarded. The condition is inherited as a homozygous recessive and is the result of the absence of the enzyme phenylalanine hydroxylase and hence the inability to transform the amino acid phenylalanine into tyrosine. The excess phenylalanine in the blood inhibits one or more chemical reactions of the brain.

Photoreceptor cell Cells located in eye retinas that carry pigment proteins, or opsins. Cells that receive light signals from external sources and that transmit these signals to optical neurons that in turn transmit them to the brain.

Pop-sociobiology A name given (by Philip Kitcher) to the extension of sociobiological "logic" that invites the human imagination to invent adaptationist explanations of all forms of human behavior, in particular outside the context of human history. For example, the behavior called rape in modern societies, they say, must have contributed to the reproductive success of the first rapist during neolithic times. These speculations are rarely supported by data or history, only by a modern imagination influenced as much by pop culture as by a desire to understand our evolutionary past.

Primary oocyte A cell that will lead directly to the formation of an egg cell, but only after a long interval of time, between birth and the onset of puberty.

Primary spermatocyte A cell within a testis that arises from a spermatogonial cell and within which the first step of the first meiotic division has taken place. From each primary spermatocytes, two secondary spermatocytes emerge, and from these four spermatids arise.

Primary structure The primary chemical structure of large biological molecules refers to the amino acid sequences of proteins, the base sequences of RNA molecules, and the base-pair sequences of DNA molecules. These large molecules are called informational molecules as a result of the key discovery that the primary structures of all of them are colinear (from the knowledge of one, the others can be predicted, exactly in the direction from DNA to protein).

Primate Among the categories used to classify all animals, Kingdom is the most inclusive and species is the most specific. The word primate is somewhat intermediate between these extremes and is called an Order, consisting of three suborders: Anthropoides (which includes humans, apes, and monkeys), Prosimii (which includes lemurs, and others), and Tarsiodea (which includes tarsiers). Primates exhibit peculiar hand movements, means of locomotion, and behavior.

Probe A DNA molecule used to locate molecules similar to itself in other genomes. Probes are made by the action of reverse transcriptase on specific mRNA molecules, are then cloned many times, and then are mixed with the restriction enzyme fragments of another genome. The probe will hybridize with fragments that are similar as determined by base pair sequences

Progeny The offspring, or children, of parents.

Protan red-green color impaired Red green colorvision-impaired persons who see red and green as green. Such persons are missing the red visual pigment proteins because of a mutant **D** gene.

Protein One of three classes of informational molecule characterized by long "chains" of building blocks called amino acids. After the chains fold into three-dimensional shapes—tertiary structure—proteins perform the biological "work" necessary for all forms of life.

Provirus The DNA molecule synthesized from an RNA viral genome that carries all of the genetic information included within the viral genome and that can become integrated within host genomes.

Pseudohermaphroditism A syndrome recognized by an automatic sex change from female to male during puberty. These individuals are 44 XY, and they produce testosterone, but they are missing a gene that programs the enzyme that transforms testosterone into dihydrotestosterone, the form needed for masculinizing early embryos. Thereby these individuals present as females at birth and remain so until puberty, at which time testosterone is an effective masculinizing hormone, and girls become boys.

Psychometrician A psychologist who designs measures of mental traits, such as tests for intelligence or psychopathic disorders.

Psychoanalytical Theory A theory that attempts to show cause-effect relationships between conscious and unconscious psychological processes. Based on this theory, attempts have been made to discover the unconscious mental processes.

Pure Lines A family, or line, of plants or animals that breed true. The progeny are always identical to the parents.

Purine One of two classes of base found in nucleic acid. Pyrimidines make up the other class. The two purines found in nucleic acids are guanine and adenine, and the three pyrimidines are cytosine, thymine, and uracil.

Pyrimidine One of two classes of base found in nucleic acid. Purines, guanine, and adenine make up the other class. The pyrimidines, cytosine and thymine, are found in DNA, and cytosine and uracil are found in RNA.

Qualitative Aspects of things that distinguish them absolutely from other things. Diamonds are qualitatively different from pearls, as are humans from hamsters. But diamonds differ from one another, as do pearls; but these kinds of differences are quantitative.

Quantitative Aspects of things that do not distinguish them absolutely from other things, for example the size of diamonds and the height of human beings.

Quantitative genetics The study of the contribution of genetic variation to phenotypic variation that is continuous (i.e., quantitative). Such phenotypes (e.g., height, skin color) are thought to be influenced by many genes and by environmental factors.

Race A concept of population biology and anthropology that assumes that within the human species there are subpopulations of individuals best described as races. There are no agreed-upon qualitative differences among races, as its usage implies, and in general the word race elicits more confusion than enlightenment. The concept of breeding population is of more use to the population biologist, and the concept of ethnic group is more useful to the anthropologist and sociologist, and certainly to the nonprofessional.

Raciation If there were such units of population as races, then raciation would include all of the processes attendant on their evolution.

Radioactivity The emission of atomic particles, usually electrons and/or neutrons, upon the spontaneous breakdown of atoms. The atomic particles released upon breakdown are capable of penetrating opaque bodies (photographic film, cells, etc.) and of producing electrical effects.

Recessive See Dominant. A recessive allele's phenotypic expression in the homozygous state is masked in the heterozygous state.

Reciprocal Altruism The relationships between individuals and groups of individuals based upon performing altruistic acts in the expectation of receiving a similar gesture of altruism in return. Often thought of in cost-accounting terms. Reduces the self-sacrificing feature of altruism.

Recombination (a) The physical exchange of chromatid segments of homologous chromosomes during meiosis. (b) The random distribution of maternal and paternal chromosomes by diploid germ line cells into haploid sex cells.

Referent An object or an event that is used as a reference point from which to evaluate a term or a symbol. For example, the population mean height is a referent by which it is determined whether an individual's height is average, below average, above average.

Reified Abstraction An abstraction or mental image which even though not real in a scientific, biological sense, can be made real by societal practices. Women and ethnic minorities, believed to be less bright than ethnic majority males, can be made to appear less bright by the practice of keeping them out of universities and in "sweat shops."

Replication DNA is characterized by self-replication in that a double-helical molecule of DNA can be propagated into two identical, double-helical molecules. This is accomplished as a result of each parent DNA strand serving as a template upon which a progeny strand is synthesized. However, DNA cannot "self"-replicate; its replication is catalyzed by enzymes called DNA polymerases.

Restriction endonucleases Endonucleases that digest (cut) nucleic acid molecules at specific base pair and base sequences. See endonucleases.

Restriction Fragment Length Polymorphism, RFLP The word *restriction* refers to endonucleases that cut DNA molecules at specific sites; the word *fragments* refers to the segments of DNA that remain after a restriction endonuclease acts on it; among similar but not identical DNA molecules, the *length* of the fragments may differ; and the word *polymorphism* refers to the fact that within large families of related individuals, there may be many (poly) different fragment sizes (fragment morphologies). The technology of RFLP analyses has greatly increased the resolving power of chromosome mapping.

Retina The inner layer of cells lining the back of the eye that intercept light directed to it from the lens of the eye. These are rod- and cone-shaped photoreceptor cells that convey the images produced by the light via the optic nerve to the brain.

Retrovirus Viruses with RNA genomes and the enzyme reverse transcriptase. The reverse transcriptase catalyzes the synthesis of a DNA complement of the RNA genome, a process that takes place in a direction opposite that predicted by the central dogma DNA $\rightarrow$ RNA $\rightarrow$ protein. Retro means going in reverse, backward.

Reverse transcriptase An enzyme found within retroviruses. In fact, reverse transcriptase is a DNA polymerase, but it differs from all other DNA polymerases in that it recognizes RNA templates upon which it catalyzes the synthesis of complementary DNA molecules.

Rhesus Blood Types Defined by their response to Rhesus monkey antiserum; blood that agglutinates is called Rh$^+$, and blood that does not is called Rh$^-$.

Rhodopsin The only visual pigment protein found on rod-shaped photoreceptor cells. Rhodopsin mediates vision in dim light and absorbs maximally in the 495-nm range.

Ribonucleic acid, RNA One of two classes of nucleic acid. The other class is DNA. RNA molecules differ from DNA molecules in that they are single-stranded molecules; they contain ribose sugar, not deoxyribose sugar, in their "backbones" and uracil, not thymine, in their primary structures. All species of RNA are synthesized upon DNA templates.

Ribosomal RNA, rRNA A species of RNA found only within the structures of ribosomes.

Ribosome Small cellular organelles upon which protein synthesis takes place. Ribosomes are composed of proteins and rRNA in a ratio close to $1:1$; each ribosome has an attachment site for mRNA, tRNA, and growing protein chains.

Risk The chance that a person whose parent(s) carries the gene in question carries the gene. If, for example, one of a child's parents is heterozygous at one gene locus, the child has a $50:50$ chance of inheriting one or the other of the two alleles. Said another way, the child is at risk of inheriting the allele that initiates the development of the genetic disease.

Sample A part of a whole, believed to represent the whole. A subset of a population selected to represent that population. For example, questions asked of 700 voting adults can be used to predict rather accurately the election of a politician within a voting district of one million adults.

Secondary Oocyte After a primary oocyte completes the first of the two meiotic divisions, two cells, unequal in size, emerge. The small cell, the first polar body, is lost; the large cell is a secondary oocyte that will initiate the second of the two meiotic divisions, leading to the formation of an egg cell and another nonfunctional polar body.

Secondary spermatocyte The products of the division of a primary spermatocyte into two cells, secondary spermatocytes; each secondary spermatocyte will, upon division, yield two spermatids.

Segregation The separation of alleles and of homologous chromosomes into different gametes during meiosis. The result is that each gamete carries one and only one member of each allelic pair and of each chromosome pair.

Selection To distinguish between types. One may select a red car over a blue car for esthetic reasons; a fish may select a mayfly over a bumblebee for reasons of taste; natural selection may have selected agouti-colored rodents over red or green ones, simply out of the circumstance that fewer agouti rodents are seen by hawks or coyotes. Darwin used the term natural selection to signify all of the forces of nature that combine to favor the survival of some types and to reduce the chances of survival of other types. Artificial selection—following the subjective choices of plant and animal breeders—accomplishes the same thing; some genotypes are encouraged to reproduce, others are discouraged.

Self-fertilization The union of female and male gametes that are produced by the same individual. In peas, all fertilized eggs result from self-fertilization, but in corn, another hermaphroditic species, most fertilized eggs result from cross-fertilization.

Semiconservative replication As progeny

DNA molecules are synthesized, each of the two strands of the parent molecule serves as a template upon which a progeny strand is synthesized. Thus progeny molecules are only half "new" in that one of their two strands is one of the parent strands and one is "new." Parent strands, not parent molecules, are "conserved."

Seminiferous tubules Testes are little more than networks of tubules, called seminiferous tubules. Each hollow tubule is lined with cells of many types, one type of which are spermatogonial cells. These cells ultimately lead to the formation of sperm cells, which flow out of the tubules and into the vas deferens.

Sense A sequence of base pairs within DNA molecules, or bases within RNA molecules, that encodes the synthesis of a functional protein. See nonsense.

Sex chromosomes Human genomes are comprised of two types of chromosome, sex chromosomes (X and Y) and autosomes. Sex chromosomes are so named because they are the only chromosomes that are distributed unequally between the two sex types.

Sexual dimorphism Two sexual morphologies (e.g., male and female morphologies). The phrase sexual dimorphism implies the processes of development of two sexual morphologies from nearly identical zygotes.

Sexuality The sum of the emotions and behaviors that lead to fantasies of and/or actual sexual interactions. It is sometimes referred to simply as sexual behavior, but it includes as well the motive forces that lead to sexual behavior, as well as the emotional consequences of the behavior.

Should In common language, should is often used as a synonym of ought, as an obligatory act called upon by duty. But, historically should implies a strict moral response, whereas ought appeals more to individual choice.

Shyness Name given to a large number of behaviors that range from fear, to retiring, to distrustful, and reluctant. In the vernacular, bashful. In pop-sociobiology, a unit of behavior assigned a genetic cause.

Sibling A brother or a sister. Siblings have the same biological father and mother. Sib is short for sibling.

Sickle Cell Anemia A fatal disease that derives its name from the fact that in oxygen-poor environments the red blood cells assume sharp, pointed, sickled shapes, as opposed to the donut shapes of normal red blood cells. The sickled cells deliver less oxygen, hence the anemia. The causative gene encodes hemoglobin S (HbS), which in turn changes the shapes of red blood cells.

Sickle Cell Trait The condition of persons heterozygous for the SCA gene, characterized by the presence of both normal and sickled red blood cells. About 60% of the red cells are normal, which is sufficient for normal health.

Social Darwinism A school of thought codified by Herbert Spencer that explains the success and failure of human societies in terms of fitness, natural selection, and biological differences. However, fitness and natural selection had a societal, not a biological meaning. Basically, Spencer used market economic language, not Darwinian language.

Societal Behavior Used to refer to human societies as opposed to animal social groups, mainly because human societal behavior is so vastly more complicated, historically determined, flexible, and dependent upon learned information. Societal behavior provides a glimpse of cultural evolution. If other animals have cultures, they are qualitatively different from human cultures.

Social Behavior Used to refer to non-human, animal, group behavior. See societal behavior.

Speciation The processes that lead to the evolution of species.

Species A population of related organisms that is biologically reproductively isolated from organisms of all other species.

Sperm Cell Male gamete, or sex cell; the sex cell that fertilizes an egg cell to form a zygote.

Spermatid Haploid cell; meiotic produce of a primary spermatocyte; cells that derive directly from secondary spermatocytes and that soon thereafter differentiate into sperm cells via spermiogenesis.

Spermatogenesis The sum of cell divisions and cell differentiation between the time of primary spermatocyte formation, from spermatogonial cells to the formation of sperm cells.

Spermatogonium Cells that line the seminiferous tubules that upon division give rise to primary spermatogonial cells and primary spermatocytes.

Spermiogenesis The name given to the processes that guide the differentiation of spermatids into sperm cells.

Standardized IQ tests Standardizing a test refers to the processes of testing its usefulness on selected populations of individuals prior to selling it to schools and other institutions for use in segregating members of the general population. For example, a test designed for six year old children must be tested on hundreds of six year old children to determine whether it works. There are many problems with standardizing test items, one of which is that a test item may appear to work on a selected population but in fact works to bias scores within the general population.

Stanford-Binet IQ Test The most widely used IQ test, developed by Lewis Terman of Stanford University from the testing procedures developed by Alfred Binet, and from the concept of intelligence quotient developed by Wilhelm Stern in 1912, i.e., that CA/MA x 100 = IQ.

Stem cell Stem cells are characterized by their retention of the ability to divide and for the ability of their progeny cells to differentiate into specialized cells. For example, all of the many specialized blood cells arise from stem cells located within the bone marrow.

Structuralism The view that experiences can be reduced to basic parts, that laws can be induced by synthesizing the parts, and that the parts can be extracted from subjects through introspection. Edward Titchener is credited with founding this school of psychology.

Symbiotic A relationship (usually) between two species, each of whose survival is enhanced, or dependent upon, the other. Human beings and the bacteria living in their guts exhibit a symbiotic relationship. Both species contribute to the survival of the other, and neither species harms the other.

Syndrome A complex of symptoms that occur together, or, we might say, a combination of phenotypic manifestations, as signs of an illness or disease (morbid state, in medicine).

Synthesis A general name given to the biochemical processes by which relatively complex molecules emerge from relatively simple ones. For example, photosynthesis is a process of transforming CO_2, H_2O, and energy into sugars, while in the liver, sugars are transformed into even larger molecules of glycogen. The opposite of digestion.

Template In biology, a molecular "mold" that determines the primary structure of molecules synthesized upon it. For example, during DNA synthesis, both strands of a prent molecule serve as templates upon which new, complementary strands are synthesized.

Teratogen Agents that induce developmental mistakes (birth defects) during embryogenesis or fetal development.

Tertiary structure As used here, a three-dimensional structure assumed by a long chain molecule after it folds into a somewhat stable state. The primary structure of a protein is a sequence of amino acids; the secondary structure is a consequence of the primary structure and reflects the fact that peptide bonds are not parallel; both the primary and secondary structures influence the shapes of the folded (tertiary) structure, but the relationships are not as yet known.

Test-cross Usually a cross between an F_1 and its homozygous recessive parent. The frequencies of phenotypes among the progeny of such a mating reveal the kinds and frequencies of gametes produced by the F_1.

Testicular feminization Persons who are discordant with respect to internal and external sex criteria. Their body types are female and their internal genitalia are male. The condition is due to an X-linked gene, *tmf*, that fails to encode receptor sites for testosterone. The testes produce testosterone, but the hormone is ineffective because receptor cells fail to bind to it.

Testosterone A hormone synthesized within

testes that appears to be essential to the processes of masculinization within male embryos. It is the male hormone necessary for sperm maturation.

Thymine One of the two pyrimidine bases found only in DNA, where it always pairs with adenine in double-helical molecules. Thymine is not found in RNA, but the closely related pyrimidine uracil is. Since the genetic dictionary includes RNA codons, T is the only DNA "letter" not found in the dictionary. U appears in its place.

Toxin Agent that inhibits cell division.

Trait See phenotypic trait.

Transcription RNA synthesis (i.e., the process of transcribing genetic information from DNA to RNA).

Transfer RNA, tRNA A species of RNA characterized by "carrying" amino acids from the cytoplasm of the cell to the sites of protein synthesis (i.e., upon ribosomes). Codons of mRNA attract anticodons of tRNA, thus selecting a specific amino acid for each site within the primary structure of proteins.

Transformation Directed modification of phenotype, usually of microorganisms, by the cell's incorporation of a gene from another organism into its own genome.

Translation Protein synthesis (i.e., the process of translating information contained within mRNA molecules into amino acid sequences of protein molecules).

Transsexual Individuals whose gender identity and biological sex are discordant, that is, they possess the biological body of one sex but "feel" as if they were a person of the opposite sex.

Trisomy A condition characterized by one extra chromosome in a genome. The diploid number of chromosomes in the human genome is 46, 23 pairs. Zygotes with 47 chromosomes may have a third chromosome of any one of the 23 types (e.g., trisomy 21, trisomy X).

Turner's syndrome Individuals with 45 chromosomes, 22 pairs of autosomes, and one X chromosome. The physical features of Turner's females are markedly different from the 44XX norm.

Unit of Evolution Units of evolution are breeding populations, the smallest populational unit within which genetic changes appear through time. While mutations occur within genomes, genomes do not evolve; gene pools made up of many genomes do evolve.

Unit of Selection It is a matter of controversy whether the forces of natural selection act directly on genes, organisms, kin groups, or breeding populations. Each of the proposed focal points of action, of the forces of selection, is referred to as a unit of selection.

Uracil One of five nucleic acid bases and one of the two pyrimidines found in RNA. U, not T, is found in the genetic dictionary as one of the four letters found within codons.

Uterus Hollow, muscular organ in females in which blastocytes become embedded and develop into fetuses. The cervix of the uterus opens into the vagina, and the fallopian tubes of the uterus open near the ovaries.

Variation Difference. There are few differences (very little variation) among purebred Boston terriers. There are many differences (a great deal of variation) among mongrel dogs that run loose on the outskirts of many small towns.

Vas deferens During early embryogenesis of male embryos, wolffian ducts become vas deferens, ducts that, in adult males, transport semen from the testes to the seminal vesicles and urethra.

Virulent As used here, virulent refers to the property of microorganisms to cause disease; the opposite of avirulent.

Virus Viruses are microorganisms that are unable to reproduce outside host cells. They are much smaller than bacteria and, from a chemical point of view, are little more than a small genome encased within a protein shell. The energy and raw materials needed for viral genome replication and protein synthesis are sequestered from host cells.

Visual pigment proteins Proteins found on cone- and rod-shaped photoreceptor cells in the retinas of the eyes; characterized by the function of absorbing specific wavelengths of visible light. See opsin.

X-Chromosome linkage All of the genes (there may be several thousand) located along the length of an X chromosome are linked, that is, they reside on one X chromosome. Such genes are referred to as X-chromosome linked or, simply, X linked.

Zygote The cell produced by fertilization, that is, by the union of an egg and a sperm cell.

Index